AF531835

Encyclopaedia of

NUMBER THEORY

Encyclopaedia of
NUMBER THEORY

Volume 2

R.K. Pandey

ANMOL PUBLICATIONS PVT. LTD.
NEW DELHI - 110 002 (INDIA)

ANMOL PUBLICATIONS PVT. LTD.
H.O.: 4374/4B, Ansari Road, Daryaganj,
New Delhi-110 002 (India)
Ph.: 23278000, 23261597
B.O.: No. 1015, Ist Main Road, BSK IIIrd Stage
IIIrd Phase, IIIrd Block,
Bangalore - 560 085 (India)
Visit us at: www.anmolpublications.com

Encyclopaedia of Number Theory

First Edition, 2009
ISBN 978-81-261-3910-1 (Set)

PRINTED IN INDIA

Printed at Mehra Offset Press, Delhi.

Contents

Preface

Natural Number is the original Mathematical concept and the most fundamental. Speculations about the nature and properties of whole numbers doubtless constitute the oldest form of mathematical thought. It is known that the sumerians and Babylonians as well as the ancient Egyptian had a fair knowledge of the properties of Natural Numbers. But first in connection with the Greeks is possible to speak of a proper theory of Numbers. Pythagoras and his pupils pursued extensive studies in the field of integers.

The Author has followed the practice of first presenting required background material, which is then used to develop the results. Relevant material is included in each chapter form other sources improves readability and makes the presentation "self-contained" to a large extent. In addition, at the end of each chapter a comprehensive Exercise is provided so that the interested reader can explore in more depth.

This volume presupposes familiarity only with the most elementary concepts of Arithmetic (Divisibility Properties, Greatest Common Divisor, etc.), is an expanded version of a series of lectures for graduate students on elementary number theory.

List of Symbols

$\pm$	$\not\subset$	$\subset$	$\subseteq$	$\in$	$\notin$	$\Re$	∂	$\neq$	$\approx$	$\angle$	∇
Σ	Π	Ω	ϕ	α	β	χ	ε	γ	η	λ	μ
ν	$\perp$	π	θ	ρ	σ	τ	$\Re$	ω	ξ	ψ	ζ
Υ		$\exists$	$\int$	$\iint$	$\iiint$	$\times$	$\sqrt{\ }$	$\because$	$\geq$	$\oint$	$\Rightarrow$

Chapter 1

Congruence Relation

In mathematics and especially in abstract algebra, a congruence relation or simply congruence is an equivalence relation that is compatible with some algebraic operation(s).

. MODULAR ARITHMETIC

The prototypical example is modular arithmetic: for n a positive integer, two integers a and b are called congruent modulo n if $a - b$ is divisible by n (or an equivalent condition is that they give the same remainder when divided by n).

For example congruent modulo 3:

$$1 \equiv 5 \pmod{3}$$

because $11 - 5$ gives 6, which is divisible by 3. Or, equally, both numbers give the same remainder when divided by 3:

$$11 = 3 \times 3 + 25 = 1 \times 3 + 2$$
$$5 = 1 \times 3 + 2$$

If $a_1 \equiv b_1 \pmod{n}$ and $a_2 \equiv b_2 \pmod{n}$ then $a_1 + a_2 \equiv b_1 + b_2 \pmod{n}$ and $a_1a_2 \equiv b_1b_2 \pmod{n}$ This turns the congruence (mod n) into an equivalence on the ring of all integers.

LINEAR ALGEBRA

Two real matrices A and B are called congruent if there is an invertible real matrix P such that

$$P^T AP = B$$

A symmetric matrix has real eigenvalues. The *inertia* of a symmetric matrix is a triple consisting of the number of positive eigenvalues, the number of zero eigenvalues, and the number of negative eigenvalues.

Sylvester's law of inertia states that two symmetric real matrices are congruent if and only if they have the same inertia. So, congruence transformations may change the eigenvalues of a matrix but they cannot change the signs of the eigenvalues.

For complex matrices, we have to distinguish between ^{T}congruency (A and B are ^{T}congruent if there is an invertible matrix P such that $P^{T}AP = B$) and *congruency (A and B are *congruent if there is an invertible matrix P such that $P^{*}AP = B$).

UNIVERSAL ALGEBRA

The idea is generalized in universal algebra: A congruence relation on an algebra A is a subset of the direct product $A \times A$ that is both an equivalence relation on A and a subalgebra of $A \times A$.

The kernel of a homomorphism is always a congruence. Indeed, every congruence arises as a kernel. For a given congruence ~ on A, the set $A/\sim$ of equivalence classes can be given the structure of an algebra in a natural fashion, the quotient algebra. The function that maps every element of A to its equivalence class is a homomorphism, and the kernel of this homomorphism is ~.

The lattice Con A of all congruence relations on an algebra A is algebraic.

CONGRUENCES OF GROUPS, AND NORMAL SUBGROUPS AND IDEALS

In the particular case of groups, congruence relations can be described in elementary terms as follows: If G is a group (with identity element e) and ~ is a binary relation on G, then ~ is a congruence whenever:

1. Given any element a of G, $a \sim a$ (reflexivity);
2. Given any elements a and b of G, if $a \sim b$, then $b \sim a$ (symmetry);
3. Given any elements a, b, and c of G, if $a \sim b$ and $b \sim c$, then $a \sim c$ (transitivity);
4. Given any elements a, a', b, and b' of G, if $a \sim a'$ and $b \sim b'$, then $a * b \sim a' * b'$.
5. Given any elements a and a' of G, if $a \sim a'$, then $a^{-1} \sim a'^{-1}$ (this can actually be proven from the other four, so is strictly redundant);

Conditions 1, 2, and 3 say that ~ is an equivalence relation.

A congruence ~ is determined entirely by the set $\{a$ " $G : a \sim e\}$ of those elements of G that are congruent to the identity element, and this set is a normal subgroup. Specifically, $a \sim b$ if and only if $b^{"1} * a \sim e$. So instead of talking about congruences on groups, people usually speak in terms of normal subgroups of them; in fact, every congruence corresponds uniquely to some normal subgroup of G.

IDEALS OF RINGS AND THE GENERAL CASE OF KERNELS

A similar trick allows one to speak of kernels in ring theory as ideals instead of congruence relations, and in module theory as submodules instead of congruence relations.

The most general situation where this trick is possible is in ideal-supporting algebras. But this cannot be done with, for example, monoids, so the study of congruence relations plays a more central role in monoid theory.

PROPERTIES OF CONGRUENCES RELATION

Let us suppose n be a positive integer. For integers a and b, we say that a is congruent to b modulo n if $n \mid (a - b)$, and we write $a \equiv b \pmod{n}$. If $n \mid (a - b)$, then we write $a \not\equiv b \pmod{n}$. Equivalently, $a \equiv b \pmod{n}$ if and only if $a = b + ny$ for some $y \in \mathbb{Z}$. The relation $a \equiv b \pmod{n}$ is called a congruence relation, or simply, a congruence. The number n appearing in such congruences is called the modulus of the congruence. If we view the modulus n as fixed, then the following theorem says that the binary relation "$. \equiv . \pmod{n}$" is an equivalence relation on the set $\mathbb{Z}$.

Theorem: Let $a, a', b, b', n \in \mathbb{Z}$ with $n > 0$. If

$$a \equiv a' \pmod{n} \text{ and } b \circ b' \pmod{n},$$

then
$$a + b \equiv a' + b' \pmod{n} \text{ and } a \,.\, b \equiv a'.b' \pmod{n}.$$

Proof: Suppose that $a \equiv a' \pmod{n}$ and $b \equiv b' \pmod{n}$. This means that there exist integers x and y such that $a = a' + nx$ and $b = b' + ny$. Therefore,

$$a + b = a' + b' + n(x + y),$$

which proves the first congruence of the theorem, and

$$ab = (a' + nx)(b' + ny) = a'b' + n(a'y + b'x + nxy),$$

which proves the second congruence. One to work with congruence relations modulo n much as one would with ordinary equalities: one can add to, subtract from, or multiply both sides of a congruence modulo n by the same integer; also, if b is congruent to a modulo n, one may substitute b for a in any simple arithmetic expression (involving addition, subtraction, and multiplication) appearing in a congruence modulo n.

Now suppose a is an arbitrary, fixed integer, and consider the set of integers z that satisfy the congruence $z \equiv a \pmod{n}$. Since z satisfies this congruence if and only if $z = a + ny$ for some $y \in \mathbb{Z}$, we may apply Theorems (with a as given, and $b := n$) to deduce that every interval of n consecutive integers contains exactly one such z.

Theorem: Let n be a positive integer. For all $a, b, c \in \mathbb{Z}$, we have:

(i) $a \equiv a \pmod{n}$;

(ii) $a \equiv b \pmod{n}$ implies $b \equiv a \pmod{n}$;

(iii) $a \equiv b \pmod{n}$ and $b \equiv c \pmod{n}$ implies $a \equiv c \pmod{n}$.

Proof:

(i) observe that n divides $0 = a - a$.

(ii) observe that if n divides $a - b$, then it also divides $-(a - b) = b - a$.

(iii) observe that if n divides $a - b$ and $b - c$, then it also divides $(a - b) + (b - c) = a - c$.

Theorem: Let $a, n \in \mathbb{Z}$ with $n > 0$. Then there exists a unique integer z such that $z \equiv a$ (mod n) and $0 \leq z < n$, namely, $z := a \bmod n$. More generally, for every $x \in \mathbb{R}$, there exists a unique integer $z \in [x, x + n)$ such that $z \equiv a$ (mod n).

Example: Let us find the set of solutions z to the congruence

$$3z + 4 \equiv 6 \pmod 7.$$

Suppose that z is a solution. Subtracting 4 from both sides, we obtain

$$3z \equiv \in \pmod 7.$$

Next, we would like to divide both sides of this congruence by 3, to get z by itself on the left-hand side. We cannot do this directly, but since $5 \,.\, 3 \equiv 1 \pmod 7$, we can achieve the same effect by multiplying both sides of by 5. If we do this, and then replace $5 \,.\, 3$ by 1, and $5 \,.\, \in$ by 3, we obtain

$$z \equiv 3 \pmod 7.$$

Thus, if z is a solution to, we must have $z \equiv 3 \pmod 7$; conversely, one can verify that if $z \equiv 3 \pmod 7$. We conclude that the integers z are precisely those integers that are congruent to 3 modulo 7, which we can list as follows:

$$\ldots, -18, -11, -4, 3, 10, 17, 24, \ldots$$

In the next section, we shall give a systematic treatment of the problem of solving linear congruences, such as the one appearing in the previous example.

EQUIVALENCE RELATIONS

Before discussing congruences, we review the definition and basic properties of equivalence relations.

Let S be a set. A binary relation 'on S is called an equivalence relation if it is

reflexive: $a \sim a$ for all $a \in S$,

symmetric: $a \sim b$ implies $b \sim a$ for all $a, b \in S$, and

transitive: $a \sim b$ and $b \sim c$ implies $a \sim c$ for all $a, b, c \in S$.

If $\sim$ is an equivalence relation on S, then for $a \in S$ one defines its equivalence class as the set $\{x \in S : x \sim a\}$.

Theorem: Let $\sim$ be an equivalence relation on a set S, and for $a \in S$, let $[a]$ denote its equivalence class. Then for all $a, b \in S$, we have:

(i) $a \in [a]$;

(ii) $a \in [b]$ implies $[a] = [b]$.

Proof:

(i) follows immediately from reflexivity.

(ii) suppose $a \in [b]$, so that $a \sim b$ by definition. We want to show that $[a] = [b]$. To this end, consider any $x \in S$. We have

$$x \in [a] \Rightarrow x \sim a \text{ (by definition)}$$
$$\Rightarrow x \sim b \text{ (by transitivity, and since } x \sim a \text{ and } a \sim b)$$
$$\Rightarrow x \in [b].$$

Thus, $[a] \subseteq [b]$. By symmetry, we also have $b \sim a$, and reversing the roles of a and b in the above argument, we see that $[b] \subseteq [a]$.

This theorem implies that each equivalence class is non-empty, and that each element of S belongs to a unique equivalence class; in other words, the distinct equivalence classes form a partition of S. A member of an equivalence class is called a representative of the class.

EXERCISE

1. Let $a, b, n, n' \in Z$ with $n > 0$, $n' > 0$, and $n' \mid n$. Show that if $a \equiv b \pmod{n}$, then $a \equiv b \pmod{n'}$.
2. Show that the equation $7y^3 + \in = z^3$ has no solutions $y, z \in \mathbb{Z}$.
3. Let $a, b, n, n' \in \mathbb{Z}$ with $n > 0$, $n' > 0$, and $\gcd(n, n') = 1$. Show that if $a \equiv b \pmod{n}$ and $a \equiv b \pmod{n'}$, then $a \equiv b \pmod{nn'}$.
4. Show that there are 14 distinct, possible, yearly (Gregorian) calendars, and show that all 14 calendars actually occur.
5. Let $a, b, n \in \mathbb{Z}$ with $n > 0$ and $a \equiv b \pmod{n}$. Also, let $c_0, c_1, ..., c_k \in \mathbb{Z}$. Show that

   $c_0 + c_1 a + ... + c_k a^k \equiv c_0 + c_1 b + ... + c_k b^k \pmod{n}$.
6. Consider the relations $=$, $\leq$, and $<$ on the set $\mathbb{R}$. Which of these are equivalence relations?
7. Let $S := (\mathbb{R} \times \mathbb{R}) \setminus \{(0, 0)\}$. For $(x, y), (x', y') \in S$, let us say $(x, y) \sim (x', y')$ if there exists a real number $\lambda > 0$ such that $(x, y) = (x', y')$. Show that $\sim$ is an equivalence relation; moreover, show that each equivalence class contains a unique representative that lies on the unit circle (i.e., the set of points (x, y) such that $x^2 + y^2 = 1$).
8. Let $a, b, n \in \mathbb{Z}$ with $n > 0$. Show that $a \equiv b \pmod{n}$ if and only if $(a \bmod n) = (b \bmod n)$.
9. Let a be a positive integer whose base-10 representation is $a = (a_{k-1} \ldots a_1 a_0)_{10}$. Let b be the sum of the decimal digits of a; that is, let $b := a_0 + a_1 + \ldots + a_{k-1}$. Show that $a \equiv b \pmod{9}$. From this, justify the usual "rules of thumb" for determining divisibility by 9 and 3: a is divisible by 9 (respectively, 3) if and only if the sum of the decimal digits of a is divisible by 9 (respectively, 3).

10. Let e be a positive integer. For $a \in \{0, ..., 2^e - 1\}$, let $\tilde{a}$ denote the integer obtained by inverting the bits in the e-bit, binary representation of a (note that $\tilde{a} \in \{0, ..., 2^e - 1\}$). Show that $\tilde{a} + 1 \equiv -a \pmod{2^e}$. This justifies the usual rule for computing negatives in 2's complement arithmetic (which is really just arithmetic modulo 2^e).
11. Let $a, b, n \in \mathbb{Z}$ with $n > 0$ and $a \equiv b \pmod n$. Show that $\gcd(a, n) = \gcd(b, n)$.

LAW OF CANCELLATION

Let us consider $a, n \in \mathbb{Z}$ with $n > 0$. Theorem gives us a cancellation law for congruences:

$$\text{if } \gcd(a, n) = 1 \text{ and } az \equiv az' \pmod n, \text{ then } z \equiv z' \pmod n.$$

More generally, if $d := \gcd(a, n)$, then we can cancel a from both sides of a congruence modulo n, as long as we replace the modulus by n/d.

Example. Let us have a look

$$5 . \in \equiv 5 . (-4) \pmod 6.$$

Theorem tells us that since $\gcd(5, 6) = 1$, we may cancel the common factor of 5 from both sides of equation, obtaining $\in \equiv -4 \pmod 6$, which one can also verify directly.

Next observe that

$$15.5 \equiv 15 . 3 \pmod 6.$$

We cannot simply cancel the common factor of 15 from both sides of equation; indeed, $5 \not\equiv 3 \pmod 6$. However, $\gcd(15, 6) = 3$, and guarantees, we do indeed have 5 3 (mod 2).

THE CHINESE REMAINDER THEOREM

Let us consider systems of linear congruences with respect to moduli that are relatively prime in pairs. The result we state here is known as the Chinese remainder theorem, and is extremely useful in a number of contexts.

Theorem: Let $\{n_i\}_{i=1}^k$ be a pairwise relatively prime family of positive integers, and let $a_1, ..., a_k$ be arbitrary integers. Then there exists a solution $a \in \mathbb{Z}$ to the system of congruences

$$a \equiv a_i \pmod{n_i} \ (i = 1, ..., k).$$

Moreover, any $a' \in \mathbb{Z}$ is a solution to this system of congruences if and only if $a \equiv a'$ (mod n), where $n := \prod_{i=1}^k n_i$.

Proof: To prove the existence of a solution a to the system of congruences, we first show how to construct integers $e_1, ..., e_k$ such that for $i, j = 1, ..., k$, we have

$$e_j \equiv \begin{cases} 1 \pmod{n_i} & \text{if } j - i, \\ 0 \pmod{n_i} & \text{if } j \neq i. \end{cases}$$

If we do this, then setting

$$a := \sum_{i=1}^{k} a_i e_i,$$

one sees that for $j = 1, ..., k$, we have

$$a \equiv \sum_{i=1}^{k} a_i e_i \equiv a_j \ (mod\ n_j),$$

since all the terms in this sum are zero modulo nj, except for the term $i = j$, which is congruent to a_j modulo n_j.

To construct $e_1, ..., e_k$ satisfying , let $n := \prod_{i=1}^{k} n_i$ as in the statement of the theorem, and for $i = 1, ..., k$, let $n_i^* := n/ni$; that is, n_i^* is the product of all the moduli n_j with $j \neq i$. From the fact that is pairwise relatively prime, it follows that for $i = 1, ..., k$, we have $\gcd(n_i, n_i^*) = 1$, and so we may define $t_i := (n_i^*)^{-1}$ mod n_i and $e_i := n_i^* t_i$. One sees that $e_i \equiv 1$ (mod n_i), while for $j \neq i$, we have $ni \mid n_j^*$, and so $e_j \equiv 0$ (mod n_i). Thus, is satisfied.

That proves the existence of a solution a to the given system of congruences. If $a \equiv a_0$ (mod n), then since $n_i \mid n$ for $i = 1, ..., k$, we see that $a' \equiv a \equiv a_i$ (mod n_i) for $i = 1, ..., k$, and so a' also solves the system of congruences.

Finally, if a' is a solution to the given system of congruences, then $a \equiv a_i \equiv a'$ (mod n_i) for $i = 1, ..., k$. Thus, $n_i \mid (a - a')$ for $i = 1, ..., k$. Since $\{n_i\}_{i=1}^{k}$ is pairwise relatively prime, this implies $n \mid (a - a')$, or equivalently, $a \equiv a'$ (mod n).

We can restate Theorem in more concrete terms, as follows. For each positive integer m, let I_m denote $\{0, ..., m - 1\}$. Suppose $\{n_i\}_{i=1}^{k}$ is a pairwise relatively prime family of positive integers, and set $n := n_1 \ldots n_k$. Then the map

$$\tau : I_n \to I_{n_1} \times \ldots \times I_{n_k}$$

$$a \to (a \bmod n_1, ..., a \bmod n_k)$$

is a bijection.

Example: The following table illustrates what Theorem says for $n_1 = 3$ and $n_2 = 5$.

a	0	1	2	3	4	5	6	7	8	9	10	11	12	13	14
$a \bmod 3$	0	1	2	0	1	2	0	1	2	0	1	2	0	1	2
$a \bmod 5$	0	1	2	3	4	0	1	2	3	4	0	1	2	3	4

We see that as a ranges from 0 to 14, the pairs (a mod 3, a mod 5) range over all pairs (a_1, a_2) with $a_1 \in \{0, 1, 2\}$ and $a_2 \in \{0, ..., 4\}$, with every pair being hit exactly once.

EXERCISE

1. Suppose n_1 and n_2 are positive integers, and let $d := \gcd(n_1, n_2)$. Let a_1 and a_2 be arbitrary integers. Show that there exists an integer a such that $a \equiv a_1 \pmod{n_1}$ and $a \equiv a_2 \pmod{n_2}$ if and only if $a_1 \equiv a_2 \pmod{d}$.

2. Let $\{n_i\}_{i=1}^{k}$ be a pairwise relatively prime family of positive integers. Let $a_1, ..., a_k$ and $b_1, ..., b_k$ be integers, and set $d_i := \gcd(a_i, n_i)$ for $i = 1, ..., k$. Show that there exists an integer z such that $a_i z \equiv b_i \pmod{n_i}$ for $i = 1, ..., k$ if and only if $d_i \mid b_i$ for $i = 1, ..., k$.

3. Compute the values e_1, e_2, e_3 in the proof of Theorem in the case where $k = 3$, $n_1 = 3$, $n_2 = 5$, and $n_3 = 7$. Also, find an integer a such that $a \equiv 1 \pmod 3$, $a \equiv 1 \pmod 5$, and $a \equiv 5 \pmod 7$.

4. If you want to show that you are a real nerd, here is an ageguessing game you might play at a party. You ask a fellow party-goer to divide his age by each of the numbers 3, 4, and 5, and tell you the remainders. Show how to use this information to determine his age.

SOLVING LINEAR CONGRUENCES

In this section, we consider the general problem of solving linear congruences. More precisely, for a given positive integer n, and arbitrary integers a and b, we wish to determine the set of integers z that satisfy the congruence

$$az \equiv b \pmod{n}.$$

If we observe the equation has a solution z, and if $z \equiv z'$ (mod n), then z' is also a solution. However, equation has may or may not have a solution, and if it does, such solutions may or may not be uniquely determined modulo n. The following theorem precisely characterizes the set of solutions; basically, it says thatequation has a solution if and only if $d := \gcd(a, n)$ divides b, in which case the solution is uniquely determined modulo n/d.

Theorem: Let $a, n \in \mathbb{Z}$ with $n > 0$, and let $d := \gcd(a, n)$.

(i) For every $b \in \mathbb{Z}$, the congruence $az \equiv b \pmod{n}$ has a solution $z \in \mathbb{Z}$ if and only if $d \mid b$.

(ii) For every $z \in \mathbb{Z}$, we have $az \equiv 0 \pmod{n}$ if and only if $z \equiv 0 \pmod{n/d}$.

(iii) For all $z, z' \in \mathbb{Z}$, we have $az \equiv az' \pmod{n}$ if and only if $z \equiv z' \pmod{n/d}$.

Proof: For

(i) let $b \in \mathbb{Z}$ be given. Then we have

$$az \equiv b \pmod{n} \text{ for some } z \in \mathbb{Z}$$

$$\Leftrightarrow \quad az = b + ny \text{ for some } z, y \in \mathbb{Z} \text{ (by definition of congruence)}$$

$$\Leftrightarrow \quad az - ny = b \text{ for some } z, y \in \mathbb{Z}$$

$$\Leftrightarrow \quad d \mid b$$

(ii) we have

$n \mid az \Leftrightarrow n/d \mid (a/d)z \Leftrightarrow n/d \mid z$.

All of these implications follow rather trivially from the definition of divisibility, except that for the implication $n/d \mid (a/d)z \Rightarrow n/d \mid z$, we use Theorem and the fact that $\gcd(a/d, n/d) = 1$.

(iii) we have

$$az \equiv az' \pmod{n} \Leftrightarrow a(z - z') \equiv 0 \pmod{n}$$

$$\Leftrightarrow \quad z - z' \equiv 0 \pmod{n/d} \text{ (by part (ii))}$$

$$\Leftrightarrow \quad z \equiv z' \pmod{n/d}.$$

We can restate Theorem in more concrete terms as follows. Let $a, n \in \mathbb{Z}$ with $n > 0$, and let $d := \gcd(a, n)$. Let $I_n := \{0, ..., n-1\}$ and consider the "multiplication by a" map

$$\tau_a : I_n \to I_n$$

$$z \to az \bmod n.$$

The image of τ_a consists of the n/d integers

$i \,.\, d \; (i = 0, ..., n/d - 1)$.

Moreover, every element b in the image of τ_a has precisely d pre-images

$z_0 + j \,.\, (n/d) \; (j = 0, ..., d - 1)$,

where $z_0 \in \{0, ..., n/d - 1\}$. In particular, τ_a is a bijection if and only if a and n are relatively prime.

Example. The following table illustrates what Theorem says for $n = 15$ and $a = 1, 2, 3, 4, 5, 6$.

z	0	1	2	3	4	5	6	7	8	9	10	11	12	13	14
$2z \bmod 15$	0	2	4	6	8	10	12	14	1	3	5	7	9	11	13
$3z \bmod 15$	0	3	6	9	12	0	3	6	9	12	0	3	6	9	12
$4z \bmod 15$	0	4	8	12	1	5	9	13	2	6	10	14	3	7	11
$5z \bmod 15$	0	5	10	0	5	10	0	5	10	0	5	10	0	5	10
$6z \bmod 15$	0	6	12	3	9	0	6	12	3	9	0	6	12	3	9

In the second row, we are looking at the values $2z$ mod 15, and we see that this row is just a permutation of the first row. So for every b, there exists a unique z such that $2z \equiv b \pmod{15}$. This is implied by the fact that gcd(2, 15) = 1.

In the third row, the only numbers hit are the multiples of 3, which follows from the fact that gcd(3, 15) = 3. Also note that the pattern in this row repeats every five columns; that is, $3z \equiv 3z' \pmod{15}$ if and only if $z \equiv z' \pmod{5}$.

In the fourth row, we again see a permutation of the first row, which follows from the fact that gcd(4, 15) = 1.

In the fifth row, the only numbers hit are the multiples of 5, which follows from the fact that gcd(5, 15) = 5. Also note that the pattern in this row repeats every three columns; that is, $5z \equiv 5z' \pmod{15}$ if and only if $z \equiv z' \pmod{3}$.

In the sixth row, since gcd (6, 15) = 3, we see a permutation of the third row.

The pattern repeats after five columns, although the pattern is a permutation of the pattern in the third row.

Modular Inverses

Again, let $a, n \in Z$ with $n > 0$. We say that $z \in \mathbb{Z}$ is a multiplicative inverse of a modulo n if az 1 (mod n). Theorem says that a has a multiplicative inverse modulo n if and only if $\gcd(a, n) = 1$. Moreover, Theorem says that the multiplicative inverse of a, if it exists, is uniquely determined modulo n; that is, if z and z' are multiplicative inverses of a modulo n, then $z \equiv z' \pmod{n}$. Note that if z is a multiplicative inverse of a modulo n, then a is a multiplicative inverse of z modulo n. Also note that if $a \equiv a' \pmod{n}$, then z is a multiplicative inverse of a modulo n if and only if z is a multiplicative inverse of a' modulo n.

Now suppose that $a, b, n \in \mathbb{Z}$ with $n > 0$, $a \neq 0$, and $\gcd(a, n) = 1$. Theorem says that there exists a unique integer z satisfying

$$az \equiv b \pmod{n} \text{ and } 0 \leq z < n.$$

Setting $s := b/a \in \mathbb{Q}$, we may generalize the "mod" operation, defining s mod n to be this value z. As the reader may easily verify, this definition of s mod n does not depend on the particular choice of fraction used to represent the rational number s. With this notation, we can simply write a^{-1} mod n to denote the unique multiplicative inverse of a modulo n that lies in the interval 0, ..., $n - 1$.

Example. Let $a, b, n \in \mathbb{Z}$ with $n > 0$. We can describe the set of solutions $z \in \mathbb{Z}$ to the congruence $az \equiv b \pmod{n}$ very succinctly in terms of modular inverses.

If $\gcd(a, n) = 1$, then setting $t := a^{-1}$ mod n, and $z_0 := tb$ mod n, we see that z_0 is the unique solution to the congruence $az \equiv b \pmod{n}$ that lies in the interval $\{0, ..., n - 1\}$.

More generally, if $d := \gcd(a, n)$, then the congruence $az \equiv b \pmod{n}$ has a solution if and only if $d \mid b$. So suppose that $d \mid b$. In this case, if we set $a' := a/d$, $b' := b/d$, and $n' := n/d$, then for each $z \in \mathbb{Z}$, we have $az \equiv b \pmod{n}$ if and only if $a'z \equiv b' \pmod{n'}$. Moreover, $\gcd(a', n') = 1$, and therefore, if we set $t := (a')^{-1} \bmod n'$ and $z' := tb' \bmod n'$, then the solutions to the congruence $az \equiv b \pmod{n}$ that lie in the interval $\{0, ..., n-1\}$ are the d integers $z_0, z_0 + n', ..., z_0 + (d-1)n'$.

EXERCISE

1. Let p be a prime, and let a, b, c, e be integers, such that $e > 0$, $a \not\equiv 0 \pmod{p^{e+1}}$, and $0 \le c < p^e$. Define N to be the number of integers $z \in \{0, ..., p^{2e} - 1\}$ such that

$$\left\lfloor \left((az+b) \bmod p^{2e}\right) / p^e \right\rfloor = c.$$

Show that $N = p^e$.

2. Let $a_1, ..., a_k, b, n$ be integers with $n > 0$. Show that the congruence

$$a_1 z_1 + ... + a_k z_k \equiv b \pmod{n}$$

has a solution $z_1, ..., z_k \in \mathbb{Z}$ if and only if $d \mid b$, where $d := \gcd(a_1, ..., a_k, n)$.

QUADRATIC RESIDUES

It is natural to study congruences of higher degree as well. In this section, we study a special case of this more general problem, namely, congruences of the form $z^2 \equiv a \pmod{n}$. The theory we develop here nicely illustrates many of the ideas we have discussed earlier, and has a number of interesting applications as well.

We begin with some general, preliminary definitions and general observations about powers in $\mathbb{Z}_n^*$. For each integer m, we define

$$(\mathbb{Z}_n^*)^m := \{\beta^m : \beta \in \mathbb{Z}_n^*\},$$

the set of mth powers in $\mathbb{Z}_n^*$. The set $(\mathbb{Z}_n^*)^m$ is non-empty, as it obviously contains [1]

Theorem: Let n be a positive integer. For each $\alpha \in \mathbb{Z}_n^*$, and all $\ell, m \in \mathbb{Z}$ with

$$\gcd(\ell, m) = 1, \text{ if } \alpha^\ell \in (\mathbb{Z}_n^*)^m, \text{ then } \alpha \in (\mathbb{Z}_n^*)^m.$$

Proof: Suppose $\alpha^\ell = \beta^m \in (\mathbb{Z}_n^*)^m$. Since $\gcd(\ell, m) = 1$, there exist integers s and t such that $\ell s + mt = 1$. We then have

$$\alpha = \alpha^{\ell s+mt} = \alpha^{\ell s}\alpha^{mt} = \beta^{ms}\alpha^{mt} = (\beta^s\alpha^t)^m \in (\mathbb{Z}_n^*)^m.$$

We now focus on the squares in $\mathbb{Z}_n^*$, rather than general powers. An integer a is called a quadratic residue modulo n if $\gcd(a, n) = 1$ and $a \equiv b^2 \pmod{n}$ for some integer b; in this case, we say that b is a square root of a modulo n. In terms of residue classes, a is a quadratic residue modulo n if and only if $[a] \in (\mathbb{Z}_n^*)^2$.

To avoid some annoying technicalities, from now on, we shall consider only the case where n is odd..

Theorem: Let n be a positive integer, let $\alpha \in \mathbb{Z}_n^*$, and let m be any integer.

(i) If $\alpha \in (\mathbb{Z}_n^*)^m$, then a–1 $(\mathbb{Z}_n^*)^m$.

(ii) If $\alpha \in (\mathbb{Z}_n^*)^m$ and $\beta \in (\mathbb{Z}_n^*)^m$, then $\alpha\beta \in (\mathbb{Z}_n^*)^m$.

(iii) If $\alpha \in (\mathbb{Z}_n^*)^m$ and $\beta \notin (\mathbb{Z}_n^*)^m$, then $\alpha\beta \notin (\mathbb{Z}_n^*)^m$.

Proof. For

(i) if $\alpha = \gamma^m$, then $\alpha^{-1} = (\gamma^{-1})^m$.

(ii) if $\alpha = \gamma^m$ and $\beta = \delta^m$, then $\alpha\beta = (\gamma\delta)^m$.

(iii) suppose that $\alpha \in (\mathbb{Z}_n^*)^m$, $\beta = 2 (\mathbb{Z}_n^*)^m$, and $\alpha\beta \in (\mathbb{Z}_n^*)^m$. Then by (i), $\alpha^{-1} \in (\mathbb{Z}_n^*)^m$, and by (ii), $\beta = \alpha^{-1}(\alpha\beta) \in (\mathbb{Z}_n^*)^m$, a contradiction.

QUADRATIC RESIDUES MODULO P^E

We next study quadratic residues modulo p^e, where p is an odd prime.

Theorem: Let p be an odd prime and e be a positive integer. Then we have

$$\prod_{\beta \in \mathbb{Z}^*_{p^e}} \beta = -1 .$$

Theorem: Let p be an odd prime, e be a positive integer, and $\gamma, \beta \in \mathbb{Z}^*_{p^e}$.

Then $\gamma^2 = \beta^2$ if and only if $\gamma = \pm\gamma$.

Theorem: Let p be an odd prime, e be a positive integer, and a be any integer.

Then a is a quadratic residue modulo pe if and only if a is a quadratic residue modulo p.

Proof: Suppose that a is a quadratic residue modulo p^e. Then a is not divisible by p and $a \equiv b^2 \pmod{p^e}$ for some integer b. It follows that $a \equiv b^2 \pmod{p}$, and so a is a quadratic residue modulo p.

Suppose that a is not a quadratic residue modulo p^e. If a is divisible by p, then by definition a is not a quadratic residue modulo p. So suppose a is not divisible by p. By Theorem, we have

$$a^{p^{e-1}(p-1)/2} \equiv -1 \pmod{p^e}.$$

This congruence holds modulo p as well, and by Fermat's little theorem (applied $e-1$ times),

$$a \equiv a^p \equiv a^{p^2} \equiv ... \equiv a^{p^{e-1}} \pmod{p},$$

and so

$$-1 \quad -1 \equiv a^{p^{e-1}(p-1)/2} \equiv a^{(p-1)/2} \pmod{p}.$$

Theorem: Let p be an odd prime and e be a positive integer. Then we have

$$|(\mathbb{Z}^*_{p^e})^2| = \varphi(p^e)/2.$$

Theorem: Let p be an odd prime, e be a positive integer, and $\beta \in \mathbb{Z}_{p^e}$. Then $\beta^2 = 1$ if and only if $\beta = \pm 1$.

Proof: Clearly, if $\beta = \pm 1$, then $\beta^2 = 1$. Conversely, suppose that $\beta^2 = 1$. Write $\beta = [b]$, where $b \in \mathbb{Z}$. Then we have $b^2 \equiv 1 \pmod{p^e}$, which means that

$$p^e \mid (b^2 - 1) = (b-1)(b+1).$$

In particular, $p \mid (b-1)(b+1)$, and so $p \mid (b \mid 1)$ or $p \mid (b+1)$. Moreover, p cannot divide both $b-1$ and $b+1$, as otherwise, it would divide their difference $(b+1)-(b-1)=2$, which is impossible (because p is odd). It follows that $p^e \mid (b-1)$ or $p^e \mid (b+1)$, which means = 1. 2

Previous theorems generalize immediately from $\mathbb{Z}^*_p$ to $\mathbb{Z}^*_{p^e}$: we really used nothing in the proofs of these theorems other than the fact that ± 1 are the only square roots of 1 modulo p. As such, we state the analogs of these theorems for $\mathbb{Z}^*_{p^e}$ without proof.

Theorem: Let p be an odd prime, e be a positive integer, and $\alpha, \beta \in \mathbb{Z}^*_{p^e}$. If

$$a \notin (\mathbb{Z}^*_{p^e})^2 \text{ and } \notin (\mathbb{Z}^*_{p^e})^2, \text{ then } \alpha\beta \in (\mathbb{Z}^*_{p^e})^2.$$

It turns out that an integer is a quadratic residue modulo p^e if and only if it is a quadratic residue modulo p.

Theorem: Let p be an odd prime, e be a positive integer, and $\alpha \in \mathbb{Z}^*_{p^e}$.

(i) $\alpha^{\varphi(p^e)/2} = 1.$

(ii) If $\alpha \in (\mathbb{Z}^*_{p^e})^2$ then $\alpha^{\varphi(p^e)/2} = 1.$

(iii) If $\alpha \notin (\mathbb{Z}^*_{p^e})^2$ then $\alpha^{\varphi(p^e)/2} = -1.$

QUADRATIC RESIDUES MODULO *P*

We first study quadratic residues modulo an odd prime p, and we begin by determining the square roots of 1 modulo p.

Theorem: Let p be an odd prime. Then $|(\mathbb{Z}_p^*)^2| = (p-1)/2$.

Proof: By the previous theorem, the "squaring map" $\sigma : \mathbb{Z}_p^* \to \mathbb{Z}_p^*$ that sends β to β^2 is a two-to-one map: every element in the image of σ has precisely two preimages. As a general principle, if we have a function $f: A \to B$, where A is a finite set and every element in $f(A)$ has exactly d pre-images, then $|f(A)| = |A|/d$. Applying this general principle to our setting, we see that the image of σ is half the size of $\mathbb{Z}_p^*$.

Thus, for every odd prime p, exactly half the elements of $\mathbb{Z}_p^*$ are squares, and half are non-squares. If we choose our representatives for the residue classes modulo p from the interval $[-p/2, p/2)$, we may list the elements of $\mathbb{Z}_p$ as

$$[-(p-1)/2], \ldots, [-1], [0], [1], \ldots, [(p-1)/2].$$

We then see that $\mathbb{Z}_p^*$ consists of the residue classes

$$[\pm 1], \ldots, [\pm(p-1)/2],$$

and so $(\mathbb{Z}_p^*)^2$ consists of the residue classes

$$[1]^2, \ldots, [(p-1)/2]^2,$$

which must be distinct, since we know that $|(\mathbb{Z}_p^*)^2| = (p-1)/2$.

Theorem: Let p be an odd prime and $\gamma, \beta \in \mathbb{Z}_p^*$. Then $\gamma^2 = \beta^2$ if and only if $\gamma = \pm\beta$.

Proof. This follows from the previous theorem:

$$\gamma^2 = \beta^2 \Leftrightarrow (\gamma/\beta)^2 = 1 \Leftrightarrow \gamma/\beta = \pm 1 \Leftrightarrow \gamma = \pm\beta.$$

This theorem says that if $\alpha = \beta^2$ for some $\beta \in \mathbb{Z}_p^*$, then has precisely two square roots: β and $-\beta$.

Theorem: Let p be an odd prime and $\beta \in \mathbb{Z}_p$. Then $\beta^2 = 1$ if and only if $\beta = \pm 1$.

Proof: Clearly, if $\beta = \pm 1$, then $\beta^2 = 1$. Conversely, suppose that $\beta^2 = 1$. Write $= [b]$, where $b \in \mathbb{Z}$. Then we have $b^2 \equiv 1 \pmod{p}$, which means that

$$p \mid (b^2 - 1) = (b-1)(b+1),$$

and since p is prime, we must have $p \mid (b-1)$ or $p \mid (b+1)$. This implies $b \pm 1 \pmod{p}$, or equivalently, $\beta = \pm 1$.

This theorem says that modulo p, the only square roots of 1 are 1 and -1, which obviously belong to distinct residue classes (since $p > 2$). From this seemingly trivial fact, a number of quite interesting and useful results may be derived.

Example. Let $p = 7$. We can list the elements of $\mathbb{Z}_p^*$ as

$$[\pm 1], [\pm 2], [\pm 3].$$

Squaring these, we see that

$$(\mathbb{Z}_p^*)^2 = \{[1]^2, [2]^2, [3]^2\} = \{[1], [4], [2]\}.$$

We next derive an extremely important characterization of quadratic residues.

Theorem: Let p be an odd prime and $\alpha, \in \mathbb{Z}_p^*$. If $\alpha = 2\, (\mathbb{Z}_p^*)^2$ and $\beta = 2\, (\mathbb{Z}_p^*)^2$, then $\alpha\beta \in (\mathbb{Z}_p^*)^2$.

Proof: Suppose $\alpha = 2\, (\mathbb{Z}_p^*)^2$ and $\beta = 2\, (\mathbb{Z}_p^*)^2$. Then by Euler's criterion, we have

$$\alpha^{(p-1)/2} = -1 \text{ and } \beta^{(p-1)/2} = -1.$$

Therefore,

$$(\alpha\beta)^{(p-1)/2} = \alpha^{(p-1)/2} \,.\, \beta^{(p-1)/2} = [-1] \,.\, [-1] = 1,$$

which again by Euler's criterion implies that $\alpha\beta \in (\mathbb{Z}_p^*)^2$.

This theorem, together with parts (ii) and (iii) of Theorem, gives us the following simple rules regarding squares in $\mathbb{Z}_p^*$:

square × square = square,

square × non-square = non-square,

non-square × non-square = square.

Euler's Criterion

Theorem: Let p be an odd prime and $\alpha \in \mathbb{Z}_p^*$.

(i) $\alpha^{(p-1)/2} = \pm 1$.

(ii) If $\in (\mathbb{Z}_p^*)^2$ then $\alpha^{(p-1)/2} = 1$.

(iii) If $\alpha \notin (\mathbb{Z}^*_p)^2$ then $\alpha^{(p-1)/2} = -1$.

Proof: For

(i) let $\gamma = \alpha^{(p-1)/2}$. By Euler's theorem, we have

$$\gamma^2 = \alpha^{p-1} = 1,$$

and hence by Theorem, we have $\gamma = \pm 1$.

(ii) suppose that $\alpha = \beta^2$. Then again by Euler's theorem, we have

$$\alpha^{(p-1)/2} = (\beta^2)^{(p-1)/2} = \beta^{p-1} = 1.$$

(iii) let $\alpha \in \mathbb{Z}^*_p \setminus (\mathbb{Z}^*_p)^2$. We study the product

$$\varepsilon := \prod_{\beta \in \mathbb{Z}^*_p} \beta.$$

We shall show that, on the one hand, $\varepsilon = \alpha^{(p-1)/2}$, while on the other hand, $\varepsilon = -1$.

To show that $\varepsilon = \alpha^{(p-1)/2}$, we group elements of $\mathbb{Z}^*_p$ into pairs of distinct elements whose product is α. More precisely, let $P := \{S \subseteq \mathbb{Z}p : |S| = 2\}$, and define $C := \{\{\kappa, \lambda\} \in P : \kappa\lambda = \alpha\}$. Note that for every $\kappa \in \mathbb{Z}^*_p$, there is a unique $\kappa \in \mathbb{Z}^*_p$ such that $\kappa\lambda = \alpha$, namely, $\alpha := \alpha/\kappa$; moreover, $\kappa \equiv \lambda$, since otherwise, we would have $\kappa^2 = \alpha$, contradicting the assumption that $\alpha \notin (\mathbb{Z}^*_p)^2$. Thus, every element of $\mathbb{Z}^*_p$ belongs to exactly one pair in C; in other words, the elements of C form a partition of $\mathbb{Z}^*_p$. It follows that

$$\varepsilon := \prod_{\{\kappa,\lambda\} \in C} (\kappa.\lambda) = \prod_{\{\kappa,\lambda\} \in C} \alpha = \alpha^{(p-1)/2}.$$

To show that $\varepsilon = -1$, we group elements of $\mathbb{Z}^*_p$ into pairs of distinct elements whose product is [1]. Define $D := \{\{\kappa, \lambda\}\} \in P : \kappa\lambda = 1\}$. For every $\kappa \in \mathbb{Z}^*_p$, there exists a unique $\kappa \in \mathbb{Z}^*_p$ such that $\kappa\lambda = 1$, namely, $\lambda := \kappa^{-1}$; moreover, $\kappa = \lambda$ if and only if $\kappa^2 = 1$, and by Theorem, this happens if and only if $\kappa = \pm 1$. Thus, every element of $\mathbb{Z}^*_p$ except for $[\pm 1]$ belongs to exactly one pair in D; in other words, the elements of D form a partition of $\mathbb{Z}^*_p \setminus \{[\pm 1]\}$. It follows that

$$\varepsilon := [1].[-1]. \prod_{\{\kappa,\lambda\} \in D} (\kappa.\lambda) = [-1]. \prod_{\{\kappa,\lambda\} \in D} [1] = -1.$$

Thus, Euler's criterion says that for every $\alpha \in \mathbb{Z}^*_p$, we have $\alpha^{(p-1)/2} = \pm 1$ and

$$\alpha \in (\mathbb{Z}p)^2 \Leftrightarrow \alpha^{(p-1)/2} = 1.$$

Wilson's theorem

Theorem: Let p be an odd prime. Then $\prod_{\beta \in \mathbb{Z}_p^*} \beta = -1$.

In the language of congruences, Wilson's theorem may be stated as follows:

$$(p-1)! \equiv -1 \pmod{p}.$$

SQUARE ROOTS OF –1 MODULO *P*

Using Euler's criterion, we can easily characterize those primes modulo which –1 is a quadratic residue. This turns out to have a number of nice applications. Consider an odd prime p. The following theorem says that the question of whether –1 is a quadratic residue modulo p is decided by the residue class of p modulo 4. Since p is odd, either $p \equiv 1 \pmod 4$ or $p \equiv 3 \pmod 4$.

Theorem: There are infinitely many primes $p \equiv 1 \pmod 4$.

Proof: Suppose there were only finitely many such primes, $p_1, ..., p_k$. Set $M := \prod_{i=1}^{k} p_i$ and $N := 4M^2 + 1$. Let p be any prime dividing N. Evidently, p is not among the p_i's, since if it were, it would divide both N and $4M^2$, and so also $N - 4M^2 = 1$. Also, p is clearly odd, since N is odd. Moreover, $(2M)^2 \equiv -1 \pmod p$; therefore, – 1 is a quadratic residue modulo p, and so $p \equiv 1 \pmod 4$, contradicting the assumption that $p_1, ..., p_k$ are the only such primes.

Theorem: Let p be an odd prime. Then –1 is a quadratic residue modulo p if and only $p \equiv 1 \pmod 4$.

Proof. By Euler's criterion, –1 is a quadratic residue modulo p if and only if $(-1)^{(p-1)/2} \equiv 1 \pmod p$. If $p \equiv 1 \pmod 4$, then $(p-1)/2$ is even, and so $(-1)^{(p-1)/2} = 1$. If $p \equiv 3 \pmod 4$, then $(p-1)/2$ is odd, and so $(-1)^{(p-1)/2} = -1$.

Theorem: There are infinitely many primes $p \equiv 3 \pmod 4$.

Proof: Suppose there were only finitely many such primes, $p_1, ..., p_k$. Set $M := \prod_{i=1}^{k} p_i$ and $N := 4M - 1$. Since $N \equiv 3 \pmod 4$, there must be some prime $p \equiv 3 \pmod 4$ dividing N (if all primes dividing N were congruent to 1 modulo 4, then so too would be their product N). Evidently, p is not among the p_i's, since if it were, it would divide both N and $4M$, and so also $4M - N = 1$. This contradicts the assumption that $p_1, ..., p_k$ are the only primes congruent to 3 modulo 4.

Theorem: Let p be a prime with $p \equiv 1 \pmod 4$, $\gamma \in \mathbb{Z}_p^* \setminus (\mathbb{Z}_p^*)^2$, and $\beta := \gamma^{(p-1)/4}$. Then $\beta^2 = -1$.

Proof. This is a simple calculation, based on Euler's criterion:

$$\beta^2 = \gamma^{(p-1)/2} = -1.$$

The fact that –1 is a quadratic residue modulo primes $p \equiv 1 \pmod 4$ can be used to prove Fermat's theorem that such primes may be written as the sum of two squares.

Fermat's two squares theorem

Theorem: Let p be an odd prime. Then

$$p = r^2 + t^2 \text{ for some } r, t \in \mathbb{Z} \text{ if and only if } p \equiv 1 \pmod 4.$$

Proof: One direction is easy. Suppose $p \equiv 3 \pmod 4$. It is easy to see that the square of every integer is congruent to either 0 or 1 modulo 4; therefore, the sum of two squares is congruent to either 0, 1, or $\in$ modulo 4, and so can not be congruent to p modulo 4 (let alone equal to p).

For the other direction, suppose $p \equiv 1 \pmod 4$. We know that -1 is a quadratic residue modulo p, so let b be an integer such that $b^2 \equiv -1 \pmod p$. Now apply Theorem with $n := p$, b as just defined, and $r^* := t^* := \lfloor\sqrt{p}\rfloor + 1$. Evidently, $\lfloor\sqrt{p}\rfloor + 1 > \sqrt{p}$, and hence $r^*t^* > p$. Also, since p is prime, $\sqrt{p}$ is not an integer, and so $\lfloor\sqrt{p}\rfloor < \sqrt{p} < p$; in particular, $r^* = \lfloor\sqrt{p}\rfloor + 1 \leq p$. Thus, the hypotheses of that theorem are satisfied, and therefore, there exist integers r and t such that

$$r \equiv bt \pmod p, |r| \leq \lfloor\sqrt{p}\rfloor < \sqrt{p}, \text{ and } 0 < |t| — \lfloor\sqrt{p}\rfloor < \sqrt{p}.$$

It follows that

$$r^2 \equiv b^2t^2 \equiv -t^2 \pmod p.$$

Thus, $r^2 + t^2$ is a multiple of p and $0 < r^2 + t^2 < 2p$. The only possibility is that $r^2 + t^2 = p$.

The fact that -1 is a quadratic residue modulo an odd prime p only if $p \equiv 1 \pmod 4$ can be used so show there are infinitely many such primes.

Thue's lemma

Theorem: Let $n, b, r^*, t^* \in \mathbb{Z}$, with $0 < r^* \leq n < r^*t^*$. Then there exist $r, t \in \mathbb{Z}$ with

$$r \equiv bt \pmod n, |r| < r^*, \text{ and } 0 < |t| < t^*.$$

Proof: For $i = 0, ..., r^* - 1$ and $j = 0, ..., t^* - 1$, we define the number $v_{ij} := i - b_j$. Since we have defined r^*t^* numbers, and $r^*t^* > n$, two of these numbers must lie in the same residue class modulo n; that is, for some $(i_1, j_1) \neq (i_2, j_2)$, we have $v_{i_1} j_i \equiv v_{i_2 j_2} \pmod n$. Setting $r \neq i_1 - i_2$ and $t := j_1 - j_2$, this implies $r \equiv bt \pmod n$, $|r| < r^*$, $|t| < t^*$, and that either $r \neq 0$ or $t \neq 0$. It only remains to show that $t \neq 0$.

Suppose to the contrary that $t = 0$. This would imply that $r \equiv 0 \pmod n$ and $r \neq 0$, which is to say that r is a non-zero multiple of n; however, this is impossible, since $|r| < r^* \leq n$.

EXERCISE

1. Let p be a prime with $p \equiv 1 \pmod 4$, and $b := ((p - 1)=2)!$. Show that $b^2 \equiv -1 \pmod p$.
2. Suppose p is an odd prime, and that $r^2 + t^2 = p$ for some integers r, t. Show that if x, y are integers such that $x^2 + y^2 = p$, then (x, y) must be $(\pm r, \pm t)$ or $(\pm t, \pm r)$.

3. Let $n := pq$, where p and q are distinct, odd primes. Show that there exist, $\alpha, \beta \in \mathbb{Z}_n^*$ such that $\alpha \notin (\mathbb{Z}_n^*)^2$, $\beta \notin (\mathbb{Z}_n^*)^2$, and $\alpha\beta \notin (\mathbb{Z}_n^*)^2$.
4. Show that if both u and v are the sum of two squares of integers, then so is their product uv.
5. Let n be an odd positive integer, and let a be any integer. Show that a is a quadratic residue modulo n if and only if a is a quadratic residue modulo p for each prime $p \mid n$.
6. Let $n, m \in \mathbb{Z}$, where $n > 0$, and let $d := \gcd(m, \varphi(n))$. Show that:

   (a) if $d = 1$, then $(\mathbb{Z}_n^*)^m = (\mathbb{Z}_n^*)$;

   (b) if $\alpha \in (\mathbb{Z}_n^*)^m$, then $\alpha^{\varphi(n)/d} = 1$.
7. Calculate the sets C and D in the proof of Theorem in the case $p = 11$ and $\alpha = -1$.
8. Calculate the square roots of 1 modulo 4, 8, and 16.
9. Show that if p is an odd prime, with $p \equiv 3 \pmod 4$, then $(\mathbb{Z}_p^*)^4 = (\mathbb{Z}_p^*)^2$. More generally, show that if n is an odd positive integer, where $p \equiv 3 \pmod 4$ for each prime $p \mid n$, then $(\mathbb{Z}_n^*)^4 = (\mathbb{Z}_n^*)^2$.
10. Suppose $r^2 + t^2 \equiv 0 \pmod n$, where n is a positive integer, and suppose p is an odd prime dividing n. Show that:

    (a) if p divides neither r nor t, then $p \equiv 1 \pmod 4$;

    (b) if p divides one of r or t, then it divides the other, and moreover, p^2 divides n, and $(r/p)^2 + (t/p)^2 \equiv 0 \pmod{n/p^2}$.
11. Let $n \in \mathbb{Z}$ with $n > 1$. Show that n is prime if and only if $(n-1)! \equiv -1 \pmod n$.
12. Let p be an odd prime, and let $e \in \mathbb{Z}$ with $e > 1$. Let a be an integer of the form $a = p^f b$, where $0 \leq f < e$ and $p \mid b$. Consider the integer solutions z to the congruence $z^2 \equiv a \pmod{p^e}$. Show that a solution exists if and only if f is even and b is a quadratic residue modulo p, in which case there are exactly $2p^f$ distinct solutions modulo p^e.
13. Let n be a positive integer, and write $n = ab^2$ where a and b are positive integers, and a is square-free. Show that n is the sum of two squares of integers if and only if no prime $p \equiv 3 \pmod 4$ divides a.

QUADRATIC RESIDUES MODULO *N*

We now study quadratic residues modulo n, where n is an arbitrary, odd integer, with $n > 1$. Let

$$n = p_1^{e_1} \ldots p_r^{e_r}$$

be the prime factorization of n. Our main tools here are the Chinese remainder map

$$\theta : \mathbb{Z}_n \to \mathbb{Z}_{p_1^{e_1}} \times \dots \times \mathbb{Z}_{p_r^{e_r}},$$

introduced in Theorem together with the results developed so far for quadratic residues modulo odd prime powers.

Let $\alpha \in \mathbb{Z}_n^*$ with $\theta(\alpha) = (\alpha_1, \dots, \alpha_r)$.

- On the one hand, suppose $\alpha = \beta^2$ for some $\beta \in \mathbb{Z}_n^*$. If $\theta(\beta) = (\beta_1, \dots, \beta_r)$, we have

$$(\alpha_1, \dots, \alpha_r) = \theta(\alpha) = \theta(\beta^2) = (\beta_1^2, \dots, \beta_r^2),$$

  where we have used part (iii.c) of Theorem. It follows that $\alpha_i = \beta_i^2$ for each i.

- On the other hand, suppose that for each i, $\alpha_i = \beta_i^2$ for some $\beta_i \in \mathbb{Z}^*_{p_i^{e_i}}$. Then setting $\beta := \theta^{-1}(\beta_1, \dots, \beta_r)$, we have

$$\theta(\beta^2) = (\beta_1^2, \dots, \beta_r^2) = (\alpha_1, \dots \alpha_r) = \theta(\alpha),$$

  where we have again used part (iii.c) of Theorem, along with the fact that θ is bijective (to define β). Thus, $\theta(\alpha) = \theta(\beta^2)$, and again since θ is bijective, it follows that $\alpha = \beta^2$.

  We have shown that

$$\alpha \in (\mathbb{Z}_n^*)^2 \Leftrightarrow \alpha_i \in -(\mathbb{Z}^*_{p_i^{e_i}})^2 \text{ for } i = 1, \dots, r.$$

  In particular, restricting θ to $(\mathbb{Z}_n^*)^2$ yields a one-to-one correspondence between $(\mathbb{Z}_n^*)^2$ and

$$\left(\mathbb{Z}^*_{p_1^{e_1}}\right)^2 \times \dots \times \left(\mathbb{Z}^*_{p_r^{e_r}}\right)^2,$$

  and therefore, by Theorem, we have

$$|(\mathbb{Z}_n^*)^2| = \prod_{i=1}^{r} (\varphi(p_i^{e_i})/2) = \varphi(n)/2^r.$$

  Now suppose that $\alpha = \beta^2$, with $\beta \in \mathbb{Z}_n^*$ and $\theta(\beta) = (\beta_1, \dots, \beta_r)$. Consider an arbitrary element $\gamma \in \mathbb{Z}_n^*$, with $\theta(\gamma) = (\gamma_1, \dots, \gamma_r)$. Then we have

$$\gamma^2 = \beta^2 \Leftrightarrow \theta(\gamma^2) = \theta(\beta^2)$$

$$\Leftrightarrow \quad (\gamma_1^2, \dots, \gamma_r^2) = (\beta_1^2, \dots, \beta_r^2)$$

$$\Leftrightarrow \quad (\gamma_1, \dots, \gamma_r) = (\pm\beta_1, \dots, \pm\beta_r).$$

Therefore, α has precisely 2^r square roots, namely, $\theta^{-1}(\pm\beta_1, \dots, \pm\beta_r)$.

SUMMATIONS OVER DIVISORS

We close this chapter with a brief treatment of summations over divisors. To this end, we introduce some terminology and notation. By an arithmetic function, we simply mean a function from the positive integers into the reals (actually, one usually considers complex-valued functions as well, but we shall not do so here).

Let f and g be arithmetic functions. The Dirichlet product of f and g, denoted $f ?\ g$, is the arithmetic function whose value at n is defined by the formula

$$(f * g)(n) := \sum_{d|n} f(d)g(n/d),$$

the sum being over all positive divisors d of n. Another, more symmetric, way to write this is

$$(f * g)(n) = \sum_{n=d_1d_2} f(d_1)g(d_2),$$

the sum being over all pairs (d_1, d_2) of positive integers with $d_1d_2 = n$.

The Dirichlet product is clearly commutative (i.e., $f * g = g * f$), and is associative as well, which one can see by checking that

$$(f * (g * h))(n) = \sum_{n=d_1d_2d_3} f(d_1)g(d_2)h(d_3) = ((f * g) * h)(n),$$

the sum being over all triples (d_1, d_2, d_3) of positive integers with $d_1d_2d_3 = n$.

We now introduce three special arithmetic functions: I, 1, and μ. The functions I and 1 are defined as follows:

$$I(n) := \begin{cases} 1 & \text{if } n = 1; \\ 0 & \text{if } n > 1; \end{cases} \quad 1\,(n) := 1.$$

The Möbius function μ is defined as follows: if $n = p_1^{e_1} \ldots p_r^{e_r}$ is the prime factorization of n, then

$$\mu(n) := \begin{cases} 0 & \text{if } e_i > 1 \text{ for } i = 1, \ldots, r; \\ (-1)^r & \text{otherwise.} \end{cases}$$

In other words, $\mu(n) = 0$ if n is not square-free; otherwise, $\mu(n)$ is $(-1)^r$ where r is the number of distinct primes dividing n. Here are some examples:

$$\mu(1) = 1,\ \mu(2) = -1,\ \mu(3) = -1,\ \mu(4) = 0,\ \mu(5) = -1,\ \mu(6) = 1.$$

It is easy to see from the definitions that for every arithmetic function f, we have

$$I * f = f \text{ and } (1 * f)(n) = \sum_{d|n} f(d).$$

Thus, I acts as a multiplicative identity with respect to the Dirichlet product, while "1 *" acts as a "summation over divisors" operator.

An arithmetic function f is called multiplicative if $f(1) = 1$ and for all positive integers n, m with $\gcd(n, m) = 1$, we have $f(nm) = f(n) f(m)$.

The reader may easily verify that I, 1, and μ are multiplicative functions. Theorem says that Euler's function φ is multiplicative. The reader may also verify the following:

Theorem: For every positive integer n, we have $\sum_{d|n} \varphi(d) = n$.

Proof: Let us define the arithmetic functions $N(n) := n$ and $M(n) := 1/n$. Our goal is to show that $N = 1 * \varphi$, and by Möbius inversion, it succees to show that

$\varphi * N = \varphi$. If $n = p_1^{e_1} \ldots p_r^{e_r}$ is the prime factorization of n, we have

$$(\mu * N)(n) = \sum_{d|n} \mu(d)(n/d) = n \sum_{d|n} \mu(d)/d$$

$$= n \prod_{i=1}^{r} (1 - 1/p_i) \text{ (applying with } f := M)$$

$$= \varphi(n).$$

Theorem: If f is a multiplicative arithmetic function, and if $n = p_1^{e_1} \ldots p_r^{e_r}$ is the prime factorization of n, then $f(n) = f(p_1^{e_1}) \ldots (p_r^{e_r})$.

Möbius inversion formula

Theorem: Let f and F be arithmetic functions.

Then $F = 1 * f$ if and only if $f = \mu * F$.

Proof: If $F = 1\ \mu f$, then

$$\mu * F = \mu * (1 * f) = (\mu * 1) * f = I * f = f,$$

and conversely, if $f = \mu * F$, then

$$1 * f = 1 * (\mu * F) = (1 * \mu) * F = I * F = F.$$

The Möbius inversion formula says this:

$$F(n) = \sum_{d|n} f(d) \text{ for all positive integers } n$$

$$\Leftrightarrow f(n) = \sum_{d|n} \mu(d) F(n/d) \text{ for all positive integers } n.$$

Theorem: Let f be a multiplicative arithmetic function. If $n = p_1^{e_1} \ldots p_r^{e_r}$ is the prime factorization of n, then

$$\sum_{d|n} \mu(d) f(d) = (1 - f(p_1)) \ldots (1 - f(p_r)).$$

Proof: The only non-zero terms appearing in the sum on the left-hand side of are those corresponding to divisors d of the form $p_{i_1} \ldots p_{i_\ell}$, where $p_{i_1} \ldots p_{i_\ell}$ are distinct; the value contributed to the sum by such a term is $(-1)^\ell f(p_{i_1} \ldots p_{i_\ell}) = (-1)^\ell f(p_{i_1}) \ldots f(p_{i_\ell})$. These are the same as the terms in the expansion of the product on the right-hand side of .

If we set $f := 1$ in the previous theorem, then we see that

$$\sum_{d|n} \mu(d) = \begin{cases} 1 & \text{if } n = 1; \\ 0 & \text{if } n > 1. \end{cases}$$

Translating this into the language of Dirichlet products, we have

$I * \text{m} = I.$

EXERCISE

1. Show that n is square-free if and only if $\sum_{d|n} \mu(d)^2 \varphi(d) = n$.
2. Show that if f and g are multiplicative, then so is $f * g$.
3. Show that for every arithmetic function f with $f(1) \neq 0$, there is a unique arithmetic function g, called the Dirichlet inverse of f, such that $f * g = I$. Also, show that if $f(1) = 0$, then f has no Dirichlet inverse.
4. In our definition of a multiplicative function f, we made the requirement that $f(1) = 1$. Show that if we dropped this requirement, the only other function that would satisfy the definition would be the zero function (i.e., the function that is everywhere zero).
5. Let $\tau(n)$ be the number of positive divisors of n. Show that:

   (a) τ is a multiplicative function;

   (b) $\tau(n) = \prod_{i=1}^{r} (e_i + 1),$, where $n = p_1^{e_1} \ldots p_r^{e_r}$ is the prime factorization of n;

   (c) $\sum_{d|n} \mu(d)\tau(n/d) = 1$;

   (d) $\sum_{d|n} \mu(d)\tau(d) = (-1)^r$, where $n = p_1^{e_1} \ldots p_r^{e_r}$ is the prime factorization of n.
6. Let f be a polynomial with integer coefficients, and for each positive integer n, define $\omega_f(n)$ to be the number of integers $x \in \{0, \ldots, n-1\}$ such that $f(x) \equiv 0 \pmod{n}$. Show that φ_f is multiplicative.
7. Show that if f is multiplicative, and if $n = p_1^{e_1} \ldots p_r^{e_r}$ is the prime factorization of n, then

   $\sum_{d|n} \mu(d)^2 f(d) = (1 + f(p_1)) \ldots (1 + f(p_r))$.

8. Show that if f is a multiplicative function, then so is its Dirichlet inverse (as defined in the previous exercise).
9. The Mangoldt function $\Lambda(n)$ is defined for all positive integers n as follows: $\Lambda(n) := \log p$, if $n = p^k$ for some prime p and positive integer k, and $\Lambda(n) := 0$, otherwise. Show that $\sum_{d|n} \Lambda(d)(d) = \log n$, and from this, deduce that $\Lambda(n) = -\sum_{d|n} \mu(d) \log d$.
10. Define $\sigma(n) = \sum_{d|n} d$. Show that:

    (a) σ is a multiplicative function;

    (b) $\sigma(n) = \sum_{i=1}^{r} \left(p_i^{e_i+1} - 1\right)/(p_i - 1)$, where $n = p_1^{e_1} ... p_r^{e_r}$ is the prime factorization of n;

    (c) $\sum_{d|n} \mu(d)\sigma(n/d) = n$;

    (d) $\sum_{d|n} \mu(d)\sigma(d) = (-1)^r \, p_1 \, ... \, p_r$, where $n = p_1^{e_1} ... p_r^{e_r}$ is the prime factorization of n.

CLASSES

We have already see that, for any fixed positive integer n, the binary relation "$\cdot \equiv \cdot \pmod{n}$" is an equivalence relation on the set $\mathbb{Z}$. As such, this relation partitions the set $\mathbb{Z}$ into equivalence classes. We denote the equivalence class containing the integer a by $[a]_n$, and when n is clear from context, we simply write $[a]$. By definition, we have

$$z \in [a] \Leftrightarrow z \equiv a \pmod{n} \Leftrightarrow z = a + ny \text{ for some } y \in \mathbb{Z},$$

and hence

$$[a] = a + n\mathbb{Z} := \{a + ny : y \in \mathbb{Z}\}.$$

These equivalence classes are called residue classes modulo n, and we shall adopt this terminology here as well. Note that a given residue class modulo n has many different "names"; for example, the residue class $[n - 1]$ is the same as the residue class $[-1]$. Any member of a residue class is called a representative of that class.

Theorem: Let us consider n be a positive integer. Then $\mathbb{Z}_n$ consists of the n distinct residue classes $[0], [1], ..., [n - 1]$. Moreover, for every $x \in \mathbb{R}$, each residue class modulo n contains a unique representative in the interval $[x, x + n)$.

When working with residue classes modulo n, one often has in mind a particular set of representatives. Typically, one works with the set of representatives $\{0, 1, ..., n - 1\}$. However, sometimes it is convenient to work with another set of representatives, such as the representatives in the interval $[-n/2, n/2)$. In this case, if n is odd, we can list the elements of $\mathbb{Z}_n$ as

$$[-(n-1)/2], ..., [-1], [0], [1], ..., [(n-1)/2],$$

and when n is even, we can list the elements of $\mathbb{Z}_n$ as

$$[-n/2], ..., [-1], [0], [1], ..., [n/2-1].$$

We can "equip" $\mathbb{Z}_n$ with binary operations defining addition and multiplication in a natural way as follows: for $a, b \in \mathbb{Z}$, we define

$$[a] + [b] := [a + b],$$

$$[a] . [b] := [a \ b].$$

Of course, one has to check that this definition is unambiguous, in the sense that the sum or product of two residue classes should not depend on which particular representatives of the classes are chosen in the above definitions. More precisely, one must check that if $[a] = [a_0]$ and $[b] = [b']$, then $[a + b] = [a' + b']$ and $[a . b] = [a' . b']$. However, this property follows immediately from Theorem.

Observe that for all $a, b, c \in \mathbb{Z}$, we have

$$[a] + [b] = [c] \Leftrightarrow a + b \equiv c \ (\text{mod } n),$$

and

$$[a] . [b] = [c] \Leftrightarrow a . b \equiv c \ (\text{mod } n),$$

Example. Consider about the residue classes modulo 6. These are as follows:

$$[0] = \{..., -12, -6, 0, 6, 12, ...\}$$

$$[1] = \{..., -11, -5, 1, 7, 13, ...\}$$

$$[2] = \{..., -10, -4, 2, 8, 14, ...\}$$

$$[3] = \{..., -9, -3, 3, 9, 15, ...\}$$

$$[4] = \{..., -8, -2, 4, 10, 16, ...\}$$

$$[5] = \{..., -7, -1, 5, 11, 17, ...\}.$$

If we write down the addition and multiplication tables for $\mathbb{Z}_6$. The addition table looks like this:

+	[0]	[1]	[2]	[3]	[4]	[5]
[0]	[0]	[1]	[2]	[3]	[4]	[5]
[1]	[1]	[2]	[3]	[4]	[5]	[0]
[2]	[2]	[3]	[4]	[5]	[0]	[1]
[3]	[3]	[4]	[5]	[0]	[1]	[2]
[4]	[4]	[5]	[0]	[1]	[2]	[3]
[5]	[5]	[0]	[1]	[2]	[3]	[4]

The multiplication table looks like this:

.	[0]	[1]	[2]	[3]	[4]	[5]
[0]	[0]	[0]	[0]	[0]	[0]	[0]
[1]	[0]	[1]	[2]	[3]	[4]	[5]
[2]	[0]	[2]	[4]	[0]	[2]	[4]
[3]	[0]	[3]	[0]	[3]	[0]	[3]
[4]	[0]	[4]	[2]	[0]	[4]	[2]
[5]	[0]	[5]	[4]	[3]	[2]	[1].

Instead of using representatives in the interval [0, 6), we could just as well use representatives from another interval, such as [–3, 3). Then, instead of naming the residue classes [0], [1], [2], [3], [4], [5], we would name them [–3], [–2], [–1], [0], [1], [2]. Observe that [–3] = [3], [–2] = [4], and [–1] = [5].

These addition and multiplication operations on $\mathbb{Z}_n$ yield a very natural algebraic structure. For example, addition and multiplication are commutative and associative; that is, for all $\alpha, \beta, \gamma \in \mathbb{Z}_n$, we have

$$\alpha + \beta = \beta + \alpha, (\alpha + \beta) + \gamma = \alpha + (\beta + \gamma),$$

$$\alpha\beta = \beta\alpha, (\alpha\beta)\gamma = (\beta\gamma).$$

Note that we have adopted here the usual convention of writing $\alpha\beta$ in place of $\alpha . \beta$ Furthermore, multiplication distributes over addition; that is, for all $\alpha, \beta, \gamma \in \mathbb{Z}_n$, we have

$$\alpha(\beta + \gamma) = \alpha\beta + \alpha\gamma.$$

All of these properties follow from the definitions, and the corresponding properties for $\mathbb{Z}$; for example, the fact that addition in $\mathbb{Z}_n$ is commutative may be seen as follows: if = $[a]$ and = $[b]$, then

$$\alpha + \beta = [a] + [b] = [a + b] = [b + a] = [b] + [a] = \beta + \alpha.$$

Because addition and multiplication in Z_n are associative, for $\alpha_1, ..., \alpha_k \in \mathbb{Z}_n$, we may write the sum $\alpha_1 + ... + \alpha_k$ and the product $\alpha_1 ... \alpha_k$ without any parenthesis, and there is no ambiguity; moreover, since both addition and multiplication are commutative, we may rearrange the terms in such sums and products without changing their values. The residue class [0] acts as an additive identity; that is, for all $\alpha \in \mathbb{Z}_n$, we have $\alpha + [0] = \alpha$; indeed, if $\alpha = [a]$, then $a + 0 \equiv a \pmod{n}$.

Moreover, [0] is the only element of $\mathbb{Z}_n$ that acts as an additive identity; indeed, if $a + z\ a \pmod{n}$ holds for all integers a, then it holds in particular for $a = 0$, which implies $z \equiv 0 \pmod{n}$. The residue class [0] also has the property that $\alpha.[0] = [0]$ for all $\alpha \in \mathbb{Z}_n$. Every $\alpha \in \mathbb{Z}_n$ has an additive inverse, that is, an element $\beta \in \mathbb{Z}_n$ such that $\alpha + \beta = [0]$; indeed, if $\alpha = [a]$, then clearly $\beta := [-a]$ does the job, since $a + (-a) \equiv 0 \pmod{n}$. Moreover, α has a unique additive inverse; indeed, if $a + z \equiv 0 \pmod{n}$, then subtracting a from both sides of this congruence yields $z \equiv -a \pmod{n}$.

We naturally denote the additive inverse of by $-\alpha$. Observe that the additive inverse of $-\alpha$ is α; that is $-(-\alpha) = \alpha$. Also, we have the identities

$$-(\alpha + \beta) = (-\alpha) + (-\beta),\ (-\alpha)\beta = -(\alpha\beta) = \alpha(-\beta),\ (-\alpha)(-\beta) = \alpha\beta.$$

For, $\alpha, \beta \in \mathbb{Z}_n$, we naturally write $\alpha - \beta$ for $\alpha + (-\beta)$.

The residue class [1] acts as a multiplicative identity; that is, for all $\alpha \in \mathbb{Z}_n$, we have $\alpha.[1] = \alpha$; indeed, if $\alpha = [a]$, then $a.1 \equiv a \pmod{n}$. Moreover, [1] is the only element of $\mathbb{Z}_n$ that acts as a multiplicative identity; indeed, if $a.z \equiv a \pmod{n}$ holds for all integers a, then in particular, it holds for $a = 1$, which implies $z \equiv 1 \pmod{n}$. For $\alpha \in \mathbb{Z}_n$, we call $\beta \in \mathbb{Z}_n$ a multiplicative inverse of α if $\alpha\beta = [1]$. Not all $\alpha \in \mathbb{Z}_n$ have multiplicative inverses. If $\alpha = [a]$ and $\beta = [b]$, then is a multiplicative inverse of if and only if $ab \equiv 1 \pmod{n}$. Theorem implies that α has a multiplicative inverse if and only if $\gcd(a, n) = 1$, and that if it exists, it is unique. When it exists, we denote the multiplicative inverse of α by α^{-1}. We define $\mathbb{Z}_n^*$ to be the set of elements of $\mathbb{Z}_n$ that have a multiplicative inverse. By the above discussion, we have

$$\mathbb{Z}_n^* = \{[a] : a = 0, ..., n-1, \gcd(a, n) = 1\}.$$

If n is prime, then $\gcd(a, n) = 1$ for $a = 1, ..., n - 1$, and we see that $\mathbb{Z}_n^* = Z_n \setminus \{[0]\}$.

If n is composite, then $\mathbb{Z}_n^* \subseteq \mathbb{Z}_n \setminus \{[0]\}$; for example, if $d \mid n$ with $1 < d < n$, we see that $[d]$ is not zero, nor does it belong to $\mathbb{Z}_n^*$. Observe that if $\alpha, \beta \in \mathbb{Z}_n^*$, then so are α^{-1} and $\alpha\beta$; indeed,

$$(\alpha^{-1})^{-1} = \alpha \text{ and } (\alpha\beta)^{-1} = \alpha^{-1}\beta^{-1}.$$

For $\alpha \in \mathbb{Z}_n$ and $\beta \in \mathbb{Z}_n^*$, we naturally write α/β for $\alpha\beta^{-1}$.

Suppose α, β, γ are elements of $\mathbb{Z}_n$ that satisfy the equation

$$\alpha\beta = \alpha\gamma.$$

If $\alpha \in \mathbb{Z}_n^*$, we may multiply both sides of this equation by α^{-1} to infer that

$$\beta = \gamma.$$

This is the cancellation law for Z_n. We stress the requirement that $\alpha \in \mathbb{Z}_n^*$, and not just $\alpha \neq [0]$. Indeed, consider any $\alpha \in \mathbb{Z}_n \setminus \mathbb{Z}_n^*$. Then we have $\alpha = [a]$ with $d := \gcd(a, n) > 1$. Setting $\beta := [n/d]$ and $\gamma := [0]$, we see that

$$\alpha\beta = \alpha\gamma \text{ and } \beta \neq \gamma.$$

Example. We list the elements of $\mathbb{Z}_{15}^*$, and for each $\alpha \in \mathbb{Z}_{15}^*$, we also give α^{-1}:

α	[1]	[2]	[4]	[7]	[8]	[11]	[13]	[14]
α^{-1}	[1]	[8]	[4]	[13]	[2]	[11]	[7]	[14]

For $\alpha_1, ..., \alpha_k \in \mathbb{Z}n$, we may naturally write their sum as $\sum_{i=1}^k \alpha_i$. By convention, this sum is [0] when $k = 0$. It is easy to see that $-\sum_{i=1}^k \alpha_i = \sum_{i=1}^k (-\alpha_i)$; that is, the additive inverse of the sum is the sum of the additive inverses. In the special case where all the α_i's have the same value α, we define $k\,.\,\alpha := \sum_{i=1}^k \alpha$; thus, $0\,.\,\alpha = [0]$, $1.\alpha = \alpha$, $\in\,.\,\alpha = \alpha + \alpha$, $3\,.\,\alpha = \alpha + \alpha + \alpha$, and so on. The additive inverse of $k\,.\,\alpha$ is $k(-\alpha)$, which we may also write as $(-k).\alpha$; thus, $(-1)\,.\,\alpha = -\alpha$, $(-2)\,.\,\alpha = (-\alpha) + (-\alpha) = -(\alpha + \alpha)$, and so on. Therefore, the notation $k\,.\,\alpha$, or more simply, k, is defined for all integers k. Note that for all integers k and a, we have $k[a] = [ka] = [k][a]$.

For all $\alpha, \beta \in \mathbb{Z}_n$ and $k, l \in \mathbb{Z}$, we have the identities:

$$k(\alpha) = (k\ell)\alpha = \ell(k\alpha),\ (k + \ell)\alpha = k\alpha + \ell\alpha,\ k(\alpha + \beta) = k\alpha + k\beta,$$
$$(k\alpha)\beta = k(\alpha\beta) = \alpha(k\beta).$$

Analogously, for $\alpha_1, ..., \alpha_k \in \mathbb{Z}_n$, we may write their product as $\prod_{i=1}^k \alpha_i$. By convention, this product is [1] when $k = 0$. It is easy to see that if all of the *αi*'s belong to $\mathbb{Z}_n^*$, then so does their product, and in particular, $\left(\prod_{i=1}^k \alpha_i\right)^{-1} = \prod_{i=1}^k \alpha_i^{-1}$; that is, the multiplicative inverse of the product is the product of the multiplicative inverses. In the special case where all the *i*'s have the same value α, we define $\alpha^k := \prod_{i=1}^k \alpha$; thus, $\alpha^0 = [1]$, $\alpha^1 = \alpha$, $\alpha^2 = \alpha\alpha$, $\alpha^3 = \alpha\alpha\alpha$, and so on. If $\alpha \in \mathbb{Z}_n^*$, then the multiplicative inverse of α^k is $(\alpha^{-1})^k$, which we may also write as α^{-k}; for example, $\alpha^{-2} = \alpha^{-1}\alpha^{-1} = (\alpha\alpha)^{-1}$. Therefore, when $\alpha \in \mathbb{Z}_n^*$, the notation α^k is defined for all integers k.

For all $\alpha, \beta \in \mathbb{Z}_n$ and all non-negative integers k and ℓ, we have the identities:

$$(\alpha^\ell)^k = \alpha^{k\ell} = (\alpha^k)^\ell, \alpha^{k+\ell} = \alpha^k\alpha^\ell, (\alpha\beta)^k = \alpha^k\beta^k.$$

If $\alpha, \beta \in \mathbb{Z}_n^*$, the identities in hold for all $k, \ell \in \mathbb{Z}$.

For all $\alpha_1, ..., \alpha_k, \beta_1, ..., \beta_\ell \in \mathbb{Z}_n$, the distributive property implies that

$$(\alpha_1 + ... + \alpha_k)(\beta_1 + .. \beta_\ell) = \sum_{\substack{1 \le i \le k \\ 1 \le j \le \ell}} \alpha_i \beta_j.$$

One last notational convention. As already mentioned, when the modulus n is clear from context, we usually write $[a]$ instead of $[a]_n$. Although we want to maintain a clear distinction between integers and their residue classes, occasionally even the notation $[a]$ is not only redundant, but distracting; in such situations, we may simply write a instead of $[a]$. For example, for every $\alpha \in \mathbb{Z}_n$, we have the identity $(\alpha + [1]_n)(\alpha - [1]_n) = \alpha^2 - [1]_n$, which we may write more simply as $(\alpha + [1])(\alpha - [1]) = \alpha^2 - [1]$, or even more simply, and hopefully more clearly, as $(\alpha + 1)(\alpha - 1) = \alpha^2 - 1$. Here, the only reasonable interpretation of the symbol "1" is $[1]$, and so there can be no confusion.

In summary, algebraic expressions involving residue classes may be manipulated in much the same way as expressions involving ordinary numbers. Extra complications arise only because when n is composite, some non-zero elements of $\mathbb{Z}_n$ do not have multiplicative inverses, and the usual cancellation law does not apply for such elements.

In general, one has a choice between working with congruences modulo n, or with the algebraic structure $\mathbb{Z}_n$; ultimately, the choice is one of taste and convenience, and it depends on what one prefers to treat as "first class objects": integers and congruence relations, or elements of $\mathbb{Z}_n$.

An alternative, and somewhat more concrete, approach to constructing $\mathbb{Z}_n$ is to directly define it as the set of n "symbols" $[0], [1], ..., [n-1]$, with addition and multiplication defined as

$$[a] + [b] := [(a + b) \bmod n], \; [a] \;\; [b] := [(a \;\; b) \bmod n],$$

for $a, b \in \{0, ..., n-1\}$. Such a definition is equivalent to the one we have given here. One should keep this alternative characterization of $\mathbb{Z}_n$ in mind; however, we prefer the characterization in terms of residue classes, as it is mathematically more elegant, and is usually more convenient to work with.

EULER'S TOTIENT FUNCTION

Euler's totient function is defined for all positive integers n as

$$\varphi(n) := \left| Z_n^* \right|.$$

Equivalently, $\varphi(n)$ is equal to the number of integers between 0 and $n - 1$ that are relatively prime to n. For example, $\varphi(1) = 1$, $\varphi(2) = 1$, $\varphi(3) = 2$, and $\varphi(4) = 2$.

It is easy to get a nice formula for $\varphi(n)$ in terms of the prime factorization of n using the Chinese remainder theorem, as we establish in the following sequence of theorems.

Theorem: Let $\{n_i\}_{i=1}^{k} n_i$. be a pairwise relatively prime family of positive integers, and let $n := \prod_{i=1}^{k} n_i$. Then

$$\varphi(n) = \prod_{i=1}^{k} \varphi(n_i).$$

Proof: Consider the map $\theta : \mathbb{Z}_n \rightarrow \mathbb{Z}_{n_1} \times ... \times \mathbb{Z}_{n_k}$ in Theorem. By parts (ii) and (iii) of that theorem, restricting θ to $\mathbb{Z}_n^*$ yields a one-to-one correspondence between $\mathbb{Z}_n^*$ and $\mathbb{Z}_{n_1}^* \times ... \times \mathbb{Z}_{n_k}^*$. The theorem now follows immediately.

We already know that $\varphi(p) = p - 1$ for every prime p, since the integers $1, ..., p - 1$ are not divisible by p, and hence are relatively prime to p. The next theorem generalizes this, giving us a formula for Euler's phi function at prime powers.

Theorem: Let p be a prime and e be a positive integer. Then

$$\varphi(p^e) = p^{e-1}(p - 1).$$

Proof: The multiples of p among $0, 1, ..., p^e - 1$ are

$$0 \cdot p, 1 \cdot p, ..., (p^{e-1} - 1) \cdot p,$$

of which there are precisely p^{e-1}. Thus, $\varphi(p^e) = p^e - p^{e-1} = p^{e-1}(p - 1)$.

If $n = p_1^{e_1} ... p_r^{e_r}$ is the factorization of n into primes, then the family of prime powers $\{p_i^{e_i}\}_{i=1}^{r}$ is pairwise relatively prime, and so Theorem implies $\varphi(n) = \varphi(p_1^{e_1}) ... \varphi(p_r^{e_r})$. Combining this with Theorem, we have:

Theorem: If $n = p_1^{e_1} ... p_r^{e_r}$ is the factorization of n into primes, then

$$\varphi(n) = \prod_{i=1}^{r} p_i^{e_i - 1}(p_i - 1) = n \prod_{i=1}^{r} (1 - 1/p_i).$$

EXERCISE

1. Define $\varphi_2(n)$ to be the number of integers $a \in \{0, ..., n - 1\}$ such that $\gcd(a, n) = \gcd(a + 1, n) = 1$. Show that if $n = p_1^{e_1} ... p_r^{e_r}$ is the factorization of n into primes, then $\varphi_2(n) = n \prod_{i=1}^{r} (1 - 2/p_i)$.
2. Show that $\varphi(nm) = \gcd(n, m) \cdot \varphi(\operatorname{lcm}(n, m))$.
3. Show that if n is divisible by r distinct odd primes, then $2^r \mid \varphi(n)$.

EULER'S THEOREM AND FERMAT'S LITTLE THEOREM

Let n be a positive integer, and let $\in \mathbb{Z}_n^*$.

Consider the sequence of powers of α:

$$1 = \alpha^0, \alpha^1, \alpha^2, \ldots.$$

Since each such power is an element of $\mathbb{Z}_n^*$, and since $\mathbb{Z}_n^*$ is a finite set, this sequence of powers must start to repeat at some point; that is, there must be a positive integer k such that $\alpha^k = \alpha^i$ for some $i = 0, \ldots, k - 1$. Let us assume that k is chosen to be the smallest such positive integer. This value k is called the multiplicative order of α.

We claim that $\alpha^k = 1$. To see this, suppose by way of contradiction that $\alpha^k = \alpha^i$, for some $i = 1, \ldots, k - 1$; we could then cancel α from both sides of the equation $\alpha^k = \alpha^i$, obtaining $\alpha^{k-1} = \alpha^{i-1}$, which would contradict the minimality of k.

Thus, we can characterize the multiplicative order of α as the smallest positive integer k such that

$$\alpha^k = 1.$$

If $\alpha = [a]$ with $a \in \mathbb{Z}$ (and $\gcd(a, n) = 1$, since $\alpha \in \mathbb{Z}_n^*$), then k is also called the multiplicative order of a modulo n, and can be characterized as the smallest positive integer k such that

$$a^k \equiv 1 \pmod{n}.$$

From the above discussion, we see that the first k powers of α, that is, $\alpha^0, \alpha^1, \ldots, \alpha^{k-1}$, are distinct. Moreover, other powers of α simply repeat this pattern. The following is an immediate consequence of this observation.

Theorem: Suppose $\alpha \in \mathbb{Z}_n^*$ has multiplicative order k. Then for every $m \in \mathbb{Z}$, the multiplicative order of α^m is $k/\gcd(m, k)$.

Proof: Applying Theorem to α^m, we see that the multiplicative order of α^m is the smallest positive integer ℓ such that $\alpha^{m\ell} = 1$. But we have

$$\alpha^{m\ell} = 1 \Leftrightarrow m\ell \equiv 0 \pmod{k}$$

$$\Leftrightarrow \ell \equiv 0 \pmod{k/\gcd(m, k)}$$

Theorem: Let n be a positive integer, and let be an element of $\mathbb{Z}_n^*$ of multiplicative order k. Then for every $i \in \mathbb{Z}$, we have $i = 1$ if and only if k divides α^i. More generally, for all $i, j \in \mathbb{Z}$, we have $\alpha^i = \alpha^j$ if and only if $i \equiv j \pmod{k}$.

Example. Let $n = 7$. For each value $a = 1, \ldots, 6$, we can compute successive powers of a modulo n to find its multiplicative order modulo n.

i	1	2	3	4	5	6
$1^i \bmod 7$	1	1	1	1	1	1
$2^i \bmod 7$	2	4	1	2	4	1
$3^i \bmod 7$	3	2	6	4	5	1
$4^i \bmod 7$	4	2	1	4	2	1
$5^i \bmod 7$	5	4	6	2	3	1
$6^i \bmod 7$	6	1	6	1	6	1

So we conclude that modulo 7: 1 has order 1; 6 has order 2; $\in$ and 4 have order 3; and 3 and 5 have order 6.

Fermat's Little Theorem

Theorem: For every prime p, and every $\alpha \in \mathbb{Z}_p$, we have $\alpha^p = \alpha$.

Proof: If $\alpha = 0$, the statement is obviously true. Otherwise, $\alpha \in \mathbb{Z}^*_p$, and by Theorem we have $\alpha^{p-1} = 1$. Multiplying this equation by α yields $\alpha^p = \alpha$.

In the language of congruences, Fermat's little theorem says that for every prime p and every integer a, we have

$$a^p \equiv a \pmod{p}.$$

For a given positive integer n, we say that $\alpha \in \mathbb{Z}$ with $\gcd(a, n) = 1$ is a primitive root modulo n if the multiplicative order of a modulo n is equal to $\varphi(n)$. If this is the case, then for $\alpha := [a] \in \mathbb{Z}^*_n$, the powers α^i range over all elements of $\mathbb{Z}^*_n$ as i ranges over the interval 0, ..., $\varphi(n) - 1$. Not all positive integers have primitive roots – we will see in §7.5 that the only positive integers n for which there exists a primitive root modulo n are

$$n = 1, 2, 4, p^e, 2p^e,$$

where p is an odd prime and e is a positive integer.

The following theorem is sometimes useful in determining the multiplicative order of an element in $\mathbb{Z}^*_n$.

Euler's Theorem

Theorem: Let n be a positive integer and $\alpha \in \mathbb{Z}^*_n$. Then $\alpha^{\varphi(n)} = 1$. In particular, the multiplicative order of divides $\varphi(n)$.

Proof: Since $\alpha \in \mathbb{Z}^*_n$, for every $\beta \in \mathbb{Z}^*_n$ we have $\alpha\beta \in \mathbb{Z}^*_n$, and so we may define the "multiplication by α" map

$$\tau_\alpha : \mathbb{Z}_n^* \to \mathbb{Z}_n^*$$
$$\beta \mapsto \alpha\beta.$$

It is easy to see that τ_α is a bijection:

Injectivity: If $\alpha\beta = \alpha\beta'$, then cancel α to obtain $\beta = \beta'$.

Surjectivity: For every $\gamma \in \mathbb{Z}_n^*$, $\alpha^{-1}\gamma$ is a pre-image of γ under τ_α.

Thus, β as ranges over the set $\mathbb{Z}_n^*$, so does $\alpha\beta$, and we have

$$\prod_{\beta\in\mathbb{Z}_n^*} \beta = \prod_{\beta\in\mathbb{Z}_n^*} (\alpha\beta) = \alpha^{\varphi(n)}\left(\prod_{\beta\in\mathbb{Z}_n^*} \beta\right).$$

Canceling the common factor $\prod_{\beta\in\mathbb{Z}_n^*} \beta \in \mathbb{Z}_n^*$ from the left- and right-hand side of , we obtain

$$1 = \alpha^{\varphi(n)}.$$

That proves the first statement of the theorem.

EXERCISE

1. Let p be any prime other than $\in$ or 5. Show that p divides infinitely many of the numbers 9, 99, 999, etc.
2. Find all elements of $\mathbb{Z}_{19}^*$ of multiplicative order 18.
3. Let $n \in \mathbb{Z}$ with $n > 1$. Show that n is prime if and only if $\alpha^{n-1} = 1$ for every non-zero $\alpha \in \mathbb{Z}_n^*$.
4. Let n be an integer greater than 1. Show that n does not divide $2^n - 1$.
5. Let $n = pq$, where p and q are distinct primes. Show that if $m := \text{lcm}(p-1, q-1)$, then $\alpha^m = 1$ for all $\alpha \in \mathbb{Z}_n^*$.
6. This exercise develops an alternative proof of Fermat's little theorem.

   (a) Using Exercise 1.14, show that for all primes p and integers a, we have
   $$(a+1)^p \equiv a^p + 1 \pmod{p}.$$
7. Prove the following generalization of Fermat's little theorem: for every positive integer n, and every $\alpha \in \mathbb{Z}_n$, we have $\alpha^n = \alpha^{n-\varphi(n)}$.

CHINESE REMAINDER MAP

Theorem: Let us consider $\{n_i\}_{i=1}^{k}$ be a pairwise relatively prime family of positive integers, and let $n := \prod_{i=1}^{k} n_i$. Define the map

$$\theta : \mathbb{Z}_n \to \mathbb{Z}_{n_1} \times \ldots \times \mathbb{Z}_{n_k}$$

$$[a]_n \mapsto ([a]_{n_1}, \ldots, [a]_{n_k}).$$

(i) The definition of θ is unambiguous.

(ii) θ is bijective.

(iii) For all $\alpha, \beta \in \mathbb{Z}_n$, if $\theta(\alpha) = (\alpha_1, \ldots, \alpha_k)$ and $\theta(\beta) = (\beta_1, \ldots, \beta_k)$, then:

(a) $\theta(\alpha + \beta) = (\alpha_1 + \beta_1, \ldots, \alpha_k + \beta_k)$;

(b) $\theta(-\beta) = (-\alpha_1, \ldots, -\alpha_k)$;

(c) $\theta(\alpha\beta) = (\alpha_1\beta_1, \ldots, \alpha_k\beta_k)$;

(d) $\alpha \in \mathbb{Z}_n^*$ if and only if $\alpha_i \in \mathbb{Z}_n^*$ for $i = 1, \ldots, k$, in which case

$$\theta(\alpha^{-1}) = (\alpha_1^{-1}, \ldots, \alpha_k^{-1}).$$

Proof:

(i) Note that $a \equiv a'$ (mod n) implies $a \equiv a'$ (mod n_i) for $i = 1, \ldots, k$, and so the definition of θ is unambiguous (it does not depend on the choice of a).

(ii) Follows directly from the statement of the Chinese remainder theorem.

(iii) Let $\alpha = [a]_n$ and $\beta = [b]_n$, so that for $i = 1, \ldots, k$, we have $\alpha_i = [a]_{n_i}$ and $\beta_i = [b]_{n_i}$. Then we have

$$\theta(\alpha + \beta) = ([a + b]_n) = ([a+b]_{n_1}, \ldots, [a+b]_{n_k}) = (\alpha_1 + \beta_1, \ldots, \alpha_k + \beta_k),$$

$$\theta(-\alpha) = \theta([-a]_n) = ([-a]_{n_1}, \ldots, [-a]_{n_k}) = (-\alpha_1, \ldots, -\alpha_k), \text{ and}$$

$$\theta(\alpha\beta) = \theta([ab]_n) = ([ab]_{n_1}, \ldots, [ab]_{n_k}) = (\alpha_1\beta_1, \ldots, \alpha_k\beta_k).$$

That proves parts (a), (b), and (c). For part (d), we have

$$\begin{aligned} \alpha \in \mathbb{Z}_n^* &\Leftrightarrow \gcd(a, n) = 1 \\ &\Leftrightarrow \gcd(a, n_i) = 1 \text{ for } i = 1, \ldots, k \\ &\Leftrightarrow \alpha_i \in \mathbb{Z}_{n_i}^* \text{ for } i = 1, \ldots, k. \end{aligned}$$

Moreover, if a $\in \mathbb{Z}_n^*$ and = α^{-1}, then

$$(\alpha\beta_1, ..., \alpha_k\beta_k) = \theta(\alpha\beta) = \theta([1]_n) = ([1]_{n_1}, ..., [1]_{n_k}),$$

and so for $i = 1, ..., k$, we have $\alpha_i\beta_i = [1]_{n_i}$, which is to say $\beta_i = \alpha_i^{-1}$.

Theorem is very powerful conceptually, and is an indispensable tool in many situations. It says that if we want to understand what happens when we add or multiply $\alpha, \beta \in \mathbb{Z}_n$, it success to understand what happens when we add or multiply their "components" $\alpha_i, \beta_i \in \mathbb{Z}_{n_i}$. Typically, we choose $n_1, ..., n_k$ to be primes or prime powers, which usually simplifies the analysis. We shall see many applications of this idea throughout the text.

EXERCISE

1. Suppose n is square-free, and let $\alpha, \beta, \gamma \in \mathbb{Z}_n$.
   Show that $\alpha^2\beta = \alpha^2\gamma$ implies $\alpha\beta = \alpha\gamma$.
2. Let p be an odd prime. Show that $\sum_{\beta\in\mathbb{Z}_p^*} \beta^{-1} = \sum_{\beta\in\mathbb{Z}_p^*} \beta = 0$.
3. Let p be an odd prime. Show that the numerator of $\sum_{i=1}^{p-1} 1/i$ is divisible by p.

Chapter 2

Euclidean Algorithm

In number theory, the Euclidean algorithm (also called Euclid's algorithm) is an algorithm to determine the greatest common divisor (GCD) of two elements of any Euclidean domain (for example, the integers). Its major significance is that it does not require factoring the two integers, and it is also significant in that it is one of the oldest algorithms known, dating back to the ancient Greeks.

HISTORY OF THE EUCLIDEAN ALGORITHM

The Euclidean algorithm is one of the oldest algorithms known, since it appeared in Euclid's *Elements* around 300 BC. Euclid originally formulated the problem geometrically, as the problem of finding the greatest common "measure" for two line lengths (a line that could be used to measure both lines without a remainder), and his algorithm proceeded by repeated subtraction of the shorter from the longer segment. However, the algorithm was probably not discovered by Euclid and it may have been known up to 200 years earlier.

DESCRIPTION OF THE ALGORITHM

Given two natural numbers a and b, not both equal to zero: check if b is zero; if yes, a is the gcd. If not, repeat the process using, respectively, b, and the remainder after dividing a by b. The remainder after dividing a by b is usually written as a mod b.

These algorithms can be used in any context where division with remainder is possible. This includes rings of polynomials over a field as well as the ring of Gaussian integers, and in general all Euclidean domains. Applying the algorithm to the more general case other than natural numbers will be discussed in more detail later in the article.

Using Recursion

Using recursion, the algorithm can be expressed:

function gcd(a, b)

```
if b = 0 return a
else return gcd(b, a mod b)
```

or in C/C++ as

```
int gcd(int a, int b)
{return ( b != 0 ? gcd(b, a % b): a );
```

USING ITERATION

An efficient, iterative method, for compilers that don't optimize tail recursion:

```
function gcd(a, b)
while    b ≠ 0
         t: = b
         b: = a mod b
         a: = t
return a
```

THE EXTENDED EUCLIDEAN ALGORITHM

By keeping track of the quotients occurring during the algorithm, one can also determine integers p and q with $ap + bq = \gcd(a, b)$. This is known as the extended Euclidean algorithm.

ORIGINAL ALGORITHM

The original algorithm as described by Euclid treated the problem geometrically, using repeated subtraction rather than mod (remainder).

```
function gcd(a, b)
if       a = 0 return b
while    b ≠ 0
if       a > b
         a: = a ≠ b
else
         b: = b ≠ a
return a
```

Proof: Suppose a and b are the natural numbers whose gcd has to be determined. Now, suppose $b > 0$, and the remainder of the division of a by b is r. Therefore $a = qb + r$ where q is the quotient of the division.

Any common divisor of a and b is also a divisor of r. To see why this is true, consider that r can be written as $r = a - qb$.

Now, if there is a common divisor d of a and b such that $a = sd$ and $b = td$, then $r = (s-qt)d$. Since all these numbers, including $s-qt$, are whole numbers, it can be seen that r is divisible by d.

The above analysis is true for any divisor d; thus, the greatest common divisor of a and b is also the greatest common divisor of b and r. Therefore it is enough if we continue searching for the greatest common divisor with the numbers b and r. Since r is smaller in absolute value than b, we will reach $r = 0$ after finitely many steps.

When analyzing the running time of Euclid's algorithm, it turns out that the inputs requiring the most divisions are two successive Fibonacci numbers (because their ratios are the convergents in the slowest continued fraction expansion to converge, that of the golden ratio) as proved by Gabriel Lamι, and the worst case requires $O(n)$ divisions, where n is the number of digits in the input. However, the divisions themselves are not constant time operations; the actual time complexity of the algorithm is $O(n^2)$.

The reason is that division of two n-bit numbers takes time $O(n(m + 1))$, where m is the length of the quotient. Consider the computation of gcd(a,b) where a and b have at most n bits, let a_0,... a_k be the sequence of numbers produced by the algorithm, and let a_0,... a_k be their lengths. Then $k = O(n)$, and the running time is bounded by

$$O\left(\sum_{i<k} n_i(n_i - n_{i+1} + 2)\right) \subseteq O\left(\sum_{i<k} n_i(n_i - n_{i+1} + 2)\right) \subseteq O(n(n_0 + 2k)) \subseteq O(n^2)$$

This is considerably better than Euclid's original algorithm, in which the modulus operation is effectively performed using repeated subtraction in $O(2^n)$ steps. Consequently, that version of the algorithm requires $O(2^n n)$ time for n-digit numbers, or $O(m \log m)$ time for the number m.

Euclid's algorithm is widely used in practice, especially for small numbers, due to its simplicity. An alternative algorithm, the binary GCD algorithm, exploits the binary representation used by computers to avoid divisions and thereby increase efficiency, although it too is $O(n^2)$; it merely shrinks the constant hidden by the big-O notation on many real machines.

There are more complex algorithms that can reduce the running time to $O(n(\log n)^2(\log\log n))$.

Relation with continued fractions

The quotients that appear when the Euclidean algorithm is applied to the inputs a and b are precisely the numbers occurring in the continued fraction representation of a/b. Take for instance the example of $a = 1071$ and $b = 1029$ used above. Here is the calculation with highlighted quotients:

$$1071 = 1029 \times 1 + 42$$
$$1029 = 42 \times 24 + 21$$
$$42 = 21 \times 2 + 0$$

Consequently,

$$\frac{1071}{1029} = 1 + \cfrac{1}{24 + \cfrac{1}{2}}.$$

This method applies to arbitrary real inputs a and nonzero b; if a/b is irrational, then the Euclidean algorithm does not terminate, but the computed sequence of quotients still represents the (now infinite) continued fraction representation of a/b.

The quotients 1,24,2 count certain squares nested within a rectangle R having length 1071 and width 1029, in the following manner:

1. There is 1 1029 × 1029 square in R whose removal leaves a 42 × 1029 rectangle, R_1;
2. There are 24 42 × 42 squares in R_1 whose removal leaves a 21 × 42 rectangle, R_2;
3. There are 2 21 × 21 squares in R_2 whose removal leaves nothing.

The "visual Euclidean algorithm" of nested squares applies to an arbitrary rectangle R. If the (length)/(width) of R is an irrational number, then the visual Euclidean algorithm extends to a visual continued fraction.

In mathematics, an abelian group, also called a commutative group, is a group $(G, *)$ with the additional property that the group operation * is commutative, so that for all a and b in G, $a * b = b * a$. Abelian groups are named after Norwegian mathematician Niels Henrik Abel. Groups in which the group operation is not commutative are called *non-abelian* (or *non-commutative*).

Since the group operation in an abelian group is commutative as well as associative, the value of a product of group elements is independent of the order in which the product is calculated. The theory of abelian groups is generally simpler than that of their non-abelian counterparts, although infinite abelian groups are the subject of current research

NOTATION

There are two main notational conventions for abelian groups — additive and multiplicative.

Convention	*Operation*	*Identity*	*Powers*	*Inverse*	*Direct/sum/product*
Addition	$x + y$	0	nx	$-x$	$G \oplus H$
Multiplication	$x * y$ or xy	e or 1	xn	x–1	G X H

The multiplicative notation is the usual notation for groups, while the additive notation is the usual notation for modules. When studying abelian groups apart from other groups, the additive notation is usually used.

Examples. Every cyclic group G is abelian, because if x, y are in G, then $xy = a^m a^n = a^{m+n} = a^{n+m} = a^n a^m = yx$. Thus the integers, Z, form an abelian group under addition, as do the integers modulo n, Z/nZ.

Every ring is an abelian group with respect to its addition operation. In a commutative ring the invertible elements, or units, form an abelian multiplicative group. In particular, the real numbers are an abelian group under addition, and the nonzero real numbers are an abelian group under multiplication.

Every subgroup of an abelian group is normal, so each subgroup gives rise to a quotient group. Subgroups, quotients, and direct sums of abelian groups are again abelian.

Matrices, even invertible matrices, do not form an abelian group under multiplication because matrix multiplication is generally not commutative.

MULTIPLICATION TABLE

To verify that a finite group is abelian, a table (matrix) - known as a Cayley table - can be constructed in a similar fashion to a multiplication table. If the group is $G = \{g_1 = e, g_2, \ldots, g_n\}$ under the operation E", the (i, j)'th entry of this table contains the product g_i E" g_j. The group is abelian if and only if this table is symmetric about the main diagonal (i.e. if the matrix is a symmetric matrix).

This is true since if the group is abelian, then g_i E" $g_j = g_j$ E" g_i. This implies that the (i, j)'th entry of the table equals the (j, i)'th entry - i.e. the table is symmetric about the main diagonal.

PROPERTIES

If n is a natural number and x is an element of an abelian group G written additively, then nx can be defined as $x + x + \ldots + x$ (n summands) and ("n)x = "(nx). In this way, G becomes a module over the ring Z of integers. In fact, the modules over Z can be identified with the abelian groups.

Theorems about abelian groups (i.e. modules over the principal ideal domain Z) can often be generalized to theorems about modules over an arbitrary principal ideal domain. A typical example is the classification of finitely generated abelian groups.

If $f, g : G \to H$ are two group homomorphisms between abelian groups, then their sum $f + g$, defined by $(f + g)(x) = f(x) + g(x)$, is again a homomorphism. (This is not true if H is a non-abelian group.) The set Hom(G, H) of all group homomorphisms from G to H thus turns into an abelian group in its own right.

Somewhat akin to the dimension of vector spaces, every abelian group has a *rank*. It is defined as the cardinality of the largest set of linearly independent elements of the group. The integers and the rational numbers have rank one, as well as every subgroup of the rationals.

Finite Abelian Groups

The fundamental theorem of finite abelian groups states that every finite abelian group can be expressed as the direct sum of cyclic subgroups of prime-power order. This is a special application of the fundamental theorem of finitely generated abelian groups in the case when G has torsion-free rank equal to 0.

$\mathbb{Z}_{mn}$ is isomorphic to the direct product of $\mathbb{Z}_m$ and $\mathbb{Z}_n$ if and only if m and n are coprime.

Therefore we can write any finite abelian group G as a direct product of the form

$$\mathbb{Z}_{m_1} \oplus \ldots \oplus \mathbb{Z}_k$$

In two unique ways:

- Where the numbers $k_1,\ldots,k_u$ are powers of primes
- Where k_1 divides k_2, which divides k_3 and so on up to k_u.

For example, $\mathbb{Z}/15\mathbb{Z} \cong \mathbb{Z}_{15}$ can be expressed as the direct sum of two cyclic subgroups of order 3 and 5: $\mathbb{Z}_{15} \cong \{0,5,10\} \oplus \{0,3,6\,9,12\}$. The same can be said for any abelian group of order 15, leading to the remarkable conclusion that all abelian groups of order 15 are isomorphic.

For another example, every abelian group of order 8 is isomorphic to either $\mathbb{Z}_8$ (the integers 0 to 7 under addition modulo 8), $\mathbb{Z}_4 \oplus \mathbb{Z}_2$ (the odd integers 1 to 15 under multiplication modulo 16), or $\mathbb{Z}_2 \oplus \mathbb{Z}_2 \oplus \mathbb{Z}_2$.

AUTOMORPHISMS OF FINITE ABELIAN GROUPS

One can apply the fundamental theorem to count (and sometimes determine) the automorphisms of a given finite abelian group G. To do this, one uses the fact (which will not be proved here) that if G splits as a direct sum $H \oplus K$ of subgroups of coprime order, then Aut $(H \oplus K)$ Aut(H) $\cong$ Aut(K).

Given this, the fundamental theorem shows that to compute the automorphism group of G it suffices to compute the automorphism groups of the Sylow p-subgroups separately (that is, all direct sums of cyclic subgroups, each with order a power of p). Fix a prime p

and suppose the exponents e_i of the cyclic factors of the Sylow p-subgroup are arranged in increasing order:

$$e_1 \le e_2 \le ... \le e_n$$

for some $n > 0$. One needs to find the automorphisms of

$$\mathbb{Z}_{p^{e1}} \oplus ... \oplus \mathbb{Z}_{p^{en}}$$

One special case is when $n = 1$, so that there is only one cyclic prime-power factor in the Sylow p-subgroup P. In this case the theory of automorphisms of a finite cyclic group can be used. Another special case is when n is arbitrary but $e_i = 1$ for $1 \le i \le n$. Here, one is considering P to be of the form

$$\mathbb{Z}_p \oplus ... \oplus \mathbb{Z}_p,$$

so elements of this subgroup can be viewed as comprising a vector space of dimension n over the finite field of p elements $\mathbb{F}_p$. The automorphisms of this subgroup are therefore given by the invertible linear transformations, so

$$\text{Aut}\ (P) \cong GL(n, \mathbb{F}_p.),$$

which is easily shown to have order

$$|\text{Aut}\ (P)| = (p^n - 1)...\ (p^n - p^{n-1}).$$

In the most general case, where the e_i and n are arbitrary, the automorphism group is more difficult to determine. It is known, however, that if one defines

$$d_k = \max\ \{r \mid e_r = e_k\}$$

and

$$c_k = \max\ \{r \mid e_r = e_k\}$$

then one has in particular $d_k \ge k$, $c_k \le k$, and

$$|\text{Aut}\ (P)| = \left(\prod_{k=1}^{n} p^{d_k} = p^{k-1}\right)\left(\prod_{j=1}^{n} (p^{e_j})^{n-d_j}\right)\left(\prod_{i=1}^{n} (p^{e_i-1})^{n-c_i+1}\right)$$

One can check that this yields the orders in the previous examples as special cases.

CYCLIC GROUP

In group theory, a cyclic group or monogenous group is a group that can be generated by a single element, in the sense that the group has an element g (called a "generator" of the group) such that, when written multiplicatively, every element of the group is a power of g (a multiple of g when the notation is additive).

Definition: A group G is called cyclic if there exists an element g in G such that

$G = \langle g \rangle = \{ g^n \mid n$ is an integer $\}$. Since any group generated by an element in a group is a subgroup of that group, showing that the only subgroup of a group G that contains g is G itself suffices to show that G is cyclic.

For example, if $G = \{ g^0, g^1, g^2, g^3, g^4, g^5 \}$ is a group, then $g^6 = g^0$, and G is cyclic. In fact, G is essentially the same as (that is, isomorphic to) the set { 0, 1, 2, 3, 4, 5 } with addition modulo 6. For example, $1 + 2 = 3 \pmod 6$ corresponds to $g^1 – g^2 = g^3$, and $2 + 5 = 1 \pmod 6$ corresponds to $g^2 \cdot g^5 = g^7 = g^1$, and so on. One can use the isomorphism φ defined by $\varphi(g^i) = i$.

For every positive integer n there is exactly one cyclic group (up to isomorphism) whose order is n, and there is exactly one infinite cyclic group (the integers under addition). Hence, the cyclic groups are the simplest groups and they are completely classified.

The name 'cyclic' may be misleading: it is possible to generate infinitely many elements and not form any literal cycles; that is, every g^n is distinct. (It can be said that it has one infinitely long cycle.) A group generated in this way is called an infinite cyclic group, and is isomorphic to the additive group of integers Z.

Since the cyclic groups are abelian, they are often written additively and denoted Z_n. However, this notation can be problematic for number theorists because it conflicts with the usual notation for p-adic number rings or localization at a prime ideal. The quotient notations Z/nZ, Z/n, and $Z/(n)$ are standard alternatives. We adopt the first of these here to avoid the collision of notation.

One may write the group multiplicatively, and denote it by C_n, where n is the order (which can be ∞). For example, $g^3g^4 = g^2$ in C_5, whereas $3 + 4 = 2$ in $Z/5Z$.

PROPERTIES

The fundamental theorem of cyclic groups states that if G is a cyclic group of order n then every subgroup of G is cyclic.

Moreover, the order of any subgroup of G is a divisor of n and for each positive divisor k of n the group G has exactly one subgroup of order k. This property characterizes finite cyclic groups: a group of order n is cyclic if and only if for every divisor d of n the group has at most one subgroup of order d. Sometimes the equivalent statement is used: a group of order n is cyclic if and only if for every divisor d of n the group has exactly one subgroup of order d.

Every finite cyclic group is isomorphic to the group { [0], [1], [2],..., $[n - 1]$ } of integers modulo n under addition, and any infinite cyclic group is isomorphic to Z (the set of all integers) under addition. Thus, one only needs to look at such groups to understand the properties of cyclic groups in general. Hence, cyclic groups are one of the simplest groups to study and a number of nice properties are known.

Given a cyclic group G of order n (n may be infinity) and for every g in G,

- G is abelian; that is, their group operation is commutative: $gh = hg$ (for all h in G). This is so since $g + h \bmod n = h + g \bmod n$.
- If n is finite, then $g^n = g^0$ is the identity element of the group, since $n \bmod n = 0$.
- If $n = \infty$, then there are exactly two generators: namely 1 and –1 for Z, and any others mapped to them under an isomorphism in other infinite cyclic groups.
- If n is finite, then there are exactly φ(n) generators where φ is the Euler phi function
- Every subgroup of G is cyclic. Indeed, each finite subgroup of G is a group of $\{0, 1, 2, 3, \ldots m - 1\}$ with addition modulo m. And each infinite subgroup of G is mZ for some m, which is bijective to (so isomorphic to) Z.
- G_n is isomorphic to Z/nZ (factor group of Z over nZ) since Z/nZ = {0 + nZ, 1 + nZ, 2 + nZ, 3 + nZ, 4 + nZ,..., n – 1 + nZ} ≅ { 0, 1, 2, 3, 4,..., n – 1} under addition modulo n.

More generally, if d is a divisor of n, then the number of elements in Z/n which have order d is φ(d). The order of the residue class of m is n / gcd(n,m).

If p is a prime number, then the only group (up to isomorphism) with p elements is the cyclic group C_p or Z/pZ.

The direct product of two cyclic groups Z/nZ and Z/mZ is cyclic if and only if n and m are coprime. Thus e.g. Z/12Z is the direct product of Z/3Z and Z/4Z, but not the direct product of Z/6Z and Z/2Z.

The definition immediately implies that cyclic groups have very simple group presentation $C = <x \mid>$ and $C_n = <x \mid x^n>$ for finite n.

A primary cyclic group is a group of the form Z/p^k where p is a prime number. The fundamental theorem of abelian groups states that every finitely generated abelian group is the direct product of finitely many finite primary cyclic and infinite cyclic groups.

Z/nZ and Z are also commutative rings. If p is a prime, then Z/pZ is a finite field, also denoted by F_p or GF(p). Every field with p elements is isomorphic to this one.

The units of the ring Z/nZ are the numbers coprime to n. They form a group under multiplication modulo n with φ(n) elements. It is written as $(Z/nZ)^X$. For example, we get $(Z/nZ)^X = \{1,5\}$ when $n = 6$, and get $(Z/nZ)^X = \{1,3,5,7\}$ when $n = 8$.

In fact, it is known that $(Z/nZ)^X$ is cyclic if and only if n is 2 or 4 or p^k or 2 p^k for an odd prime number p and $k \geq 1$, in which case every generator of $(Z/nZ)^X$ is called a primitive root modulo n. Thus, $(Z/nZ)^X$ is cyclic for $n = 6$, but not for $n = 8$, where it is instead isomorphic to the Klein four-group.

The group $(Z/pZ)^X$ is cyclic with $p - 1$ elements for every prime p, and is also written $(Z/pZ)^*$ because it consists of the non-zero elements. More generally, every *finite* subgroup of the multiplicative group of any field is cyclic.

In mathematics, ring theory is the study of rings, algebraic structures in which addition and multiplication are defined and have similar properties to those familiar from the integers. Ring theory studies the structure of rings, their representations, or, in different language, modules, special classes of rings (group rings, division rings, universal enveloping algebras), as well as an array of properties that proved to be of interest both within the theory itself and for its applications, such as homological properties and polynomial identities.

Commutative rings are much better understood than noncommutative ones. Due to its intimate connections with algebraic geometry and algebraic number theory, which provide many natural examples of commutative rings, their theory, which is considered to be part of commutative algebra and field theory rather than of general ring theory, is quite different in flavour from the theory of their noncommutative counterparts.

A fairly recent trend, started in the 1980s with the development of noncommutative geometry and with the discovery of quantum groups, attempts to turn the situation around and build the theory of certain classes of noncommutative rings in a geometric fashion as if they were rings of functions on (non-existent) 'noncommutative spaces'.

ELEMENTARY INTRODUCTION

Definition: Formally, a ring is an Abelian group $(R, +)$, together with a second binary operation * such that for all a, b and c in R,

$$a * (b * c) = (a * b) * c$$

$$a * (b + c) = (a * b) + (a * c)$$

$$(a + b) * c = (a * c) + (b * c)$$

also, if there exists a *multiplicative identity* in the ring, that is, an element e such that for all a in R,

$$a * e = e * a = a$$

then it is said to be a *ring with unity*. The number 1 is a common example of a unity.

It is simple to show that any ring in which $1 = 0$ must have just one element; any such ring is called a zero ring.

Rings that sit inside other rings are called subrings. Maps between rings which respect the ring operations are called ring homomorphisms. Rings, together with ring homomorphisms, form a category (the category of rings). Closely related is the notion of ideals, certain subsets of rings which arise as kernels of homomorphisms and can serve to define factor rings. Basic facts about ideals, homomorphisms and factor rings are recorded in the isomorphism theorems and in the Chinese remainder theorem.

A ring is called *commutative* if its multiplication is commutative. Commutative rings resemble familiar number systems, and various definitions for commutative rings are designed to recover properties known from the integers.

Commutative rings are also important in algebraic geometry. In commutative ring theory, numbers are often replaced by ideals, and the definition of prime ideal tries to capture the essence of prime numbers. Integral domains, non-trivial commutative rings where no two non-zero elements multiply to give zero, generalize another property of the integers and serve as the proper realm to study divisibility.

Principal ideal domains are integral domains in which every ideal can be generated by a single element, another property shared by the integers.

Euclidean domains are integral domains in which the Euclidean algorithm can be carried out. Important examples of commutative rings can be constructed as rings of polynomials and their factor rings.

Summary: Euclidean domain => principal ideal domain $\Rightarrow$ unique factorization domain $\Rightarrow$ integral domain $\Rightarrow$ Commutative ring.

Non-commutative rings resemble rings of matrices in many respects. Following the model of algebraic geometry, attempts have been made recently at defining non-commutative geometry based on non-commutative rings.

Non-commutative rings and associative algebras (rings that are also vector spaces) are often studied via their categories of modules. A module over a ring is an Abelian group that the ring acts on as a ring of endomorphisms, very much akin to the way fields (integral domains in which every non-zero element is invertible) act on vector spaces. Examples of non-commutative rings are given by rings of square matrices or more generally by rings of endomorphisms of Abelian groups or modules, and by monoid rings.

Finite fields are the general starting point for the constructions of many combinatorial structures. It will be important to know the fundamentals concerning these fields in order to investigate combinatorial structures and related areas of combinatorial interest. Unfortunately, the area of field theory is rather large and it would be impossible for us to cover it in detail and still have time to work with the results.

In the interest of conserving time, we will present the elements of general field theory without proofs and only prove statements when we turn our attention specifically to finite fields. The material that we will gloss over is usually covered in a graduate (or good undergraduate) course in abstract algebra.

It will be assumed that you are familiar with the definition of a field, the definition and basic properties of vector spaces and the fact that the integers modulo a prime p form a field (a finite one), which we will denote by GF(p) (standing for the Galois Field of order p).

GENERAL FIELD THEORY

Let L be a field containing a subset K, which is itself a field under the operations inherited from L. Then L is called an *extension* of K, and K a *subfield* of L.

Every field has a smallest subfield, called the *prime subfield*, which is isomorphic to either the field of rationals Q, in which case we say that it has *characteristic zero*, or to a GF(p) for some prime p, in which case we say that it has *characteristic p*. We shall denote the characteristic of an arbitrary field K by char K.

If L is an extension of K (denoted L > K), and þ is an element of L but not an element of K, then the smallest field containing both K and þ will be denoted by K(þ) and is a subfield of L. Similarly, the smallest field containing K and the elements $þ_1,þ_2,...,þ_n$ in L > K will be written as $K(þ_1,þ_2,...,þ_n)$. Any extension field L of K can be viewed as a vector space over K and the dimension of this vector space is called the *degree* of L over K, and is denoted by [L:K]. If the vector space is finite dimensional we say that L is a *finite extension* of K. If L is a finite extension of K and M is a finite extension of L, then [M:K] = [M:L][L:K].

Denote by K[x] the ring of polynomials over K in the variable x. K[x] is a principal ideal domain (an ideal is a subring that is closed under multiplication, a principal ideal is an ideal generated by a single element, and a principal ideal domain is a commutative ring with unity all of whose ideals are principal). A polynomial in K[x] is said to be *monic* if the coefficient of the highest power in x is unity. It is *irreducible* if it is not the product of two nonscalar polynomials in K[x].

If f(x) is an irreducible polynomial in K[x], then any zero (i.e., root) of f(x) is not in K and so there is a smallest extension field L of K that contains it. Furthermore, L is isomorphic to the quotient field K[x]/<f(x)>, where <f(x)> denotes the principal ideal of K[x] generated by f(x). If the irreducible polynomial f(x) is of degree n, then [L:K] = n. Furthermore, if þ is the zero of f(x) in question, then L = K(þ) and the elements $1,þ,þ^2,...,$ $þ^{n-1}$ form a basis of L over K.

Before we continue, let us illustrate these ideas with a well-known example. The field C of complex numbers is usually described in one of two ways, either as the set of numbers {a + bi} where a and b are real numbers and i = sqrt(-1), or if you prefer not to mention the imaginary number i, C can be described as the set of all pairs (a,b) of real numbers where addition of two pairs is the usual componentwise addition and multiplication of two pairs is defined by (a,b)(c,d) = (ac - bd, ad + bc).

The second version is of course viewing C as a vector space of dimension 2 over the real numbers R. Let us see how we would construct C starting with the subfield R. Now K = R and K[x] is the ring of all polynomials with real coefficients. The polynomial $x^2 + 1$ is monic since the coefficient of x^2 is 1 and it is irreducible over R since it cannot be factored into polynomials with only real coefficients.

The principal ideal <x^2+1> consists of all polynomials that have x^2+1 as a factor. The quotient field structure, R[x]/<x^2+1> is obtained by taking each polynomial in R[x] and dividing it by x^2+1. The polynomials that have the same remainder after division form equivalence classes, which are the elements of the quotient field. The different possible remainders are the polynomials a + bx, where a,b in R, and we identify the equivalence classes with these remainders.

The association a + bx iff a + bi clearly shows that the quotient field and C are isomorphic. Now, let þ be a zero of $x^2 + 1$, i.e. $þ^2 + 1 = 0$, so þ = sqrt(-1) = i which is not an element of R. {1,i} forms a basis for C over R since every element of C can be written as a(1) + b(i) with a,b in R and 1 and i are easily seen to be linearly independent. Using this basis, we can identify the elements of C with their coefficient vectors, i.e. a + bi iff (a,b) to get the second representation as a vector space of dimension 2 (notice the highest power of x in the irreducible polynomial).

Let $f(x)$ be in K[x], then the smallest field containing K and all the zeros of $f(x)$ is called the *splitting field* of f(x) over K. We may call it "the" splitting field due to the following theorem:

Theorem: Let $f(x)$ be an irreducible polynomial in K[x], then a splitting field for $f(x)$ over K exists and any two such splitting fields are isomorphic.

If $f(x)$ is a polynomial in K[x] of degree n, then its splitting field over K is at most of degree n! and this bound may or may not be obtained, depending on the polynomial and the field. We now consider some elementary properties of polynomials over a field. If f(x) in K[x] is given by

$$f(x) = \beta_n x^n + \beta_{n-1} x^{n-1} + \ldots + \beta_1 x + \beta_0, \text{ where } \beta_i \text{ in K}$$

then the formal derivative of f(x), denoted by f'(x) is the polynomial

$$f'(x) = n\beta_n x^{n-1} + (n-1)\beta_{n-1} x^{n-2} + \ldots + \beta_1$$

which is a polynomial of degree at most n - 1. Note that the formal derivative of a polynomial may be zero even though the polynomial is not a constant, for example if K = GF(3), where 3 = 0, the polynomial f(x) = x^3 + 2 has a zero formal derivative. A polynomial f(x) in K[x] is said to have a zero þ of *multiplicity* m in some extension field L of K if m is the largest positive integer for which

$$(x - \Rightarrow)^m \mid f(x))$$

In L[x], where the vertical bar indicates division with no remainder. The zeros of an irreducible polynomial $f(x)$ in K[x] in the splitting field for $f(x)$ over K are called *conjugates*. Let $f(x)$, g(x) in K[x] be two polynomials such that deg f> deg g. By the Euclidean algorithm there exist two polynomials $q(x)$ and r(x) such that

$$f(x) = q(x)g(x) + r(x), \text{ where deg r} < \text{deg g}.$$

By repeated application of the algorithm, the greatest common divisor (or gcd) d(x) of $f(x)$ and $g(x)$ (denoted by ($f(x)$, g(x))) can be expressed as

$$d(x) = (f(x),g(x)) = a(x)f(x) + b(x)g(x), \text{ for some } a(x),b(x) \text{ in } K[x].$$

For further reference we collect some elementary properties of polynomials in the following theorem.

Theorem: *Let f(x), g(x) in K[x] and let L be any extension of K. Then:*

- If ($f(x)$, g(x)) = $d(x)$ in K[x] then ($f(x)$, $g(x)$) = $d(x)$ in L[x].
- If $f(x)$ | g(x) in K[x] iff $f(x)$ | $g(x)$ in L[x].
- *f(x) has a multiple zero iff (f(x), f′(x)) is not 1.*

It is significant that two polynomials have a common root in some extension field if they have a common divisor over the original field. The question of multiplicities of roots of irreducible polynomials can be settled precisely. If f(x) in K[x], char K = 0 and f(x) is irreducible, then f(x) cannot have multiple zeros.

If char K = p, then an irreducible polynomial f(x) having multiple zeros must be of the form f(x) = g(x^p) for some polynomial g(x) in K[x]. In this case each zero of f(x) has the same multiplicity. It can be shown that if f(x) is an irreducible polynomial over a finite field, then it has only simple (multiplicity one) zeros.

Theorem: A polynomial $f(x)$ in $K[x]$ of degree n has at most n zeros in any extension of K.

We consider now the concept of field isomorphism, which will be useful in the investigation of finite fields. An *isomorphism* of the field K_1 onto the field K_2 is a one-to-one onto map that preserves both field operations, i.e.,

$$\mu(þ + \beta) = \mu(þ) + \mu(\beta),\ \mu(þ\beta) = \mu(þ)\mu(\beta) \text{ for all } þ, \beta \text{ in K1}.$$

An *automorphism* of K is an isomorphism of K onto itself. The set of all automorphisms of a field forms a group under composition. If $\mu_1,\mu_2,...,\mu_n$ are distinct isomorphisms of K_1onto K_2, then these isomorphisms are linearly independent over K_2 in the sense that if

Sum a_i μ_i (b) = 0 with a_i in K_2

for all b in K_1, then all the a_i must = 0. In particular, distinct automorphisms of a field are linearly independent.

A field automorphism leaves the elements of the prime subfield fixed. The automorphism may leave a larger subfield fixed. More generally, if

$$K' = \{ \text{a in K} \mid \mu_i \text{ (a)} = \text{a, i} = 1,...,\text{n} \}$$

then K′ is a subfield of K, called the *fixed field* of K with respect to the automorphisms μ_i. It can be shown that [K:K'] >= n. We will be interested in subgroups of the group of all automorphisms that fix certain subfields.

We denote the group of all automorphisms of a field L by G(L) and the subgroup of G(L) that fixes all elements of the subfield K of L by G(L/K). It is important to note that the fixed field of G(L/K) may properly contain K. It is easily shown that G(L/K) is a subgroup of G(L), conversely, if H is any subgroup of G(L), then the set of elements of L fixed by H is a subfield of L.

FUNDAMENTAL PROPERTIES OF FINITE FIELDS

We now restrict our attention to finite fields, i.e., fields with a finite number of elements. A finite field, since it cannot contain Q, must have a prime subfield of the form GF(p) for some prime p, also:

Theorem: *Any finite field with characteristic p has p^n elements for some positive integer n.*

Proof: Let L be the finite field and K the prime subfield of L. The vector space of L over K is of some finite dimension, say n, and there exists a basis $þ_1, þ_2, ..., þ_n$ of L over K. Since every element of L can be expressed uniquely as a linear combination of the $þ_i$ over K, i.e., every a in L can be written as, a = Sum β_i $þ_i$, with β_i in K, and since K has p elements, L must have p^n elements.

This theorem, while it does restrict the size of a finite field, does not say that one will exist for a particular power of a prime, nor does it specify how many finite fields can exist of a particular order. The answers to these questions can be deduced from the following theorem.

Theorem: *Let L be a field with characteristic p and prime subfield K. Then L is the splitting field for $f(x) = xp^n - x$ iff L has p^n elements.*

Proof: Suppose that L is the splitting field for $f(x) = xp^n - x$ over *K*. Since $(f(x), f'(x)) = 1$, the roots of $f(x)$ are distinct and so L has at least p^n elements. Consider the subset

$$E = \{ þ \text{ in } L \mid þ^{pn} = þ \}$$

of L. Clearly E contains p^n elements since it consists of the roots of f(x). Suppose that þ,í in E; then $(þ\beta)^{pn} = (þ)^{pn} (\beta)^{pn} = þ\beta$ and hence, þβ in *E*. Also,

$$(Þ + \beta)^{p^n} = \sum þ^i \beta^{p^n - 1} = þ^{p^n} + \beta^{p^n} = Þ + \beta$$

since p | C(p^n, i) for $0 < i < p^n$, and hence, (þ + í) in E. The existence of additive and multiplicative inverses is easy to show, so E is a subfield of L and also a splitting field for $f(x)$. Thus by Thm II.1.1 E = L and L contains p^n elements.

Suppose now that L contains p^n elements. The multiplicative group of L, which we will denote by L*, forms a group of order p^n - 1 and hence the order of any element of L* divides p^n - 1. Thus $þ^{pn} = þ$ for all þ in L* and the relation is trivially true for þ = 0. Thus $f(x)$ splits in L.

The following important theorem is useful in establishing the subfield structure of the Galois Fields among other things.

Theorem: $GF(p^n)^*$ *is cyclic.*

Proof: The multiplicative group $GF(p^n)^*$ is, by definition, abelian and of order $p^n - 1$. If $p^n - 1 = p_1^{n_1} \dots p_k^{e_k}$, then, factoring $GF(p^n)^*$ into a direct product of its Sylow subgroups, we have

$$GF(p^n)^* = S(p_1) \text{ X} \dots \text{ X } S(p_k)$$

where $S(p_i)$ is the Sylow subgroup of order $(p_i)^{e_i}$. The order of every element in $S(p_i)$ is a power of p_i and let a_i in $S(p_i)$ have the maximal order, say $(p_i)^{e'_i}$, $e'_i <= e_i$, for $i = 1,\dots,k$. Since $(p_i, p_j) = 1$, i not equal j, the element $a = a_1 a_2 \dots a_k$ has maximal order $m = (p_1)^{e'-1} \dots (p_k)^{e'_k}$ in $GF(p^n)^*$. Furthermore every element of $GF(p^n)^*$ satisfies the polynomial $x^m - 1$, implying that $m >= p^n - 1$. Since a in $GF(p^n)^*$ has order m, m divides $p^n - 1$ and so, $m = p^n - 1$. Thus the element a is a generator and $GF(p^n)^*$ is cyclic.

A generator of $GF(p^n)^*$ is called a *primitive element* of $GF(p^n)$.

The following theorem has some useful consequences.

Theorem: *Over any field K, $(x^m - 1) \mid (x^n - 1)$ iff $m \mid n$.*

Proof: If $n = qm + r$, with $r < m$, then by direct computation

$$x^n + 1 = x^r \left(\sum_{i=0}^{q-1} x^{im} \right) (x^m - 1) + (x^r - 1)$$

It follows that $(x^m - 1) \mid (x^n - 1)$ iff $x^r - 1 = 0$, i.e., $r = 0$.

Corollary: *For any prime integer p, $(p^m - 1) \mid (p^n - 1)$ iff $m \mid n$.*

Proof: Basically the same as that of the theorem, do for homework.

Theorem: $GF(p^m)$ is a subfield of $GF(p^n)$ *iff $m \mid n$.*

Proof: Suppose $GF(p^m)$ is a subfield of $GF(p^n)$; then $GF(p^n)$ may be interpreted as a vector space over $GF(p^m)$ with dimension, say, k. Hence, $p^n = p^{km}$ and $m \mid n$.

Now suppose $m \mid n$, which from the previous theorem and its corollary implies that $(x^{pm-1} - 1) \mid (x^{pn-1} - 1)$. Thus every zero of $x^{pm} - x$ that is in $GF(p^m)$ is also a zero of $x^{pn} - x$ and hence in $GF(p^n)$. It follows that $GF(p^m)$ is contained in $GF(p^n)$. Notice that there is precisely one subfield of $GF(p^n)$ of order p^m, otherwise $x^{pm} - x$ would have more than p^m roots.

Although we will not prove it, the automorphism group of a finite field is cyclic. The standard generator of this group is the so-called *Frobenius automorphism* defined for a finite field of characteristic p as the map $x \rightarrow x^p$ for all x in $GF(p^n)$.

In mathematics, plural matrices is a rectangular table of *elements* (or *entries*), which may be numbers or, more generally, any abstract quantities that can be added and multiplied. Matrices are used to describe linear equations, keep track of the coefficients of linear transformations and to record data that depend on multiple parameters. Matrices are described by the field of matrix theory. Matrices can be added, multiplied, and decomposed in various ways, which also makes them a key concept in the field of linear algebra.

The horizontal lines in a matrix are called rows and the vertical lines are called columns. A matrix with m rows and n columns is called an m-by-n matrix (written $m \times n$) and m and n are called its dimensions.

The dimensions of a matrix are always given with the number of rows first, then the number of columns. It is commonly said that an m-by-n matrix has an order of $m \times n$ ("order" meaning size). Two matrices of the same order whose corresponding entries are equivalent are considered equal.

Almost always capital letters denote matrices with the corresponding lower-case letters with two indices representing the entries. For example, the entry of a matrix A that lies in the i-th row and the j-th column is written as $a_{i,j}$ and called the i,j entry or (i,j)-th entry of A. Alternative notations for that entry are A[i,j] or $A_{i,j}$.

The row is always noted first, then the column. In this example, A (with no subscripts) would symbolize the entire matrix. In addition to using uppercase letters as symbols representing matrices, many authors use a special typographical style, commonly boldface upright (non-italic), to further distinguish matrices from other variables. Following this convention, A is a matrix, distinguished from A, a scalar. An alternate convention is to annotate matrices with their dimensions in small type underneath the symbol, for example, $\underset{r \times c}{A}$ for an r-by-c matrix.

We often write A: $(a_{i,j})_{i=1,m; j=1,\ldots n}$ or A: = $(a_{i,j})$ to define an $m \times n$ matrix A. In this case, the entries $a_{i,j}$ are defined separately for all integers $1 \le i \le m$ and $1 \le j \le n$. In some programming languages, the numbering of rows and columns starts at zero. Texts which use any such language extensively frequently follow that convention, so we have $0 \le i \le m-1$ and $0 \le j \le n-1$.

A matrix where one of the dimensions equals one is often called a *vector*, and interpreted as an element of real coordinate space. An $m \times 1$ matrix (one column and m rows) is called a column vector and a 1 X n matrix (one row and n columns) is called a row vector.

MATHEMATICAL DEFINITION

An $m \times n$ $(m, n \in \mathbb{N})$ matrix A is a function $A: \{1, 2, \ldots., m\} \times \{1, 2, \ldots, n\} \to S$ where S is any non-empty set.

$(\{1, 2, ..., m\} \times \{1, 2, ..., n\}$ is the Cartesian product of sets $\{1, 2, ..., m\}$ and $(\{1, 2, ..., n\})$.

We say that matrix A is a matrix over the set S. Important thing to note is that, if we want to have matrix algebra, the set S must be a ring and matrix A must be a square matrix. Since the set of all square matrices over a ring is also a ring, matrix algebra is usually called matrix ring.

Since this article mainly considers matrices over real numbers, matrices shown here are actually functions $A : (\{1, 2, ..., m\} \times (\{1, 2, ..., n\} \to \mathbb{R}$.

Example: The matrix

$$A = \begin{bmatrix} 9 & 8 & 6 \\ 1 & 2 & 7 \\ 4 & 9 & 2 \\ 6 & 0 & 5 \end{bmatrix} \text{ or } A = \begin{pmatrix} 9 & 8 & 6 \\ 1 & 2 & 7 \\ 4 & 9 & 2 \\ 6 & 0 & 5 \end{pmatrix}$$

is a matrix 4 X 3. The element $a_{2,3}$ or A [2, 3] is 7. In terms of the mathematical definition given above, this matrix is a function $A : \{1, 2, 3, 4\} \times (\{1, 2, 3\} \to \mathbb{R}$ and, for example, A ((2, 3)) = 7 and A ((3, 1)) = 4.

The matrix

$$\text{R} = \begin{bmatrix} 1 & 2 & 3 & 4 & 5 & 6 & 7 & 8 & 9 \end{bmatrix}$$

is a 1 X 9 matrix, or 9-element row vector.

BASIC OPERATIONS

Sum

Two or more matrices of identical dimensions *m* and *n* can be added. Given *m*-by-*n* matrices A and B, their sum A+B is the *m*-by-*n* matrix computed by adding corresponding elements:

For example:

$$\begin{bmatrix} 1 & 3 & 1 \\ 1 & 0 & 0 \\ 1 & 2 & 2 \end{bmatrix} + \begin{bmatrix} 0 & 0 & 5 \\ 7 & 5 & 0 \\ 2 & 1 & 1 \end{bmatrix} = \begin{bmatrix} 1+0 & 3+0 & 1+5 \\ 1+7 & 0+5 & 0+0 \\ 1+2 & 2+1 & 2+1 \end{bmatrix} = \begin{bmatrix} 1 & 3 & 6 \\ 8 & 5 & 0 \\ 3 & 3 & 3 \end{bmatrix}$$

Another, much less often used notion of matrix addition is the direct sum.

SCALAR MULTIPLICATION

Given a matrix A and a number c, the scalar multiplication cA is computed by multiplying every element of A by the scalar c (i.e.) $(cA)_{i,j} = c \cdot a_{ij}$). For example:

$$2 \cdot \begin{bmatrix} 1 & 8 & -3 \\ 4 & -2 & 5 \end{bmatrix} = \begin{bmatrix} 2\cdot 1 & 2\cdot 8 & 2\cdot -3 \\ 2\cdot 4 & 2\cdot -2 & 2\cdot 5 \end{bmatrix} = \begin{bmatrix} 2 & 16 & -6 \\ 8 & -4 & 10 \end{bmatrix}.$$

Matrix addition and scalar multiplication turn the set $M(m, n, \mathbb{R})$ of all m-by-n matrices with real entries into a real vector space of dimension $m \cdot n$.

MATRIX MULTIPLICATION

Multiplication of two matrices is well-defined only if the number of columns of the left matrix is the same as the number of rows of the right matrix. If A is an m-by-n matrix and B is an n-by-p matrix, then their matrix product AB is the m-by-p matrix given by:

$$(AB)_{ij} = a_{i,1}b_{1,j} + a_{i,2}b_{2,j} + \ldots + a_{i,n}b_{n,j}$$

for each pair (i,j). For example:

$$\begin{bmatrix} 1 & 0 & 2 \\ -1 & 3 & 1 \end{bmatrix} \times \begin{bmatrix} 3 & 1 \\ 2 & 1 \\ 1 & 0 \end{bmatrix} = \begin{bmatrix} (1\times 3 + 0\times 2 + 2\times 1) & (1\times 1 + 0\times 1 + 2\times 0) \\ (-1\times 3 + 3\times 2 + 1\times 1) & (-1\times 1 + 3\times 1 + 1\times 0) \end{bmatrix}$$

$$= \begin{bmatrix} 5 & 1 \\ 4 & 2 \end{bmatrix}.$$

Matrix multiplication has the following properties:

- (AB)C = A(BC) for all k-by-m matrices A, m-by-n matrices B and n-by-p matrices C ("associativity").
- (A + B)C = AC + BC for all m-by-n matrices A and B and n-by-k matrices C ("right distributivity").
- C(A + B) = CA + CB for all m-by-n matrices A and B and k-by-m matrices C ("left distributivity").

Matrix multiplication is not commutative; that is, given matrices A and B and their product defined, then generally AB $\neq$ BA. It may also happen that AB is defined but BA is not defined.

Besides the ordinary matrix multiplication just described, there exist other operations on matrices that can be considered forms of multiplication, such as the Hadamard product and the Kronecker product.

LINEAR TRANSFORMATIONS

Matrices can conveniently represent linear transformations because matrix multiplication neatly corresponds to the composition of maps, as will be described next.

This same property makes them powerful data structures in high-level programming languages.

Here and in the sequel we identify R^n with the set of "columns" or *n*-by-1 matrices. For every linear map $f : R^n \rightarrow R^m$ there exists a unique *m*-by-*n* matrix A such that $f(x) = Ax$ for all x in R^n. We say that the matrix A "represents" the linear map *f*. Now if the *k*-by-*m* matrix B represents another linear map $g : R^m \rightarrow R^k$, then the linear map $g \circ f$ is represented by BA. This follows from the above-mentioned associativity of matrix multiplication.

More generally, a linear map from an *n*-dimensional vector space to an *m*-dimensional vector space is represented by an *m*-by-*n* matrix, provided that bases have been chosen for each.

RANKS

The rank of a matrix A is the dimension of the image of the linear map represented by A; this is the same as the dimension of the space generated by the rows of A, and also the same as the dimension of the space generated by the columns of A.

It can also be defined without reference to linear algebra as follows: the rank of an *m*-by-*n* matrix A is the smallest number *k* such that A can be written as a product BC where B is an *m*-by-*k* matrix and C is a *k*-by-*n* matrix (although this is not a practical way to compute the rank).

TRANSPOSE

The transpose of an *m*-by-*n* matrix A is the *n*-by-*m* matrix A^{tr} (also sometimes written as A^T or ${}^tA$) formed by turning rows into columns and columns into rows, i.e. $A^{tr}[i, j] = A[j, i]$ for all indices *i* and *j*. If A describes a linear map with respect to two bases, then the matrix A^{tr} describes the transpose of the linear map with respect to the dual bases.

We have $(A + B)^{tr} = A^{tr} + B^{tr}$ and $(AB)^{tr} = B^{tr} A^{tr}$.

SQUARE ROOT

Given two bounded matrices T and B, B is a square root of T if T = B*B.

SQUARE MATRICES AND RELATED DEFINITIONS

A square matrix is a matrix which has the same number of rows and columns. The set of all square *n*-by-*n* matrices, together with matrix addition and matrix multiplication is a ring. Unless $n = 1$, this ring is not commutative.

M(*n*, R), the ring of real square matrices, is a real unitary associative algebra. M(*n*, C), the ring of complex square matrices, is a complex associative algebra.

The unit matrix or identity matrix I_n, with elements on the main diagonal set to 1 and all other elements set to 0, satisfies $MI_n = M$ and $I_nN = N$ for any *m*-by-*n* matrix M and *n*-by-*k* matrix N. For example, if $n = 3$:

$$I_3 = \begin{bmatrix} 1 & 0 & 0 \\ 0 & 1 & 0 \\ 0 & 0 & 1 \end{bmatrix}.$$

The identity matrix is the identity element in the ring of square matrices.

Invertible elements in this ring are called invertible matrices or non-singular matrices. An *n* by *n* matrix A is invertible if and only if there exists a matrix B such that

$$AB = I_n \; (= BA).$$

In this case, B is the inverse matrix of A, denoted by A^{-1}. The set of all invertible *n*-by-*n* matrices forms a group (specifically a Lie group) under matrix multiplication, the general linear group.

If λ is a number and v is a non-zero vector such that $Av = \lambda v$, then we call v an eigenvector of A and λ the associated eigenvalue. (*Eigen* means "own" in German and in Dutch.) The number λ is an eigenvalue of A if and only if $A - \lambda I_n$ is not invertible, which happens if and only if $p_A(\lambda) = 0$. Here $p_A(x)$ is the characteristic polynomial of A. This is a polynomial of degree *n* and has therefore *n* complex roots (counting multiple roots according to their multiplicity). In this sense, every square matrix has *n* complex eigenvalues.

The determinant of a square matrix A is the product of its *n* eigenvalues, but it can also be defined by the *Leibniz formula*. Invertible matrices are precisely those matrices with a nonzero determinant.

The Gaussian elimination algorithm is of central importance: it can be used to compute determinants, ranks and inverses of matrices and to solve systems of linear equations.

SPECIAL TYPES OF MATRICES

In many areas in mathematics, matrices with certain structure arise. A few important examples are

- Symmetric matrices are such that elements symmetric about the *main diagonal* (from the upper left to the lower right) are equal, that is, $a_{i,j} = a_{j,i} \Leftrightarrow A^T = A$.
- Skew-symmetric matrices are such that elements symmetric about the *main diagonal* are the negative of each other, that is $a_{i,j} = a_{j,j} \Leftrightarrow A^T = -A$. In a skew-symmetric matrix, all diagonal elements are zero, that is, $a_{i,i} = -a_{i,i} = a_{i,i} \Rightarrow = 0$.
- Hermitian (or self-adjoint) matrices are such that elements symmetric about the diagonal are each others complex conjugates, that is, $a_{i,j} = \overline{a}_{j,i} \Leftrightarrow A^H = A,$, where $\overline{z}$ signifies the complex conjugate of a complex number *z* and A^H the conjugate transpose of *A*.

- Toeplitz matrices have common elements on their diagonals, that is, $a_{i,j} = a_{j+1\, j+1}$.
- Stochastic matrices are square matrices whose rows are probability vectors; they are used to define Markov chains.
- A square matrix A is called idempotent if $A^2 = AA = A$.

Probability is the likelihood or chance that something is the case or will happen. Probability theory is used extensively in areas such as statistics, mathematics, science and philosophy to draw conclusions about the likelihood of potential events and the underlying mechanics of complex systems.

INTERPRETATIONS

The word *probability* does not have a consistent direct definition. In fact, there are two broad categories of probability interpretations:

- Frequentists talk about probabilities only when dealing with well defined random experiments. The probability of a random event denotes the *relative frequency of occurrence* of an experiment's outcome, when repeating the experiment.
- Bayesians, however, assign probabilities to any statement whatsoever, even when no random process is involved. Probability, for a Bayesian, is a way to represent an individual's *degree of belief* in a statement, given the evidence.

MATHEMATICAL TREATMENT

In mathematics a probability of an event, A is represented by a real number in the range from 0 to 1 and written as P(A), p(A) or Pr(A). An impossible event has a probability of 0, and a certain event has a probability of 1. However, the converses are not always true: probability 0 events are not always impossible, nor probability 1 events certain. The rather subtle distinction between "certain" and "probability 1" is treated at greater length in the article on "almost surely".

The *opposite* or *complement* of an event A is the event [not A] (that is, the event of A not occurring); its probability is given by P(not A) = 1 – P(A). As an example, the chance of not rolling a six on a six-sided die is 1 – (chance of rolling a six) $= 1 - \frac{1}{6} = \frac{5}{6}$.

If two events, A and B are independent then the joint probability is

$$P(A \text{ and } B) = P(A \cap B) = P(A)\,P(B),$$

for example if two coins are flipped the chance of both being heads is $\frac{1}{2} \times \frac{1}{2} = \frac{1}{4}$.

If two events are mutually exclusive then the probability of either occurring is

$$P(A \text{ or } B) = P(A \cup B) = P(A) + P(B).$$

For example, the chance of rolling a 1 or 2 on a six-sided die is $P(1 \text{ or } 2) = P(1) + P(2) = \frac{1}{6} + \frac{1}{6} = \frac{1}{3}$.

If the events are not mutually exclusive then

$$P(A \text{ or } B) = P(A) + P(B) - P(A \text{ and } B).$$

For example, when drawing a single card at random from a regular deck of cards, the chance of getting a heart or a face card (J,Q,K) (or one that is both) is $\frac{13}{52} + \frac{12}{52} - \frac{3}{52} = \frac{11}{26}$, because of the 52 cards of a deck 13 are hearts, 12 are face cards, and 3 are both: here the possibilities included in the "3 that are both" are included in each of the "13 hearts" and the "12 face cards" but should only be counted once.

Conditional probability is the probability of some event A, given the occurrence of some other event B. Conditional probability is written $P(A|B)$, and is read "the probability of A, given B". It is defined by

$$P(A \mid B) = \frac{P(A \cap B)}{P(B)}.$$

If $P(B) = 0$ then $P(A \mid B)$ is undefined.

Summary of Probabilities

Event	**Probability**
A	$P(A) \in [0, 1]$
not A	$P(A') = 1 - P(A)$
A or B	$P(A \cap B) = P(A) + P(B) - P(P(A \cap B)$ $= P(A) + P(B)$ if A and B are mutually exclusive
A and B	$P(A \cap B) = P(A/B)\, P(B)$ $= P(A)\, P(B)$ if A and B are independent
A given B	$P(A|B)$

THEORY

Like other theories, the theory of probability is a representation of probabilistic concepts in formal terms-that is, in terms that can be considered separately from their meaning. These formal terms are manipulated by the rules of mathematics and logic, and any results are then interpreted or translated back into the problem domain.

There have been at least two successful attempts to formalize probability, namely the Kolmogorov formulation and the Cox formulation.

In Kolmogorov's formulation, sets are interpreted as events and probability itself as a measure on a class of sets. In Cox's theorem, probability is taken as a primitive (that is, not further analyzed) and the emphasis is on constructing a consistent assignment of probability values to propositions. In both cases, the laws of probability are the same, except for technical details.

There are other methods for quantifying uncertainty, such as the Dempster-Shafer theory and possibility theory, but those are essentially different and not compatible with the laws of probability as they are usually understood.

APPLICATIONS

Two major applications of probability theory in everyday life are in risk assessment and in trade on commodity markets.

Governments typically apply probabilistic methods in environmental regulation where it is called "pathway analysis", often measuring well-being using methods that are stochastic in nature, and choosing projects to undertake based on statistical analyses of their probable effect on the population as a whole.

It is not correct to say that statistics are involved in the modelling itself, as typically the assessments of risk are one-time and thus require more fundamental probability models, e.g. "the probability of another 9/11". A law of small numbers tends to apply to all such choices and perception of the effect of such choices, which makes probability measures a political matter.

A good example is the effect of the perceived probability of any widespread Middle East conflict on oil prices - which have ripple effects in the economy as a whole. An assessment by a commodity trader that a war is more likely vs. less likely sends prices up or down, and signals other traders of that opinion.

Accordingly, the probabilities are not assessed independently nor necessarily very rationally. The theory of behavioral finance emerged to describe the effect of such groupthink on pricing, on policy, and on peace and conflict.

It can reasonably be said that the discovery of rigorous methods to assess and combine probability assessments has had a profound effect on modern society. Accordingly, it may be of some importance to most citizens to understand how odds and probability assessments are made, and how they contribute to reputations and to decisions, especially in a democracy.

Another significant application of probability theory in everyday life is reliability. Many consumer products, such as automobiles and consumer electronics, utilize reliability theory in the design of the product in order to reduce the probability of failure. The probability of failure is also closely associated with the product's warranty.

RELATION TO RANDOMNESS

In a deterministic universe, based on Newtonian concepts, there is no probability if all conditions are known. In the case of a roulette wheel, if the force of the hand and the period of that force are known, then the number on which the ball will stop would be a certainty.

Of course, this also assumes knowledge of inertia and friction of the wheel, weight, smoothness and roundness of the ball, variations in hand speed during the turning and so forth. A probabilistic description can thus be more useful than Newtonian mechanics for analysing the pattern of outcomes of repeated rolls of roulette wheel. Physicists face the same situation in kinetic theory of gases, where the system, while deterministic *in principle*, is so complex (with the number of molecules typically the order of magnitude of Avogadro constant ($6{\cdot}10^{23}$) that only statistical description of its properties is feasible.

A revolutionary discovery of 20th century physics was the random character of all physical processes that occur at microscopic scales and are governed by the laws of quantum mechanics. The wave function itself evolves deterministically as long as no observation is made, but, according to the prevailing Copenhagen interpretation, the randomness caused by the wave function collapsing when an observation is made, is fundamental.

PROBABILITY DISTRIBUTION

A probability distribution describes the *values* and *probabilities* associated with a random event. The values must cover all of the possible outcomes of the event, while the total probabilities must sum to exactly 1, or 100%. For example, a single coin flip can take values *Heads* or *Tails* with a probability of exactly 1/2 for each; these two values and two probabilities make up the probability distribution of the single coin flipping event.

This distribution is called a *discrete distribution* because there are a countable number of discrete outcomes with positive probabilities.

A *continuous distribution* describes events over a continuous range, where the probability of a specific outcome is zero. For example, a dart thrown at a dartboard has essentially zero probability of landing at a specific point, since a point is vanishingly small, but it has some probability of landing within a given area.

The probability of landing within the small area of the bullseye would (hopefully) be greater than landing on an equivalent area elsewhere on the board. A smooth function that describes the probability of landing anywhere on the dartboard is the probability distribution of the dart throwing event. The integral of the probability density function (pdf) over the entire area of the dartboard (and, perhaps, the wall surrounding it) must be equal to 1, since each dart must land somewhere.

The concept of the probability distribution and the random variables which they describe underlies the mathematical discipline of probability theory, and the science of statistics. There is spread or variability in almost any value that can be measured in a population (e.g. height of people, durability of a metal, etc.); almost all measurements are made with some intrinsic error; in physics many processes are described probabilistically, from the kinetic properties of gases to the quantum mechanical description of fundamental particles.

For these and many other reasons, simple numbers are often inadequate for describing a quantity, while probability distributions are often more appropriate models. There are, however, considerable mathematical complications in manipulating probability distributions, since most standard arithmetic and algebraic manipulations cannot be applied.

RIGOROUS DEFINITIONS

In probability theory, every random variable may be attributed to a function defined on a state space equipped with a probability distribution that assigns a probability to every subset (more precisely every measurable subset) of its state space in such a way that the probability axioms are satisfied.

That is, probability distributions are probability measures defined over a state space instead of the sample space.

A random variable then defines a probability measure on the sample space by assigning a subset of the sample space the probability of its inverse image in the state space. In other words the probability distribution of a random variable is the push forward measure of the probability distribution on the state space.

Probability Distributions of Real-valued Random Variables

Because a probability distribution Pr on the real line is determined by the probability of being in a half-open interval Pr$(a, b]$, the probability distribution of a real-valued random variable X is completely characterized by its cumulative distribution function:

$$F(x) = \Pr[X \leq x] \quad \forall x \in \mathbb{R}$$

Discrete Probability Distribution

A probability distribution is called *discrete* if its cumulative distribution function only increases in jumps.

The set of all values that a discrete random variable can assume with non-zero probability is either finite or countably infinite because the sum of uncountably many positive real numbers (which is the smallest upper bound of the set of all finite partial sums) always diverges to infinity. Typically, the set of possible values is topologically discrete in the sense that all its points are isolated points. But, there are discrete random variables for which this countable set is dense on the real line.

Discrete distributions are characterized by a probability mass function, p such that

$$F(x) = \Pr[X \le x] = \sum_{x_i \le x} p(x_i).$$

Continuous Probability Distribution

By one convention, a probability distribution is called *continuous* if its cumulative distribution function is continuous, which means that it belongs to a random variable X for which $\Pr[X = x] = 0$ for all x in R.

Another convention reserves the term *continuous probability distribution* for absolutely continuous distributions. These distributions can be characterized by a probability density function: a non-negative Lebesgue integrable function f defined on the real numbers such that

$$F(x) = \Pr[X \le x] = \int_{-\infty}^{x} f(t)dt$$

Discrete distributions and some continuous distributions (like the devil's staircase) do not admit such a density.

Terminology

The support of a distribution is the smallest closed set whose complement has probability zero.

The probability density function of the sum of two independent random variables is the convolution of each of their density functions.

The probability density function of the difference of two random variables is the cross-correlation of each of their density functions.

A discrete random variable is a random variable whose probability distribution is discrete. Similarly, a continuous random variable is a random variable whose probability distribution is continuous.

FUNDAMENTALS OF EUCLIDEAN ALGORITHM

We consider the following problem: given two non-negative integers a and b, compute their greatest common divisor, gcd(a, b). We can do this using the well-known Euclidean algorithm, also called Euclid's algorithm.

The basic idea is the following. Without loss of generality, we may assume that $a \ge b \ge 0$. If $b = 0$, then there is nothing to do, since in this case, gcd(a, 0) = a. Otherwise, $b > 0$, and we can compute the integer quotient $q := \lfloor a/b \rfloor$ and remainder $r := a \bmod b$, where $0 \le r < b$. From the equation

$$a = bq + r,$$

it is easy to see that if an integer d divides both b and r, then it also divides a; likewise, if an integer d divides a and b, then it also divides r. From this observation, it follows that $\gcd(a, b) = \gcd(b, r)$, and so by performing a division, we reduce the problem of computing $\gcd(a, b)$ to the "smaller" problem of computing $\gcd(b, r)$.

Theorem: Euclid's algorithm runs in time $O(\text{len}(a)\ \text{len}(b))$.

Proof: We may assume that $b > 0$. The running time is $O(T)$, where

$$T = \sum_{i=1}^{\lambda} len(r_i) len(q_i) \leq len(b) \sum_{i=1}^{\lambda} len(q_i)$$

$$\leq len(b) \sum_{i=1}^{\lambda} (len(r_{i-1}) - len(r_i) + 1$$

$$= \text{len}(b)(\text{len}(r_0) - \text{len}(r_\lambda) + \lambda)$$

$$\leq \text{len}(b)(\text{len}(a) + \log b/\log \Phi + 1)$$

$$= O(\text{len}(a)\ \text{len}(b)).$$

Theorem: Let a, b be integers, with $a \geq b \geq 0$. Using the division with remainder property, define the integers $r_0, r_1, \ldots, r_{\lambda+1}$ and $q_1, \ldots, q_\lambda$, where $\lambda \geq 0$, as follows:

$$\begin{aligned}
a &= r_0, \\
b &= r_1, \\
r_0 &= r_1 q_1 + r_2\ (0 < r_2 < r_1), \\
&\vdots \\
r_{i-1} &= r_i q_i + r_{i+1}\ (0 < r_{i+1} < r_i), \\
&\vdots \\
r_{\lambda-2} &= r_{\lambda-1} q_{\lambda-1} + r_\lambda\ (0 < r_\lambda < r_{\lambda-1}), \\
r_{\lambda-1} &= r_\lambda q_\lambda\ (r_{\lambda+1} = 0).
\end{aligned}$$

Note that by definition, $\lambda = 0$ if $b = 0$, and $\lambda > 0$, otherwise. Then we have $r_\lambda = \gcd(a, b)$. Moreover, if $b > 0$, then $\lambda \leq \log b = \log \phi + 1$, where $\phi := (1 + \sqrt{5})/2 \approx 1.62$.

Proof: For the first statement, one sees that for $i = 1, \ldots, \lambda$, we have $r_{i-1} = r_i q_i + r_{i+1}$, from which it follows that the common divisors of r_{i-1} and r_i are the same as the common divisors of r_i and r_{i+1}, and hence $\gcd(r_{i-1}, r_i) = \gcd(r_i, r_{i+1})$.

From this, it follows that

$$\gcd(a, b) = \gcd(r_0, r_1) = \ldots = \gcd(r_\lambda, r_{\lambda+1}) = \gcd(r_\lambda, 0) = r_\lambda.$$

To prove the second statement, assume that $b > 0$, and hence $\lambda > 0$. If $\lambda = 1$, the statement is obviously true, so assume $\lambda > 1$. We claim that for $i = 0,..., \lambda - 1$, we have $r_{\lambda-i} \geq \phi^i$. The statement will then follow by setting $i = \lambda - 1$ and taking logarithms.

We now prove the above claim. For $i = 0$ and $i = 1$, we have

$$r_\lambda \geq 1 = \phi^0 \text{ and } r_{\lambda-1} \geq r_\lambda + 1 \geq \in \geq \phi^1.$$

For $i = 2,..., \lambda - 1$, using induction and applying the fact that $\phi^2 = \phi + 1$, we have

$$r_{\lambda-i} \geq r_{\lambda-(i-1)} + r_{\lambda-(i-2)} \geq \Phi^{i-1} + \Phi^{i-2} = \Phi^{i-2}(1+\Phi) = \Phi^i,$$

which proves the claim.

Example. Suppose $a = 100$ and $b = 35$. Then the numbers appearing in Theorem are easily computed as follows:

i	0	1	2	3	4
r_i	100	35	30	5	0
q_i		2	1	6	

So we have $\gcd(a, b) = r_3 = 5$.

EUCLID'S ALGORITHM

On input a, b, where a and b are integers such that $a \geq b \geq 0$, compute $d = \gcd(a, b)$ as follows:

$$r \leftarrow a,\ r' \leftarrow b$$

while $r' \neq 0$ do

$$r'' \leftarrow r \bmod r'$$

$$(r, r') \leftarrow (r', r'')$$

$$d \leftarrow r$$

output d

We now consider the running time of Euclid's algorithm. Naively, one could estimate this as follows. Suppose a and b are ℓ-bit numbers. The number of divisions performed by the algorithm is the number in Theorem, which is $O(\ell)$.

Moreover, each division involves numbers of ℓ bits or fewer in length, and so takes time $O(\ell^2)$. This leads to a bound on the running time of $O(\ell^3)$. However, as the following theorem shows, this cubic running time bound is well off the mark. Intuitively, this is because the cost of performing a division depends on the length of the quotient: the larger the quotient, the more expensive the division, but also, the more progress the algorithm makes towards termination.

EXERCISE

1. Show that if we run Euclid's algorithm on input a, b, where $a \geq b > 0$, then its running time is $O(\text{len}(a/d)\ \text{len}(b))$, where $d := \gcd(a, b)$.
2. With notation as in Theorem, give a direct and simple proof that for each $i = 1,\ldots, \Phi$, we have $r_{i+1} \leq r_{i-1}/2$. Thus, with every two division steps, the bit length of the remainder drops by at least 1. Based on this, give an alternative proof that the number of divisions is $O(\text{len}(b))$.

   Show that on these inputs, the number of division steps performed is also λ, the remainder sequence is $\{r_i / d\}_{i=0}^{\lambda+1}$, and the quotient sequence is $\{q_i\}_{i=1}^{\lambda}$
3. Let λ be a positive integer. Show that there exist integers a, b with $a > b > 0$ and $\geq \log b / \log \Phi$, such that Euclid's algorithm on input a, b performs at least λ divisions.
4. Show how to compute $\text{lcm}(a, b)$ in time $O(\text{len}(a)\ \text{len}(b))$.
5. Let $\alpha, \beta \in \mathbb{Z}$ with $a \geq b \geq 0$, let $d := \gcd(a, b)$, and assume $d > 0$. Suppose that on input a, b, Euclid's algorithm performs λ division steps, and computes the remainder sequence $\{r_i\}_{i=0}^{\lambda+1}$ and the quotient sequence $\{q_i\}_{i=1}^{\lambda}$. Now suppose we run Euclid's algorithm on input a/d, b/d.

MODULAR COMPUTATION

An important practical application of the above "computational" version of the Chinese remainder theorem is a general algorithmic technique that can significantly speed up certain types of computations involving long integers. Instead of trying to describe the technique in some general form, we simply illustrate the technique by means of a specific example: integer matrix multiplication.

Suppose we have two m X m matrices A and B whose entries are large integers, and we want to compute the product matrix $C := AB$. Suppose that for $r, s = 1,\ldots, m$, the entry of A at row r and column s is a_{rs}, and that for $s, t = 1,\ldots, m$, the entry of B at row s and column t is b_{st}. Then for $r, t = 1,\ldots, m$, the entry of C at row r and column t is c_{rt}, which is given by the usual rule for matrix multiplication:

$$c_{rt} = \sum_{s=1}^{m} a_{rs} b_{st}.$$

Suppose further that M is the maximum absolute value of the entries in A and B, so that the entries in C are bounded in absolute value by $M' := M^2 m$. Let $\ell := \text{len}(M)$. To simplify calculations, let us also assume that $m \leq M$ (this is reasonable, as we want to consider large values of M, greater than say 2^{100}, and certainly, we cannot expect to work with 2^{100} X 2^{100} matrices).

By just applying the formula, we can compute the entries of C using m^3 multiplications of numbers of length at most ℓ, and m^3 additions of numbers of length at most len(M'), where len(M') $\leq \ell \in +$ len(m) = $O(\ell)$. This yields a running time of

$$O(m^3 \ell^2).$$

Using the Chinese remainder theorem, we can actually do much better than this, as follows.

For every integer $n > 1$, and for all $r, t = 1,..., m$, we have

$$c_{rt} \equiv \sum_{s=1}^{m} a_{rs} b_{st} \pmod{n}.$$

Moreover, if we compute integers c'_{rt} such that

$$c'_{rt} \equiv \sum_{s=1}^{m} a_{rs} b_{st} \pmod{n}$$

and if we also have

$$-n/2 \leq c'_{rt} < n/2 \text{ and } n > 2M',$$

then we must have

$$c_{rt} = c'_{rt}.$$

that $c_{rt} \equiv c'_{rt}$ (mod n), which means that n divides ($c_{rt} - c'_{rt}$). Then from the bound $|c_{rt}| \leq M'$ and from, we obtain

$$|c_{rt} - c'_{rt}| \leq |c_{rt}| + |c'_{rt}| \leq M' + n/2 < n/2 + n/2 = n.$$

So we see that the quantity ($c_{rt} - c'_{rt}$) is a multiple of n, while at the same time this quantity is strictly less than n in absolute value; hence, this quantity must be zero.

So from the above discussion, to compute C, it suffices to compute the entries of C modulo n, where we have to make sure that we compute "balanced" remainders in the interval $[-n/2, n/2)$, rather than the more usual "least non-negative" remainders.

To compute C modulo n, we choose a number of small integers $n_1,..., n_k$, such that the family $\{n_i\}_{i=1}^{k}$ is pairwise relatively prime, and the product $n := \prod_{i=1}^{k} n_i$ is just a bit larger than $2M'$. In practice, one would choose the ni's to be small primes, and a table of such primes could easily be computed in advance, so that all problems up to a given size could be handled. For example, the product of all primes of at most 16 bits is a number that has more than 90,000 bits. Thus, by simply pre-computing and storing a table of small primes, we can handle input matrices with quite large entries (up to about 45,000 bits).

Let us assume that we have pre-computed appropriate small primes $n_1,..., n_k$. Further, we shall assume that addition and multiplication modulo each n_i can be done in constant time. This is reasonable from a practical (and theoretical) point of view, since such primes easily "fit" into a machine word, and we can perform modular addi. on and multiplication using a constant number of built-in machine operations. Finally, we assume that we do not use more n_i's than are necessary, so that $\text{len}(n) = O(\ell)$ and $k = O(\ell)$.

To compute C, we execute the following steps:

1. For each $i = 1,..., k$, do the following:

   (*a*) Compute $\hat{a}_{rs}^{(i)} \leftarrow a_{rs} \bmod n_i$ for $r, s = 1,..., m$,

   (*b*) Compute $\hat{b}_{rs}^{(i)} \leftarrow b_{st} \bmod n_i$ for $s, t = 1,..., m$,

   (*c*) For $r, t = 1,..., m$, compute

   $$\hat{c}_{rs}^{(i)} \leftarrow \sum_{s=1}^{m} \hat{a}_{rs}^{(i)} \hat{b}_{st}^{(i)} \bmod n_i.$$

2. For each $r, t = 1,..., m$, apply the Chinese remainder theorem to $\hat{c}_{rt}^{(1)}, \hat{c}_{rt}^{(2)},..., \hat{c}_{rt}^{(k)}$, obtaining an integer c_{rt}, which should be computed as a balanced remainder modulo n, so that $-n/2 \leq c_{rt} < n/2$.

3. Output the matrix C, whose entry in row r and column t is c_{rt}.

Note that in step 2, if our Chinese remainder algorithm happens to be implemented to return an integer a with $0 \leq a < n$, we can easily get a balanced remainder by just subtracting n from a if $a \geq n/2$.

The correctness of the above algorithm has already been established. Let us now analyze its running time. The running time of steps 1a and 1b is easily seen to be $O(m^2 \ell^2)$. Under our assumption about the cost of arithmetic modulo small primes, the cost of step 1c is $O(m^3 k)$, and since $k = O(\ell)$, the cost of this step is $O(m^3 \ell)$. Finally, by Theorem, the cost of step $\in$ is $O(m^2 \ell^2)$. Thus, the total running time of this algorithm is

$$O(m^2 \ell^2 + m^3 \ell).$$

This is a significant improvement over; for example, if $\ell \approx m$, then the running time of the original algorithm is $O(m^5)$, while the running time of the modular algorithm is $O(m^4)$.

EXTENSION OF EUCLIDEAN ALGORITHM

Let a and b be integers, and let $d:= \gcd(a, b)$. By Theorem there exist integers s and t such that $as + bt = d$.

The extended Euclidean algorithm allows us to efficiently compute s and t. The following theorem defines the quantities computed by this algorithm, and states a number of important facts about them – these will play a crucial role, both in the analysis of the running time of the algorithm, as well as in applications of the algorithm that we will discuss later.

Theorem: The extended Euclidean algorithm runs in time $O(\text{len}(a)\ \text{len}(b))$.

Proof. We may assume that $b > 0$. It suffices to analyze the cost of computing the coefficient sequences $\{s_i\}$ and $\{t_i\}$. Consider first the cost of computing all of the t_i's, which is $O(T)$, where $T = \sum_{i=1}^{\lambda} len(t_i)\ \text{len}(qi)$. We have $t_1 = 1$ and, by part (vi) of Theorem, we have $|t_i| \leq a$ for $i = 2,..., \lambda$. Arguing as in the proof of Theorem, we have

$$T \leq \text{len}(q_1) + \text{len}(a) \sum_{i=2}^{\lambda} len(q_i)$$

$$\leq \text{len}(a) + \text{len}(a)(\text{len}(r_1) - \text{len}(r_\lambda) + \lambda - 1) = O(\text{len}(a)\ \text{len}(b)).$$

An analogous argument shows that one can also compute all of the s_i's in time $O(\text{len}(a)$ $\text{len}(b))$, and in fact, in time $O(\text{len}(b)^2)$.

For the reader familiar with the basics of the theory of matrices and determinants, it is instructive to view Theorem as follows. For $i = 1,..., \lambda$, we have

$$\begin{pmatrix} r_i \\ r_{i+1} \end{pmatrix} = \begin{pmatrix} 0 & 1 \\ 1 & -q_i \end{pmatrix} \begin{pmatrix} r_{i-1} \\ r_i \end{pmatrix}.$$

Recursively expanding the right-hand side of this equation, we have

$$\begin{pmatrix} r_i \\ r_{i+1} \end{pmatrix} = \overbrace{\begin{pmatrix} 0 & 1 \\ 1 & -q_i \end{pmatrix} \cdots \begin{pmatrix} 0 & 1 \\ 1 & -q_1 \end{pmatrix}}^{M_i :=} \begin{pmatrix} a \\ b \end{pmatrix}.$$

This defines the $\in X \in$ matrix M_i for $i = 1,..., \lambda$. If we additionally define M_0 to be the $\in X \in$ identity matrix, then it is easy to see that for $i = 0,..., \lambda$, we have

$$M_i = \begin{pmatrix} s_i & t_i \\ s_{i+1} & t_{i+1} \end{pmatrix}.$$

From these observations, part (i) of Theorem is immediate, and part (ii) follows from the fact that M_i is the product of i matrices, each of determinant –1, and the determinant of M_i is evidently $s_it_{i+1} - t_is_{i+1}$.

Theorem: Let a, b, $r_0,..., r_{\lambda+1}$ and $q_1,..., q_\lambda$ be as in Theorem. Define integers $s_0,..., s_{\lambda+1}$ and $t_0,..., t_{\lambda+1}$ as follows:

$s_0 := 1,\quad t_0 := 0,$

$s_1 := 0,\quad t_1 := 1,$

$s_{i+1} := s_{i-1} - s_iq_i, t_{i+1} := t_{i-1} - t_iq_i \ (i = 1,..., \lambda).$

Then:

- For $i = 0,..., \lambda + 1$, we have $as_i + bt_i = r_i$; in particular, $as_\lambda + bt_\lambda = \gcd(a, b)$;
- For $i = 0,..., \lambda$, we have $s_it_{i+1} - t_is_{i+1} = (-1)^i$;
- For $i = 0,..., \lambda + 1$, we have $\gcd(s_i, t_i) = 1$;
- For $i = 0,..., \lambda$, we have $t_it_{i+1} \leq 0$ and $| t_i | \leq | t_{i+1} |$; for $i = 1,..., \lambda$, we have $s_is_{i+1} \leq 0$ and $| s_i | \leq | s_{i+1} |$;
- For $i = 1,..., \lambda + 1$, we have $r_{i-1} | t_i | \leq a$ and $r_{i-1} | s_i | \leq b$;
- If $a > 0$, then for $i = 1,..., \lambda + 1$, we have $| t_i | \leq a$ and $| s_i | \leq b$; if $a > 1$ and $b > 0$, then $| t_\lambda | \leq a/2$ and $| s_\lambda | \leq b/2$.

Proof: (*i*) Is easily proved by induction on i. For $i = 0, 1$, the statement is clear.

For $i = 2,..., \lambda + 1$, we have

$$\begin{aligned} as_i + bt_i &= a(s_{i-2} - s_{i-1}q_{i-1}) + b(t_{i-2} - t_{i-1}q_{i-1}) \\ &= (as_{i-2} + bt_{i-2}) - (as_{i-1} + bt_{i-1})q_{i-1} \\ &= r_{i-2} - r_{i-1}q_{i-1} \text{ (by induction)} \\ &= r_i. \end{aligned}$$

(*ii*) Is also easily proved by induction on i. For $i = 0$, the statement is clear. For $i = 1,..., \lambda$, we have

$$\begin{aligned} s_it_{i+1} - t_is_{i+1} &= s_i(t_{i-1} - t_iq_i) - ti(si-1 - siqi) \\ &= -(s_{i-1}t_i - t_{i-1}s_i) \text{ (after expanding and simplifying)} \\ &= -(-1)^{i-1} \text{ (by induction)} \\ &= (-1)^i. \end{aligned}$$

(*iii*) Follows directly from (ii).

For (iv), one can easily prove both statements by induction on i. The statement involving the t_i's is clearly true for $i = 0$. For $i = 1,..., \lambda$, we have

$t_{i+1} = t_{i-1} - t_i q_i$; moreover, by the induction hypothesis, t_{i-1} and t_i have opposite signs and $|t_i| \geq |t_{i-1}|$; it follows that $|t_{i+1}| = |t_{i-1}| + |t_i| q_i \geq |t_i|$, and that the sign of t_{i+1} is the opposite of that of t_i. The proof of the statement involving the s_i's is the same, except that we start the induction at $i = 1$.

For (v), one considers the two equations:

$$as_{i-1} + bt_{i-1} = r_{i-1},$$

$$as_i + bt_i = r_i.$$

Subtracting t_{i-1} times the second equation from t_i times the first, and applying (ii), we get $a = t_i r_{i-1} - t_{i-1} r_i$; consequently, using the fact that t_i and t_{i-1} have opposite sign, we obtain

$$a = |t_i r_{i-1} - t_{i-1} r_i| = |t_i| r_{i-1} + |t_{i-1}| r_i \geq |t_i| r_{i-1}.$$

The inequality involving s_i follows similarly, subtracting s_{i-1} times the second equation from s_i times the first.

(vi) follows from (v) and the following observations: if $a > 0$, then $r_{i-1} > 0$ for $i = 1,\ldots, \lambda + 1$; if $a > 1$ and $b > 0$, then $\lambda > 0$ and $r_{\lambda-1} \geq 2$.

Example. The *si*'s and *ti*'s are easily computed from the q_i's:

i	0	1	2	3	4
r_i	100	35	30	5	0
q_i		2	1	6	
s_i	1	0	1	−1	7
t_i	0	1	−2	3	−20

So we have $\gcd(a, b) = 5 = -a + 3b$.

THE EXTENDED EUCLIDEAN ALGORITHM

On input a, b, where a and b are integers such that $a \geq b \geq 0$, compute integers d, s, and t, such that $d = \gcd(a, b)$ and $as + bt = d$, as follows:

$r \leftarrow a, r' \leftarrow b$

$s \leftarrow 1, s' \leftarrow 0$

$t \leftarrow 0, t' \leftarrow 1$

while $r' \neq 0$ do

$q \leftarrow \lfloor r/r' \rfloor, r'' \leftarrow r \bmod r'$

$(r, s, t, r', s', t') \leftarrow (r', s', t', r'', s - s'q, t - t'q)$

$d \leftarrow r$

output d, s, t

EXERCISE

1. Suppose that on input a, b, the extended Euclidean algorithm computes the coefficient sequences $\{s_i\}_{i=0}^{\lambda+1}$ and $\{t_i\}_{i=0}^{\lambda+1}$. Show that the extended Euclidean algorithm on input a/d, b/d computes the same coefficient sequences.
2. Suppose we modify the extended Euclidean algorithm so that it computes balanced remainders; that is, for $i = 1,\dots, \lambda$, the values q_i and r_{i+1} are computed so that $r_{i-1} = r_i q_i + r_{i+1}$ and $-|r_i|/2 \leq r_{i+1} < |r_i|/2$. Give a detailed analysis of the running time of this algorithm, which should include an analysis of the number of division steps, and the sizes of the s_i's and t_i's.
3. In our description of the extended Euclidean algorithm, we made the restriction that the inputs a and b satisfy $a \geq b \geq 0$. Using this restricted algorithm as a subroutine, give an algorithm that works without any restrictions on its input.

FERMAT'S TWO SQUARES THEOREM

We have already that every prime $p \equiv 1 \pmod 4$ can be expressed as a sum of two squares of integers. In this section, we make this theorem computationally effective; that is, we develop an effcient algorithm that takes as input a prime $p \equiv 1 \pmod 4$, and outputs integers r and t such that $p = r^2 + t^2$.

This lemma asserts the existence of certain numbers, and we proved it using the "pigeonhole principle," which unfortunately does not translate directly into an efficient algorithm to actually find these numbers. However, we can show that these numbers arise as a "natural by-product" of the extended Euclidean algorithm. To make this more precise, let us introduce some notation. For integers a, b, with $a \geq b \geq 0$, let us define

$$\text{EEA}(a, b) := \{(r_i, s_i, t_i)\}_{i=0}^{\lambda+1},$$

where r_i, s_i, and t_i, for $i = 0,\dots, \lambda + 1$.

Thue's lemma

Theorem: Let $n, b, r^*, t^* \in \mathbb{Z}$, with $0 \leq b < n$ and $0 < r^* \leq n < r^* t^*$. Further, let $\text{EEA}(n, b) = \{(r_i, s_i, t_i)\}_{i=0}^{\lambda+1}$, and let j be the smallest index (among $0,\dots, \lambda + 1$) such that $r_j < r^*$. Then, setting $r := r_j$ and $t := t_j$, we have

$$r \equiv bt \pmod n,\ 0 \leq r < r^*, \text{ and } 0 < |t| < t^*.$$

Proof: Since $r_0 = n \geq r^* > 0 = r_{\lambda+1}$, the value of the index j is well defined; moreover, $j \geq 1$ and $r_{j-1} \geq r^*$. It follows that

$$
\begin{aligned}
|t| &\le n/r_{j-1} \\
&\le n/r^* \\
&< t^* \text{ (since } n < r^*t^*).
\end{aligned}
$$

Since $j \geq 1$, by part (iv) of Theorem, we have $|t_j| \geq |t_1| > 0$. Finally, since $r_j = ns_j + bt_j$, we have $r_j \equiv bt_j \pmod n$.

What this theorem says is that given n, b, r^*, t^*, to find the desired values r and t, we run the extended Euclidean algorithm on input n, b.

This generates a sequence of remainders $r_0 > r_1 > r_2 > \ldots$, where $r_0 = n$ and $r_1 = b$. If r_j is the first remainder in this sequence that falls below r^*, and if s_j and t_j are the corresponding numbers computed by the extended Euclidean algorithm, then $r := r_j$ and $t := t_j$ do the job.

The essential ingredient, which guarantees the existence of a square root of -1 modulo p when p is a prime congruent to 1 modulo 4. We need an effective version of this result as well.

Assume we are given a prime $p \equiv 1 \pmod 4$, and we want to compute $\beta \in \mathbb{Z}_p^*$ such that $\beta^2 = -1$. By Theorem, it suffices to find $\gamma \in \mathbb{Z}_p^* \setminus (\mathbb{Z}_p^*)^2$, since then $\beta := \gamma^{(p-1)/4}$ (which we can efficiently compute via repeated squaring) satisfies $\beta^2 = -1$.

While there is no known efficient, deterministic algorithm to find such a γ, we do know that half the elements of $\mathbb{Z}_p^*$ are squares and half are not, which suggests the following simple "trial and error" strategy to compute β:

repeat
choose $\gamma \in \mathbb{Z}_p^*$
compute $\beta \leftarrow \gamma^{(p-1)/4}$
until $\beta^2 = -1$
output β

As an algorithm, this is not fully specified, as we have to specify a procedure for selecting γ in each loop iteration.

A reasonable approach is to simply choose γ at random: this would be an example of a probabilistic algorithm. Let us assume for the moment that this makes sense from a mathematical and algorithmic point of view, so that with each loop iteration, we have a 50% chance of picking a "good" γ, that is, one that is not in $(\mathbb{Z}_p^*)^2$. From this, it follows that with high probability, we should find a "good" in just a few loop iterations (the probability that after k loop iterations we still have not found one is $1/2^k$), and that the expected number of loop iterations is just 2.

The running time of each loop iteration is dominated by the cost of repeated squaring, which is $O(\text{len}(p)^3)$. It follows that the expected running time of this algorithm is $O(\text{len}(p)^3)$.

Let us now put all the ingredients together to get an algorithm to find r, t such that $p = r^2 + t^2$.

1. Find $\beta \in \mathbb{Z}_p^*$ such that $\beta^2 = -1$, using the above "trial and error" strategy.
2. Set $b \leftarrow \text{rep}(\beta)$ (so that $\beta = [b]$ and $b \in \{0,\ldots, p-1\}$).
3. Run the extended Euclidean algorithm on input p, b to obtain EEA(p, b), and then apply Theorem with $n := p$, b, and $r^* := t^* := \lfloor\sqrt{p}\rfloor + 1$, to obtain the values r and t.
4. Output r, t.

When this algorithm terminates, we have $r^2 + t^2 = p$, as required: as we argued in the proof of Theorem4, since $r \equiv bt \pmod p$ and $b^2 \equiv -1 \pmod p$, it follows that $r^2 + t^2 \equiv 0 \pmod p$, and since $0 < r^2 + t^2 < 2p$, we must have $r^2 + t^2 = p$. The (expected) running time of step 1 is $O(\text{len}(p)^3)$. The running time of step 3 is $O(\text{len}(p)^2)$. Thus, the total (expected) running time is $O(\text{len}(p)^3)$.

Example. One can check that $p := 1009$ is prime and $p \equiv 1 \pmod 4$. Let us express p as a sum of squares using the above algorithm.

First, we need to find a square root of -1 modulo p. Let us just try a random number, say 17, and raise this to the power $(p-1)/4 = 252$.

One can calculate that $17^{252} \equiv 469 \pmod{1009}$, and $469^2 \equiv -1 \pmod{1009}$. So we were lucky with our first try. Now we run the extended Euclidean algorithm on input $p = 1009$ and $b = 469$, obtaining the following data:

i	r_i	q_i	s_i	t_i
0	1009		1	0
1	469	2	0	1
2	71	6	1	−2
3	43	1	−6	13
4	28	1	7	−15
5	15	1	−13	28
6	13	1	20	−43
7	2	6	−33	71
8	1	2	218	−469
9	0		−469	1009

The first r_j that falls below the threshold $r^* = \lfloor\sqrt{1009}\rfloor + 1 = 32$ is at $j = 4$, and so we set $r := 28$ and $t := -15$. One verifies that $r^2 + t^2 = 28^2 + 15^2 = 1009 = p$.

It is natural to ask whether one can solve this problem without resorting to randomization. The answer is "yes", but the only known deterministic algorithms for this problem are quite impractical (albeit polynomial time). This example illustrates the utility of randomization as an algorithm design technique, one that has proved to be invaluable in solving numerous algorithmic problems in number theory; we already mentioned its use in connection with primality testing, and we will explore many other applications as well.

CHINESE REMAINDERING

An important application of the extended Euclidean algorithm is to the problem of computing multiplicative inverses in $\mathbb{Z}_n$.

Theorem: Suppose we are given integers $n_1,\ldots, n_k$ and $a_1,\ldots, a_k$, where the family $\{n_i\}_{i=1}^k$ is pairwise relatively prime, and where $n_i > 1$ and $0 \leq a_i < n_i$ for $i = 1,\ldots, k$. Let $n := \prod_{i=1}^k n_i$. Then in time $O(\text{len}(n)^2)$, we can compute the unique integer a satisfying $0 \leq a < n$ and $a \equiv a_i \pmod{n_i}$ for $i = 1,\ldots, k$.

Proof. The algorithm is a straightforward implementation of the proof of Theorem, and runs as follows:

$n \leftarrow \prod_{i=1}^k n_i$

for $i \leftarrow 1$ to k do

$n_i^* \leftarrow n/n_i$, $b_i \leftarrow n_i^* \bmod n_i$, $t_i \leftarrow b_i^{-1} \bmod n_i$, $e_i \leftarrow n_i^* t_i$

$a \leftarrow \left(\sum_{i=1}^k a_i e_i\right) \bmod n$

We leave it to the reader to verify the running time bound.

Theorem: Suppose we are given integers a, n, where $0 \leq a < n$. Then in time $O(\text{len}(n)^2)$, we can determine if a is relatively prime to n, and if so, compute $a^{-1} \bmod n$.

Proof: We may assume $n > 1$, since when $n = 1$, we have $a = 0$ and $a^{-1} \bmod n = 0$. We run the extended Euclidean algorithm on input n, a, obtaining integers d, s, and t, such that $d = \gcd(n, a)$ and $ns + at = d$. If $d \neq 1$, then a does not have a multiplicative inverse modulo n. Otherwise, if $d = 1$, then t is a multiplicative inverse of a modulo n; however, it may not lie in the range $\{0,\ldots, n-1\}$, as required.

We have $|t| \leq n/2 < n$. Thus, if $t \geq 0$, then $a^{-1} \bmod n$ is equal to t; otherwise, a^{-1} mod

n is equal to $t + n$. Based on Theorem, it is clear that all the computations can be performed in time $O(\text{len}(n)^2)$.

Example. Suppose we are given integers a, b, n, where $0 \leq a < n$, and $0 \leq b < n$, and we want to compute a solution z to the congruence $az \equiv b \pmod{n}$, or determine that no such solution exists. Based on the discussion in Example, the following algorithm does the job:

$d \leftarrow \gcd(a, n)$

if $d \mid b$ then

output "no solution"

else

$a' \leftarrow a/d$, $b' \leftarrow b/d$, $n' \leftarrow n/d$

$t \leftarrow (a')^{-1} \bmod n'$

$z \leftarrow tb' \bmod n'$

output z

Using Euclid's algorithm to compute d, and the extended Euclidean algorithm to compute t, the running time of this algorithm is clearly $O(\text{len}(n)^2)$.

EXERCISE

1. Suppose we are given $\alpha_1, \ldots, \alpha_k \in \mathbb{Z}_n$. Show how to compute $a_1^{-1}, \ldots, a_k^{*1}$ k by computing one multiplicative inverse modulo n, and performing fewer than $3k$ multiplications modulo n. This result is useful, as in practice, if n is several hundred bits long, it may take 10–20 times longer to compute multiplicative inverses modulo n than to multiply modulo n.
2. Suppose that we are given a positive integer n, two elements $\alpha, \beta \in \mathbb{Z}_n^*$, and integers ℓ and m, such that $\alpha^{\ell} = \beta^m$ and $\gcd(\ell, m) = 1$. Show how to compute $\alpha \in \mathbb{Z}_n^*$ such that $\alpha = \gamma^m$ in time $O(\text{len}(\ell)\,\text{len}(m) + (\text{len}(\ell) + \text{len}(m))\,\text{len}(n)^2)$.
3. Using the algorithm in the previous exercise as a subroutine, give a simple $O(\text{len}(n)^2)$ algorithm that takes as input integers $n_1, \ldots, n_k$ and $a_1, \ldots, a_k$, where the family $\{n_i\}_{i=1}^k$ is pairwise relatively prime, and where $n_i > 1$ and $0 \leq a_i < n_i$ for $i = 1, \ldots, k$, and outputs integers a and n such that $0 \leq a < n$, $n = \sum_{i=1}^k n_i$, and $a \equiv a_i \pmod{n_i}$ for $i = 1, \ldots, k$. The algorithm should be "incremental," in that it processes the pairs (a_i, n_i) one at a time, using time $O(\text{len}(n)\,\text{len}(n_i))$ per pair.
4. In this exercise and the next, you are to analyze an "incremental Chinese remaindering algorithm." Consider the following algorithm, which takes as input integers a_1, n_1, a_2, n_2 satisfying

$$0 \leq a_1 < n_1, 0 \leq a_2 < n_2, \text{ and } \gcd(n_1, n_2) = 1.$$

It outputs integers a, n satisfying

$n = n_1 n_2$, $0 \leq a < n$, $a \equiv a_1 \pmod{n_1}$, and $a \equiv a_2 \pmod{n_2}$, and runs as follows:

$b \leftarrow n_1 \bmod n_2$, $t \leftarrow b^{-1} \bmod n_2$, $h \leftarrow (a_2 - a_1)t \bmod n_2$

$a \leftarrow a_1 + n_1 h$, $n \leftarrow n_1 n_2$

output a, n

Show that the algorithm correctly computes a and n as specified, and runs in time $O(\text{len}(n)\ \text{len}(n_2))$.

RATIONAL RECONSTRUCTION

This lemma asserts that for given integers n and b, there exists a pair of integers (r, t) satisfying $r \equiv bt \pmod{n}$, and contained in a prescribed rectangle, provided the area of the rectangle is large enough, relative to n.

In this section, we first prove a corresponding uniqueness theorem, under the assumption that the area of the rectangle is not too large; of course, if $r \equiv bt \pmod{n}$, then for any non-zero integer q, we also have $rq \equiv b(tq) \pmod{n}$, and so we can only hope to guarantee that the ratio r/t is unique. After proving this uniqueness theorem, we show how to make this theorem computationally effective, and then develop several ver,' neat applications.

Theorem: Let $n, b, r^*, t^* \in Z$ witn $0 \leq b < n$, $0 \leq r^* < n$, and $t^* > 0$. Further, let EEA (n, b) $=\{(r_i, s_i, t_i)\}_{i=0}^{\lambda+1}$, and let j be the smallest index (among $0, \ldots, \lambda + 1$) such that $r_j \leq r^*$, and set

$$r' := r_j,\ s' := s_j, \text{ and } t' := t_j.$$

Finally, suppose that there exist $r, s, t \in \mathbb{Z}$ such that

$$r = ns + bt,\ |r| \leq r^*, \text{ and } 0 < |t| \leq t^*.$$

Then we have:

- $0 < |t'| \leq t$;
- If $n > 2r^*t^*$, then for some non-zero integer q,
  $r = r'q$, $s = s'q$, and $t = t'q$.

Proof: Since $r' = n > r^* \geq 0 = r_{\lambda+1}$, the value of j is well defined, and moreover, $j \geq 1$, and we have the inequalities

$$0 \leq r_j \leq r^* < r_{j-1},\ 0 < |t_j|,\ |r| \leq r^*, \text{ and } 0 < |t| \leq t^*,$$

along with the identities

$$r_{j-1} = ns_{j-1} + bt_{j-1},$$
$$r_j = ns_j + bt_j,$$
$$r = ns + bt.$$

We now turn to part (i) of the theorem. Our goal is to prove that

$$| t_j | \le t.$$

This is the hardest part of the proof. To this end, let

$$\varepsilon: = s_j t_{j-1} - s_{j-1} t_j,\ \mu: = (t_{j-1}s - s_{j-1}t)/\varepsilon,\ v:= (s_j t - t_j s)/\varepsilon.$$

Since $\varepsilon = \pm 1$, the numbers μ and v are integers; moreover, one may easily verify that they satisfy the equations

$$s_j\mu + s_{j-1} = s,$$
$$t_j\mu + t_{j-1} = t.$$

We now use these identities. We consider three cases:

- Suppose $v = 0$. In this case, implies $t_j \mid t$, and since $t \neq 0$, this implies $| t_j | \le | t | \le t$.
- Suppose $\mu v < 0$. In this case, since t_j and t_{j-1} have opposite sign, implies $| t | = | t_j\mu | + | t_{j-1}v | \ge | t_j |$, and so again, we have $| t_j | \le | t | \le t^*$.
- The only remaining possibility is that $v \neq 0$ and $\mu v \ge 0$. We argue that this is impossible. Adding n times to b times and using the identities, we get

$$r_j\mu + r_{j-1} = r.$$

If $v \neq 0$ and μ and v had the same sign, we would have $| r | = | r_j | + | r_{j-1}v | \ge r_{j-1}$, and hence $r_{j-1} \le | r | \le r^*$; however, this contradicts the fact that $r_{j-1} > r^*$.

That proves the inequality. We now turn to the proof of part (ii) of the theorem, which relies critically on this inequality. Assume that

$$n > 2r^*t^*.$$

we have

$$r_j \equiv bt_j \pmod{n} \text{ and } r \equiv bt \pmod{n}.$$

Combining this with the inequalities, we see that the hypotheses of Theorem are satisfied, and so we may conclude that

$$rt_j - r_j t = 0.$$

Subtracting t_j times from t times, and using the identity, we obtain $n(stj - s_j t) = 0$, and hence

$$st_j - s_j t = 0.$$

We see that $t_j \mid s_j t$, and since $\gcd(s_j, t_j) = 1$, we must have $t_j \mid t$. So $t = t_j q$ for some q, and we must have $q \neq 0$ since $t \neq 0$. Substituting $t_j q$ for t in equations yields $r = r_j q$ and $s = s_j q$. That proves part (ii) of the theorem.

Theorem: Let $n, b, r^*, t^* \in \mathbb{Z}$ with r^* 0, $t^* > 0$, and $n > 2r^*t^*$. Further,

suppose that $r, t, r', t' \in \mathbb{Z}$ satisfy

$$r \in bt \pmod{n}, |r| \leq r^*, 0 < |t| \leq t^*,$$

$$r' \in bt' \pmod{n}, |r'| \leq r^*, 0 < |t'| \leq t^*.$$

Then $r/t = r'/t'$.

Proof. Consider the two congruences

$$r \equiv bt \pmod{n},$$

$$r' \equiv bt' \pmod{n}.$$

Subtracting t times the second from t' times the first, we obtain

$$rt' - r't \equiv 0 \pmod{n}.$$

However, we also have

$$|rt' - r't| \leq |r||t'| + |r'||t| \leq 2r^*t^* < n.$$

Thus, $rt' - r't$ is a multiple of n, but less than n in absolute value; the only possibility is that $rt' - r't = 0$, which means $r/t = r'/t'$.

Now suppose that we are given $n, b, r^*, t^* \in \mathbb{Z}$ as in the above theorem; moreover, suppose that there exist $r, t \in \mathbb{Z}$ satisfying, but that these values are not given to us.

Note that under the hypothesis of Theorem, Thue's lemma cannot be used to ensure the existence of such r and t, but in our eventual applications, we will have other reasons that will guarantee this. We would like to find $r', t' \in \mathbb{Z}$ satisfying, and if we do this, then by the theorem, we know that $r/t = r'/t'$.

We call this the rational reconstruction problem. We can solve this problem efficiently using the extended Euclidean algorithm; indeed, just as in the case of our effective version of Thue's lemma, the desired values of r' and t' appear as "natural by-products" of that algorithm. To state the result precisely, let us recall the notation we introduced in the last section: for integers a, b, with $a \geq b \geq 0$, we defined

$$\text{EEA}(a, b) := \{(r_i, s_i, t_i)\}_{i=0}^{\lambda+1},$$

where r_i, s_i, and t_i, for $i = 0, \ldots, \lambda + 1$.

APPLICATION

RECOVERING FRACTIONS FROM THEIR DECIMAL EXPANSIONS

A familiar fact to the reader that every real number has a decimal expansion, and that this decimal expansion is unique, provided one rules out those expansions that end in an infinite run of 9's (e.g., 1/10 = 0.1000... = 0.0999...).

Now suppose that Alice and Bob play a game. Alice thinks of a rational number $z := s/t$, where s and t are integers with $0 \leq s < t$, and tells Bob some of the highorder digits in the decimal expansion of z. Bob's goal in the game is to determine z. Can he do this?

The answer is "yes," provided Bob knows an upper bound$M$ on t, and provided Alice gives Bob enough digits.

Of course, Bob probably remembers from grade school that the decimal expansion of z is ultimately periodic, and that given enough digits of z so that the periodic part is included, he can recover z; however, this technique is quite useless in practice, as the length of the period can be huge – $\Theta(M)$ in the worst case. The method we discuss here requires only $O(\text{len}(M))$ digits.

Suppose Alice gives Bob the high-order k digits of z, for some $k \geq 1$. That is, if

$$z = 0.\, z_1 z_2 z_3 \ldots$$

is the decimal expansion of z, then Alice gives Bob $z_1, \ldots, z_k$. Now, if 10^k is much smaller than M^2, the number z is not even uniquely determined by these digits, since there are $\Omega(M^2)$ distinct rational numbers of the form s/t, with $0 \leq s < t \leq M$. However, if $10^k > 2M^2$, then not only is z uniquely determined by $z_1, \ldots, z_k$, but using Theorem, Bob can efficiently compute it.

We shall presently describe efficient algorithms for both Alice and Bob, but before doing so, we make a few general observations about the decimal expansion of z. Let e be an arbitrary non-negative integer, and suppose that the decimal expansion of z. Observe that

$$10^e z = z_1 \ldots z_e .\, z_{e+1} z_{e+2} \ldots$$

It follows that

$$\lfloor 10^e z \rfloor = z_1 \ldots z_e .0.$$

Since $z = s/t$, if we set $r := 10^e s \bmod t$, then $10^e s = \lfloor 10^e z \rfloor t + r$, and dividing this by t,

we have $10^e z = \lfloor 10^e z \rfloor + r/t$, where $r/t \in [0, 1)$. Therefore,

$$\frac{10^e s \bmod t}{t} = 0.z_{e+1} z_{e+2} z_{e+3 \ldots}.$$

Next, consider Alice. Based on the above discussion, Alice may use the following simple, iterative algorithm to compute $z_1,..., z_k$, for arbitrary $k \geq 1$, after she chooses s and t:

$x_1 \leftarrow s$

for $i \leftarrow 1$ to k do

$y_i \leftarrow 10x_i$

$z_i \leftarrow \lfloor y_i / t \rfloor$

$x_{i+1} \leftarrow y_i \bmod t$

output $z_1,..., z_k$

Correctness follows easily from the observation that for each $i = 1, 2,...$, we have $x_i = 10^{i-1} s \bmod t$; indeed, applying with $e = i - 1$, we have $x_i/t = 0.z_i z_{i+1} z_{i+2}...$, and consequently, by with $e = 1$ and x_i/t in the role of z, we have $\lfloor 10x_i / t \rfloor = z_i$. The total time for Alice's computation is $O(k \text{ len}(M))$, since each loop iteration takes time $O(\text{len}(M))$.

Finally, consider Bob. Given the high-order digits $z_1,..., z_k$ of $z = s/t$, along with the upper bound M on t, he can compute z as follows:

1. Compute $n \leftarrow 10^k$ and $b \leftarrow \sum_{i=1}^{k} z_i 10^{k-i}$.
2. Run the extended Euclidean algorithm on input n, b to obtain EEA(n, b), and then apply Theorem with n, b, and $r^* := t^* := M$, to obtain the values r', s', t'.
3. Output the rational number $-s'/t'$.

Let us analyze this algorithm, assuming that $10^k > 2M^2$.

For correctness, we must show that $z = -s'/t'$. To prove this, observe that by with $e = k$, we have $b = \lfloor nz \rfloor = \lfloor nz / t \rfloor$. Moreover, if we set $r := ns \bmod t$, then we have

$$r = ns - bt,\ 0 \leq r < t \leq r^*,\ 0 < t \leq t^*, \text{ and } n > 2r^*t^*.$$

It follows that the integers s', t' from Theorem satisfy $s = s'q$ and $-t = t'q$ for some non-zero integer q. Thus, $s/t = -s'/t'$, as required. As a bonus, since the extended Euclidean algorithm guarantees that $\gcd(s', t') = 1$, not only do we obtain z, but we obtain z expressed as a fraction in lowest terms.

We leave it to the reader to verify that Bob's computation may be performed in time $O(k^2)$.

We conclude that both Alice and Bob can successfully play this game with k chosen so that $k = O(\text{len}(M))$, in which case, their algorithms run in time $O(\text{len}(M)^2)$.

One interpretation of the Chinese remainder theorem is that if we "encode" an integer a, with $0 \leq a < n$, as the sequence $(a_1, \ldots, a_k)$, where $a_i = a \bmod n_i$ for $i = 1, \ldots, k$, then we can efficiently recover a from this encoding. Here, of course, $n = n_1 \ldots n_k$, and the family $\{n_i\}_{i=1}^{k}$ is pairwise relatively prime.

Suppose that Alice encodes a as $(a_1, \ldots, a_k)$, and sends this encoding to Bob over some communication network; however, because the network is not perfect, during the transmission of the encoding, some (but hopefully not too many) of the values $a_1, \ldots, a_k$ may be corrupted. The question is, can Bob still efficiently recover the original a from its corrupted encoding?

To make the problem more precise, suppose that the original, correct encoding of a is $(a_1, \ldots, a_k)$, and the corrupted encoding is $(b_1, \ldots, b_k)$. Let us define $G \subseteq \{1, \ldots, k\}$ to be the set of "good" positions i with $a_i = b_i$, and $B \subseteq \{1, \ldots, k\}$ to be the set of "bad" positions i with $a_i \neq b_i$. We shall assume that $|B| \leq \ell$, where ℓ is some specified parameter.

Of course, if Bob hopes to recover a, we need to build some redundancy into the system; that is, we must require that $0 \leq a \leq M$ for some bound M that is somewhat smaller than n. Now, if Bob knew the location of bad positions, and if the product of the n_i's at the good positions exceeds M, then Bob could simply discard the errors, and reconstruct a by applying the Chinese remainder theorem to the a_i's and n_i's at the good positions. However, in general, Bob will not know a priori the locations of the bad positions, and so this approach will not work.

Despite these apparent difficulties, Theorem may be used to solve the problem quite easily, as follows. Let P be an upper bound on the product of any ℓ of the integers $n_1, \ldots, n_k$ (e.g., we could take P to be the product of the ℓ largest numbers among $n_1, \ldots, n_k$). Further, let us assume that $n > 2MP^2$.

Now, suppose Bob obtains the corrupted encoding $(b_1, \ldots, b_k)$. Here is what Bob does to recover a:

1. Apply the Chinese remainder theorem, obtaining the integer b satisfying $0 \leq b < n$ and $b \equiv b_i \pmod{n_i}$ for $i = 1, \ldots, k$.
2. Run the extended Euclidean algorithm on input n, b to obtain EEA(n, b), and then apply Theorem with n, b, $r^* := MP$ and $t^* := P$, to obtain values r', s', t'.
3. If $t' \mid r'$, output the integer r'/t'; otherwise, output "error."

We claim that the above procedure outputs a, under our assumption that the set B of bad positions is of size at most ℓ. To see this, let $t := \prod_{i \in B} n_i$. By construction, we have $1 \leq t \leq P$. Also, let $r := at$, and note that $0 \leq r \leq r^*$ and $0 < t \leq t^*$. We claim that

$$r \equiv bt \pmod{n}.$$

To show that holds, it suffices to show that

$$at \equiv bt \pmod{n_i}$$

for all $i = 1,..., k$. To show this, for each index i we consider two cases:

Case 1. $i \in G$. In this case, we have $a_i = b_i$, and therefore,

$$at \equiv a_i t \equiv b_i t \equiv bt \pmod{n_i}.$$

Case 2. $i \in B$. In this case, we have $n_i \mid t$, and therefore,

$$at \equiv 0 \equiv bt \pmod{n_i}.$$

Thus, holds for all $i = 1,..., k$, and so it follows that holds. Therefore, the values r', t' obtained from Theorem satisfy

$$\frac{r'}{t'} = \frac{r}{t} = \frac{at}{t} = a.$$

One easily checks that both the procedures to encode and decode a value a run in time $O(\text{len}(n)^2)$.

The above scheme is an example of an error correcting code, and is actually the integer analog of a Reed–Solomon code.

Example. Suppose we want to encode a 1024-bit message as a sequence of 16- bit blocks, so that the above scheme can correct up to 3 corrupted blocks. Without any error correction, we would need just 1024/16 = 64 blocks. However, to correct this many errors, we need a few extra blocks; in fact, 7 will do.

Of course, a 1024-bit message can naturally be viewed as an integer a in the set $\{0,..., 2^{1024} - 1\}$, and the ith 16-bit block in the encoding can be viewed as an integer a_i in the set $\{0,..., 2^{16} - 1\}$. Setting $k := 71$, we select k primes, $n_1,..., n_k$, each 16-bits in length. In fact, let us choose $n_1,..., n_k$ to be the largest k primes under 2^{16}. If we do this, then the smallest prime among the n_i's turns out to be 64717, which is greater than $2^{15.98}$. We may set $M := 2^{1024}$, and since we want to correct up to 3 errors, we may set $P := 2^{3.16}$. Then with $n := \prod_i n_i$, we have

$$n > 2^{71.15.98} = 2^{1134.58} > 2^{1121} = 2^{1+1024+6.16} = 2MP^2.$$

Thus, with these parameter settings, the above scheme will correct up to 3 corrupted blocks. This comes at a cost of increasing the length of the message from 1024 bits to 71.16 = 1136 bits, an increase of about 11%.

APPLICATIONS TO SYMBOLIC ALGEBRA

Rational reconstruction also has a number of applications in symbolic algebra. We briefly sketch one such application here. Suppose that we want to find the solution v to the equation $vA = w$, where we are given as input a non-singular square integer matrix A and an integer vector w. The solution vector v will, in general, have rational entries.

We stress that we want to compute the exact solution v, and not some floating point approximation to it. Now, we could solve for v directly using Gaussian elimination; however, the intermediate quantities computed by that algorithm would be rational numbers whose numerators and denominators might get quite large, leading to a rather lengthy computation (however, it is possible to show that the overall running time is still polynomial in the input length).

Another approach is to compute a solution vector modulo n, where n is a power of a prime that does not divide the determinant of A. Provided n is large enough, one can then recover the solution vector v using rational reconstruction. With this approach, all of the computations can be carried out using arithmetic on integers not too much larger than n, leading to a more e—cient algorithm.

EXERCISE

1. Let $z = s/t \in \mathbb{Q}$, where s and t are relatively prime integers with $0 \leq s < t$. Show that:
   (*a*) if $\gcd(10, t) = 1$, then the decimal expansion of z is purely periodic with period equal to the multiplicative order of 10 modulo t;
   (*b*) more generally, if k is the smallest non-negative integer such that 10 and $t' := t/\gcd(10^k, t)$ are relatively prime, then the decimal expansion of z is ultimately periodic with pre-period k and period equal to the multiplicative order of 10 modulo t'.
   (c) there exist integers k, k' such that $0 \leq k < k'$ and $s10^k \equiv s10^{k'} \pmod t$;
   (d) for all integers k, k' with $0 \leq k < k'$, the decimal expansion of z is $(k, k' - k)$-periodic if and only if $s10^k \equiv s10^{k'} \pmod t$;

2. Let $n, b \in \mathbb{Z}$ with $0 \leq b < n$, and let $\text{EEA}(n, b) = \{(r_i, s_i, t_i)\}_{i=0}^{\lambda+1}$. This exercise develops some key properties of the fractions $-s_i/t_i$ as approximations to b/n. For $i = 1, \ldots, \lambda + 1$, let $\varepsilon_i := b/n + s_i/t_i$.
   (a) Show that $\varepsilon_i = r_i/t_i n$ for $i = 1, \ldots, \lambda + 1$.
   (b) Show that successive ε_i's strictly decrease in absolute value, and alternate in sign.
   (c) Show that $|\varepsilon_i| < 1/t_i^2$ for $i = 1, \ldots, \lambda$, and $\varepsilon_{\lambda+1} = 0$.
   (d) Show that for all $s, t \in \mathbb{Z}$ with $t \neq 0$, if $|b/n - s/t| < 1/2t^2$, then $s/t = -s_i/t_i$ for some $i = 1, \ldots, \lambda + 1$.

(e) Consider a fixed index $i \in \{2,..., \lambda + 1\}$. Show that for all $s, t \in \mathbb{Z}$, if $0 < | t | \leq | t_i |$ and $| b/n - s/t | \leq | \varepsilon_i |$, then $s/t = - s_i/t_i$. In this sense, $-s_i/t_i$ is the unique, best approximation to b/n among all fractions of denominator at most $| t_i |$.

3. Using the decimal approximation $\pi \approx 3.141592654$, apply Euclid's algorithm to calculate a rational number of denominator less than 1000 that is within 10^{-6} of π.
4. Show that if a sequence Ψ is ultimately periodic, then it is (k^*, ℓ^*)-periodic for some uniquely determined pair (k^*, ℓ^*) for which the following holds: for every pair (k, ℓ) such that Ψ is (k, ℓ)-periodic, we have $k^* \leq k^*$ and $\ell^* | \ell$.

   The value ℓ^* in the above exercise is called the period of Ψ, and k^* is called the pre-period of Ψ. If its pre-period is zero, then Ψ is called purely periodic.
5. Show that given integers s, t, k, with $0 \leq s < t$, and $k > 0$, we can compute the kth digit in the decimal expansion of s/t in time $O(\text{len}(k)\ \text{len}(t)^2)$.

For the following exercises, we need a definition. Let $\Psi = \{z_i\}_{i=1}^{\infty}$ be a sequence of elements drawn from some arbitrary set. For integers $k \geq 0$ and $\ell \geq 1$, we say that Ψ is (k, ℓ)-periodic if $z_i = z_{i+\ell}$ for all $i > k$; in addition, we say that Ψ is ultimately periodic if it is (k, ℓ)-periodic for some (k, ℓ).

6. Let z be a real number whose decimal expansion is an ultimately periodic sequence. Show that z is rational.

A famous conjecture of Artin postulates that for every integer d, not equal to -1 or to the square of an integer, there are infinitely many primes t such that d has multiplicative order $t - 1$ modulo t. If Artin's conjecture is true, then by part (c) of the previous exercise, there are infinitely many primes t such that the decimal expansion of s/t, for every s with $0 < s < t$, is a purely periodic sequence of period $t - 1$.

In light of these observations, the "grade school" method of computing a fraction from its decimal expansion using the period is hopelessly impractical.

Chapter 3

Abelian Groups

In mathematics, an abelian group, also called a commutative group, is a group $(G, *)$ with the additional property that the group operation * is commutative, so that for all a and b in G, $a * b = b * a$. Abelian groups are named after Norwegian mathematician Niels Henrik Abel. Groups in which the group operation is not commutative are called *non-abelian* (or *non-commutative*).

Since the group operation in an abelian group is commutative as well as associative, the value of a product of group elements is independent of the order in which the product is calculated. The theory of abelian groups is generally simpler than that of their non-abelian counterparts, although infinite abelian groups are the subject of current research

NOTATION

There are two main notational conventions for abelian groups — additive and multiplicative.

Convention	*Operation*	*Identity*	*Powers*	*Inverse*	*Direct/sum/product*
Addition	$x + y$	0	nx	$-x$	$G \oplus H$
Multiplication	$x * y$ or xy	e or 1	xn	$x-1$	$G \times H$

The multiplicative notation is the usual notation for groups, while the additive notation is the usual notation for modules. When studying abelian groups apart from other groups, the additive notation is usually used.

Examples. Every cyclic group G is abelian, because if x, y are in G, then $xy = a^m a^n = a^{m+n} = a^{n+m} = a^n a^m = yx$. Thus the integers, Z, form an abelian group under addition, as do the integers modulo n, Z/nZ.

Every ring is an abelian group with respect to its addition operation. In a commutative ring the invertible elements, or units, form an abelian multiplicative group. In particular, the real numbers are an abelian group under addition, and the nonzero real numbers are an abelian group under multiplication.

Every subgroup of an abelian group is normal, so each subgroup gives rise to a quotient group. Subgroups, quotients, and direct sums of abelian groups are again abelian.

Matrices, even invertible matrices, do not form an abelian group under multiplication because matrix multiplication is generally not commutative.

MULTIPLICATION TABLE

To verify that a finite group is abelian, a table (matrix) - known as a Cayley table - can be constructed in a similar fashion to a multiplication table. If the group is $G = \{g_1 = e, g_2, \ldots, g_n\}$ under the operation E", the (i, j)'th entry of this table contains the product g_i E" g_j. The group is abelian if and only if this table is symmetric about the main diagonal (i.e. if the matrix is a symmetric matrix).

This is true since if the group is abelian, then g_i E" $g_j = g_j$ E" g_i. This implies that the (i, j)'th entry of the table equals the (j, i)'th entry - i.e. the table is symmetric about the main diagonal.

PROPERTIES

If n is a natural number and x is an element of an abelian group G written additively, then nx can be defined as $x + x + \ldots + x$ (n summands) and ("n)x = "(nx). In this way, G becomes a module over the ring Z of integers. In fact, the modules over Z can be identified with the abelian groups.

Theorems about abelian groups (i.e. modules over the principal ideal domain Z) can often be generalized to theorems about modules over an arbitrary principal ideal domain. A typical example is the classification of finitely generated abelian groups.

If $f, g : G \to H$ are two group homomorphisms between abelian groups, then their sum $f + g$, defined by $(f + g)(x) = f(x) + g(x)$, is again a homomorphism. (This is not true if H is a non-abelian group.) The set Hom(G, H) of all group homomorphisms from G to H thus turns into an abelian group in its own right.

Somewhat akin to the dimension of vector spaces, every abelian group has a *rank*. It is defined as the cardinality of the largest set of linearly independent elements of the group. The integers and the rational numbers have rank one, as well as every subgroup of the rationals.

Finite abelian Groups

The fundamental theorem of finite abelian groups states that every finite abelian group can be expressed as the direct sum of cyclic subgroups of prime-power order. This is a special application of the fundamental theorem of finitely generated abelian groups in the case when G has torsion-free rank equal to 0.

$\mathbb{Z}_{mn}$ is isomorphic to the direct product of $\mathbb{Z}_m$ and $\mathbb{Z}_n$ if and only if m and n are coprime. Therefore we can write any finite abelian group G as a direct product of the form

$$\mathbb{Z}_{m_1} \oplus \ldots \oplus \mathbb{Z}_k$$

in two unique ways:

- Where the numbers $k_1,\ldots,k_u$ are powers of primes
- Where k_1 divides k_2, which divides k_3 and so on up to k_u.

For example, $\mathbb{Z}/15\mathbb{Z} \cong \mathbb{Z}_{15}$ can be expressed as the direct sum of two cyclic subgroups of order 3 and 5: $\mathbb{Z}_{15} \cong \{0,5,10\} \oplus \{0,3,6\,9,12\}$. The same can be said for any abelian group of order 15, leading to the remarkable conclusion that all abelian groups of order 15 are isomorphic.

For another example, every abelian group of order 8 is isomorphic to either $\mathbb{Z}_8$ (the integers 0 to 7 under addition modulo 8), $\mathbb{Z}_4 \oplus \mathbb{Z}_2$ (the odd integers 1 to 15 under multiplication modulo 16), or $\mathbb{Z}_2 \oplus \mathbb{Z}_2 \oplus \mathbb{Z}_2$.

AUTOMORPHISMS OF FINITE ABELIAN GROUPS

One can apply the fundamental theorem to count (and sometimes determine) the automorphisms of a given finite abelian group G. To do this, one uses the fact (which will not be proved here) that if G splits as a direct sum $H \oplus K$ of subgroups of coprime order, then Aut $(H \oplus K)$ Aut$(H) \cong$ Aut(K).

Given this, the fundamental theorem shows that to compute the automorphism group of G it suffices to compute the automorphism groups of the Sylow p-subgroups separately (that is, all direct sums of cyclic subgroups, each with order a power of p). Fix a prime p and suppose the exponents e_i of the cyclic factors of the Sylow p-subgroup are arranged in increasing order:

$$e_1 \leq e_2 \leq \ldots \leq e_n$$

for some $n > 0$. One needs to find the automorphisms of

$$\mathbb{Z}_{p^{e1}} \oplus \ldots \oplus \mathbb{Z}_{p^{en}}$$

One special case is when $n = 1$, so that there is only one cyclic prime-power factor in the Sylow p-subgroup P. In this case the theory of automorphisms of a finite cyclic group can be used. Another special case is when n is arbitrary but $e_i = 1$ for $1 \leq i \leq n$. Here, one is considering P to be of the form

$$\mathbb{Z}_p \oplus \ldots \oplus \mathbb{Z}_p,$$

so elements of this subgroup can be viewed as comprising a vector space of dimension n over the finite field of p elements $\mathbb{F}_p$. The automorphisms of this subgroup are therefore given by the invertible linear transformations, so

$$\text{Aut}\ (P) \cong GL(n, \mathbb{F}_p.),$$

which is easily shown to have order

$$|\text{Aut}\ (P)| = (p^n - 1) \ldots (p^n - p^{n-1}).$$

In the most general case, where the e_i and n are arbitrary, the automorphism group is more difficult to determine. It is known, however, that if one defines

$$d_k = \max \{r \mid e_r = e_k\}$$

and

$$c_k = \max \{r \mid e_r = e_k\}$$

then one has in particular $d_k \geq k$, $c_k \leq k$, and

$$|\text{Aut}\ (P)| = \left(\prod_{k=1}^{n} p^{d_k} = p^{k-1}\right)\left(\prod_{j=1}^{n} (p^{e_j})^{n-d_j}\right)\left(\prod_{i=1}^{n} (p^{e_i-1})^{n-c_i+1}\right)$$

One can check that this yields the orders in the previous examples as special cases.

CYCLIC GROUP

In group theory, a cyclic group or monogenous group is a group that can be generated by a single element, in the sense that the group has an element g (called a "generator" of the group) such that, when written multiplicatively, every element of the group is a power of g (a multiple of g when the notation is additive).

Definition: A group G is called cyclic if there exists an element g in G such that $G = \langle g \rangle = \{ g^n \mid n \text{ is an integer} \}$. Since any group generated by an element in a group is a subgroup of that group, showing that the only subgroup of a group G that contains g is G itself suffices to show that G is cyclic.

For example, if $G = \{ g^0, g^1, g^2, g^3, g^4, g^5 \}$ is a group, then $g^6 = g^0$, and G is cyclic. In fact, G is essentially the same as (that is, isomorphic to) the set { 0, 1, 2, 3, 4, 5 } with addition modulo 6. For example, 1 + 2 = 3 (mod 6) corresponds to $g^1 – g^2 = g^3$, and 2 + 5 = 1 (mod 6) corresponds to $g^2 \cdot g^5 = g^7 = g^1$, and so on. One can use the isomorphism φ defined by $\varphi(g^i) = \text{i}$.

For every positive integer n there is exactly one cyclic group (up to isomorphism) whose order is n, and there is exactly one infinite cyclic group (the integers under addition).

Hence, the cyclic groups are the simplest groups and they are completely classified.

The name 'cyclic' may be misleading: it is possible to generate infinitely many elements and not form any literal cycles; that is, every g^n is distinct. (It can be said that it has one infinitely long cycle.) A group generated in this way is called an infinite cyclic group, and is isomorphic to the additive group of integers Z.

Since the cyclic groups are abelian, they are often written additively and denoted Z_n. However, this notation can be problematic for number theorists because it conflicts with the usual notation for *p*-adic number rings or localization at a prime ideal. The quotient notations Z/*n*Z, Z/*n*, and Z/(*n*) are standard alternatives. We adopt the first of these here to avoid the collision of notation.

One may write the group multiplicatively, and denote it by C_n, where *n* is the order (which can be "). For example, $g^3g^4 = g^2$ in C_5, whereas 3 + 4 = 2 in Z/5Z.

PROPERTIES

The fundamental theorem of cyclic groups states that if *G* is a cyclic group of order *n* then every subgroup of *G* is cyclic. Moreover, the order of any subgroup of *G* is a divisor of *n* and for each positive divisor *k* of *n* the group *G* has exactly one subgroup of order *k*. This property characterizes finite cyclic groups: a group of order *n* is cyclic if and only if for every divisor *d* of *n* the group has at most one subgroup of order *d*. Sometimes the equivalent statement is used: a group of order *n* is cyclic if and only if for every divisor *d* of *n* the group has exactly one subgroup of order *d*.

Every finite cyclic group is isomorphic to the group { [0], [1], [2], ..., [*n* " 1] } of integers modulo *n* under addition, and any infinite cyclic group is isomorphic to Z (the set of all integers) under addition. Thus, one only needs to look at such groups to understand the properties of cyclic groups in general. Hence, cyclic groups are one of the simplest groups to study and a number of nice properties are known.

Given a cyclic group *G* of order *n* (*n* may be infinity) and for every *g* in *G*,

- G is abelian; that is, their group operation is commutative: $gh = hg$ (for all *h* in *G*). This is so since $g + h \bmod n = h + g \bmod n$.
- If *n* is finite, then $g^n = g^0$ is the identity element of the group, since $n \bmod n = 0$.
- If $n = \infty$, then there are exactly two generators: namely 1 and –1 for Z, and any others mapped to them under an isomorphism in other infinite cyclic groups.
- If *n* is finite, then there are exactly φ(n) generators where φ is the Euler phi function
- Every subgroup of *G* is cyclic. Indeed, each finite subgroup of G is a group of { 0, 1, 2, 3, ... $m - 1$} with addition modulo m. And each infinite subgroup of G is *m*Z for some m, which is bijective to (so isomorphic to) Z.

- G_n is isomorphic to Z/nZ (factor group of Z over nZ) since Z/nZ = {0 + nZ, 1 + nZ, 2 + nZ, 3 + nZ, 4 + nZ, ..., n – 1 + nZ} ≅ { 0, 1, 2, 3, 4, ..., n – 1} under addition modulo *n*.

More generally, if *d* is a divisor of *n*, then the number of elements in Z/*n* which have order *d* is φ(*d*). The order of the residue class of *m* is *n* / gcd(*n*,*m*).

If *p* is a prime number, then the only group (up to isomorphism) with *p* elements is the cyclic group C_p or Z/*p*Z.

The direct product of two cyclic groups Z/*n*Z and Z/*m*Z is cyclic if and only if *n* and *m* are coprime. Thus e.g. Z/12Z is the direct product of Z/3Z and Z/4Z, but not the direct product of Z/6Z and Z/2Z.

The definition immediately implies that cyclic groups have very simple group presentation $C = < x \mid >$ and $C_n = < x \mid x^n >$ for finite *n*.

A primary cyclic group is a group of the form Z/p^k where *p* is a prime number. The fundamental theorem of abelian groups states that every finitely generated abelian group is the direct product of finitely many finite primary cyclic and infinite cyclic groups.

Z/*n*Z and Z are also commutative rings. If *p* is a prime, then Z/*p*Z is a finite field, also denoted by F_p or GF(*p*). Every field with *p* elements is isomorphic to this one.

The units of the ring Z/*n*Z are the numbers coprime to *n*. They form a group under multiplication modulo *n* with φ(*n*) elements. It is written as $(Z/nZ)^\times$. For example, we get $(Z/nZ)^\times = \{1,5\}$ when $n = 6$, and get $(Z/nZ)^\times = \{1,3,5,7\}$ when $n = 8$.

In fact, it is known that $(Z/nZ)^\times$ is cyclic if and only if *n* is 2 or 4 or p^k or $2\,p^k$ for an odd prime number *p* and $k \geq 1$, in which case every generator of $(Z/nZ)^\times$ is called a primitive root modulo *n*. Thus, $(Z/nZ)^\times$ is cyclic for $n = 6$, but not for $n = 8$, where it is instead isomorphic to the Klein four-group.

The group $(Z/pZ)^\times$ is cyclic with $p - 1$ elements for every prime *p*, and is also written $(Z/pZ)^*$ because it consists of the non-zero elements. More generally, every *finite* subgroup of the multiplicative group of any field is cyclic.

PROPERTIES OF ABELIAN GROUPS

Definition: An abelian group is a set *G* together with a binary operation * on *G* such that

- For all $a, b, c \in G$, $a * (b * c) = (a * b)$? c (i.e., * is associative),
- There exists $e \in G$ (called the identity element) such that for all $a \in G$, $a * e = a = e * a$,
- For all $a \in G$ there exists $a' \in G$ (called the inverse of *a*) such that $a * a' = e = a' * a$,
- For all $a, b \in G$, $a * b = b * a$ (i.e., * is commutative).

The restriction to abelian groups helps to simplify the discussion significantly. Because we will only be dealing with abelian groups, we may occasionally simply say "group" instead of "abelian group."

Before looking at examples, let us state some very basic properties of abelian groups that follow directly from the definition:

Theorem: Let G be an abelian group. Then for all $a, b, c \in G$, we have:

- If $a + b = a + c$, then $b = c$;
- The equation $a + x = b$ has a unique solution $x \in G$;
- $-(a + b) = (-a) + (-b)$;
- $-(-a) = a$.

Proof: For (i), just add $-a$ to both sides of the equation $a + b = a + c$. For (ii), the solution is $x = b - a$. For (iii), we have

$$(a + b) + ((-a) + (-b)) = (a + (-a)) + (b + (-b)) = 0_G + 0_G = 0_G,$$

which shows that $(-a) + (-b)$ is indeed the inverse of $a + b$. For (iv), we have $(-a) + a = 0_G$, which means that a is the inverse of $-a$.

Part (i) of the above theorem is the cancellation law for abelian groups.

If $a_1,..., a_k$ are elements of an abelian group G, we naturally write $\sum_{i=1}^{k} a_i$ for their sum. By convention, the sum is 0_G when $k = 0$. Part (iii) of previous theorem obviously generalizes, so that $-\sum_{i=1}^{k} a_i = \sum_{i=1}^{k} (-a_i)$. In the special case where all the a_i's have the same value a, we define $k. a: = \sum_{i=1}^{k} a$, whose inverse is $k. (-a)$, which we may write as $(-k).a$. Thus, the notation $k.a$, or more simply, ka, is defined for all integers k. Observe that by definition, $1a = a$ and $(-1)a = -a$.

Theorem: Let G be an abelian group with binary operation ?. Then we have:

- G contains only one identity element;
- Every element of G has only one inverse.

Proof: Suppose e, e' are both identities. Then we have

$$e = e * e' = e',$$

where we have used part (ii) of Definition 6.1, once with e' as the identity, and once with e as the identity. That proves part (i) of the theorem.

To prove part (ii) of the theorem, let $a \in G$, and suppose that a has two inverses, a' and a''. Then using parts (i) – (iii) of Definition 6.1, we have

$$a' = a' * e \text{ (by part (ii))}$$
$$= a' * (a * a'') \text{ (by part (iii) with inverse } a'' \text{ of } a)$$
$$= (a' * a) * a'' \text{ (by part (i))}$$
$$= e * a'' \text{ (by part (iii) with inverse } a' \text{ of } a)$$
$$= a'' \text{ (by part (ii)).}$$

These uniqueness properties justify use of the definite article in Definition 6.1 in conjunction with the terms "identity element" and "inverse." Note that we never used part (iv) of the definition in the proof of the above theorem.

Abelian groups are lurking everywhere, as the following examples illustrate.

Examples.

1. The set of rational numbers $\mathbb{Q} = \{a/b: a, b \in \mathbb{Z}, b \neq 0\}$ under addition forms an abelian group, with 0 being the identity, and $(-a)/b$ being the inverse of a/b.
2. The set F^* of all arithmetic functions f, such that $f(1)$ 'H 0, and with the Dirichlet product as the binary operation forms an abelian group.
3. The set $\mathbb{Z}_n^*$ of residue classes $[a]_n$ with $\gcd(a, n) = 1$ under multiplication forms an abelian group, where $[1]n$ is the identity, and if b is a multiplicative inverse of a modulo n, then $[b]_n$ is the inverse of $[a]_n$.
4. For every positive integer n, the set of n-bit strings under the "exclusive or" operation forms an abelian group, where the "all zero" bit string is the identity, and every bit string is its own inverse.
5. For each integer n, the set $n\mathbb{Z} = \{nz: z \in \mathbb{Z}\}$ under addition forms an abelian group, again, with 0 being the identity, and $n(-z)$ being the inverse of nz.
6. The set of non-negative integers under addition does not form an abelian group, since additive inverses do not exist for any positive integers.
7. The set of all finite bit strings under concatenation does not form an abelian group. Although concatenation is associative and the empty string acts as an identity element, inverses do not exist (except for the empty string), nor is concatenation commutative.
8. The set of non-zero rational numbers $\mathbb{Q}^*$ under multiplication forms an abelian group, with 1 being the identity, and b/a being the inverse of a/b.
9. The set $\mathbb{Z}_n$ under addition forms an abelian group, where $[0]_n$ is the identity, and where $[-a]_n$ is the inverse of $[a]_n$.
10. The set of $\in$ X $\in$ integer matrices with determinant ± 1, together with the binary

operation of matrix multiplication, is an example of a non-abelian group; that is, it satisfies properties (i) – (iii) of Definition 6.1, but not property (iv).

11. The set of all permutations on a given set of size $n \geq 3$, together with the binary operation of function composition, is another example of a nonabelian group (for $n = 1, 2$, it is an abelian group).
12. The set of integers $\{\pm 1\}$ under multiplication forms an abelian group, with 1 being the identity, and –1 its own inverse.
13. The set of integers $\mathbb{Z}$ under addition forms an abelian group, with 0 being the identity, and $-a$ being the inverse of $a \in \mathbb{Z}$.
14. The set of integers under multiplication does not form an abelian group, since inverses do not exist for any integers other than ± 1.

Consider an abelian group G with binary operation $*$. Since the group operation is associative, for all $a_1, \ldots, a_k \in G$, we may write $a_1 * \ldots * a_k$ without parentheses, and there can be no ambiguity as to the value of such an expression: any explicit parenthesization of this expression yields the same value.

Furthermore, since the group operation is commutative, reordering the *ai*'s does not change this value.

Note that in specifying a group, one must specify both the underlying set G as well as the binary operation; however, in practice, the binary operation is often implicit from context, and by abuse of notation, one often refers to G itself as the group.

For example, when talking about the abelian groups $\mathbb{Z}$ and $\mathbb{Z}_n$, it is understood that the group operation is addition, while when talking about the abelian group $\mathbb{Z}^*_n$, it is understood that the group operation is multiplication.

Typically, instead of using a special symbol like "*" for the group operation, one uses the usual addition ("+") or multiplication ("*") operations.

ADDITIVE NOTATION

If an abelian group G is written additively, using "+" as the group operation, then the identity element is denoted by 0_G (or just 0 if G is clear from context), and the inverse of an element $a \in G$ is denoted by $-a$. For $a, b \in G$, $a - b$ denotes $a + (-b)$.

MULTIPLICATIVE NOTATION

If an abelian group G is written multiplicatively, using "." as the group operation, then the identity element is denoted by 1_G (or just 1 if G is clear from context), and the inverse of an element $a \in G$ is denoted by a^{-1}. As usual, one may write ab in place of $a.\ b$. Also, one may write a/b for ab^{-1}.

For any particular, concrete abelian group, the most natural choice of notation is clear (e.g., addition for $\mathbb{Z}$ and $\mathbb{Z}_n$, multiplication for $\mathbb{Z}_n^*$); however, for a "generic" group, the choice is largely a matter of taste. By convention, whenever we consider a "generic" abelian group, we shall use additive notation for the group operation, unless otherwise specified.

The next theorem states a few simple but useful properties of abelian groups (stated using our default, additive notation).

Theorem: Let G be an abelian group. Then for all $a, b \in G$ and $k, \ell \in \mathbb{Z}$, we Have:

(*i*) $k(\ell a) = (k\ell)a = \ell(ka)$;

(*ii*) $(k + \ell)a = ka + \ell a$;

(*iii*) $k(a + b) = ka + kb$.

Proof: The proof of this is easy, but tedious. We leave the details as an exercise to the reader.

Multiplicative notation: It is perhaps helpful to translate the above discussion from additive to multiplicative notation. If a group G is written using multiplicative notation, then Theorem says that (i) $ab = ac$ implies $b = c$, (ii) $ax = b$ has a unique solution, (iii) $(ab)^{-1} = a^{-1} b^{-1}$, and (iv) $(a^{-1})^{-1} = a$. If $a_1,..., a_k \in G$, we write their product as $\sum_{i=1}^{k} a_i$, which is 1_G when $k = 0$. We have $(\sum_{i=1}^{k} a_i)^{-1} = \sum_{i=1}^{k} a_i^{-1}$. We also define $a^k := \sum_{i=1}^{k} a$, and we have $(a^k)^{-1} = (a^{-1})^k$, which we may write as a^{-k}. Theorem says that (i) $(a^{\ell})^k = a^{k\ell} = (a^k)^{\ell}$, (ii) $a^{k+\ell} = a^k a^{\ell}$, and (iii) $(ab)^k = a^k b^k$.

The definition of an abelian group includes the trivial group, which consists of just the zero element $\{0_G\}$, with $0_G + 0_G = 0_G$. An abelian group G may be infinite or finite. If the group is finite, we define its order to be the number of elements in the underlying set G; otherwise, we say that the group has infinite order.

Example.

1. The additive group $\mathbb{Z}$ has infinite order.
2. The order of the additive group $\mathbb{Z}_n$ is n. If $n = 1$, then $\mathbb{Z}_n$ is the trivial group.
3. The order of the multiplicative group $\mathbb{Z}_n^*$ is $\varphi(n)$, where φ is Euler's phi function.
4. Let G be an abelian group. An element $(a_1,..., a_k)$ of G^{Xk} may be identified with the function f: $\{1,..., k\} \to G$ given by $f(i) = a_i$ for $i = 1,..., k$.

We can generalize this, replacing $\{1,..., k\}$ by an arbitrary set I. We define Map(I, G)

to be the set of all functions $f: I \to G$, which we naturally view as a group by defining the group operation point-wise: for $f, g \in \text{Map}(I, G)$, we define

$$(f + g)(i) := f(i) + g(i) \text{ for all } i \in I.$$

Again, we leave it to the reader to verify that Map(I, G) is an abelian group, where the identity element is the function that maps each $i \in I$ to 0_G, and for $f \in \text{Map}(I, G)$, we have $(-f)(i) = -(f(i))$ for all $i \in I$.

5. If $G_1, \ldots, G_k$ are abelian groups, we can form the direct product $H := G_1 \text{ X} \ldots, G_k$, which consists of all k-tuples $(a_1, \ldots, a_k)$ with $a_1 \in G_1, \ldots, a_k \in G_k$. We can view H in a natural way as an abelian group if we define the group operation component-wise:

$$(a_1, \ldots, a_k) + (b_1, \ldots, b_k) := (a_1 + b_1, \ldots, a_k + b_k).$$

Of course, the groups $G_1, \ldots, G_k$ may be different, and the group operation applied in the ith component corresponds to the group operation associated with G_i. We leave it to the reader to verify that H is in fact an abelian group, where $0_H = (0_{G_1}, \ldots, 0_{G_k})$ and $-(a_1, \ldots, a_k) = (-a_1, \ldots, -a_k)$. As a special case, if $G = G_1 = \ldots = G_k$, then the k-wise direct product of G is denoted G^{Xk}.

EXERCISE

1. Let G be a finite abelian group of even order. Show that there exists $a \in G$ with $a \neq 0_G$ and $2a = 0_G$.
2. Let G be an abelian group, and let g be an arbitrary, fixed element of G. Assume that the group operation of G is written additively. We define a new binary operation $\odot$ on G, as follows: for $a, b \in G$, let $a \odot b := a + b + g$. Show that the set G under $\odot$ forms an abelian group.
3. Let ? be a binary operation on a finite set G. Assume that * is associative, commutative, and satisfies the cancellation law: $a * b = a * c$ implies $b = c$. Show that G under * forms an abelian group.
4. Let $G := \{x \in \mathbb{R} : x > 1\}$, and define $a * b := ab - a - b + \in$ for all $a, b \in \mathbb{R}$. Show that:

   (*a*) G is closed under *;

   (*b*) The set G under the operation * forms an abelian group.
5. Show that the result of the previous exercise need not hold if G is infinite.
6. For a finite abelian group, one can completely specify the group by writing down the group operation table. For instance, Example presented an addition table for $\mathbb{Z}_6$.

(*a*) Write down group operation tables for the following finite abelian groups: $\mathbb{Z}_5, \mathbb{Z}_5^*$, and $\mathbb{Z}_3 \times \mathbb{Z}_4^*$.

(*b*) Show that the group operation table for every finite abelian group is a Latin square; that is, each element of the group appears exactly once in each row and column.

(*c*) Below is an addition table for an abelian group that consists of the elements $\{a, b, c, d\}$; however, some entries are missing. Fill in the missing entries.

+	*a*	*b*	*c*	*d*
a	*a*			
b	*b*	*a*		
c			*a*	
d				

CYCLIC GROUPS

Let G be an abelian group. For $a \in G$, define $\langle a \rangle := \{za: z \in \mathbb{Z}\}$. It is easy to theorem that $\langle a \rangle$ is a subgroup of G; indeed, it is the image of the group homomorphism already discussed. Moreover, $\langle a \rangle$ is the smallest subgroup of G containing a; that is, $\langle a \rangle$ contains a, and every subgroup of G that contains a must contain everything in h*a*i.

Indeed, if a subgroup contains a, it must contain $a + a = 2a$, $a + a + a = 3a$, and so on; it must also contain $0_G = 0a$, $-a = (-1)a$, $(-a) + (-a) = (-2)a$, and so on. The subgroup $\langle a \rangle$ is called the subgroup (of G) generated by a. Also, one defines the order of a to be the order of the subgroup $\langle a \rangle$.

More generally, for $a_1, ..., a_k \in G$, we define

$$\langle a_1, ..., a_k \rangle := \{z_1 a_1 + ... + z_k a_k : z_1, ..., z_k \in \mathbb{Z}\}.$$

It is easy to see that $\langle a_1, ..., a_k \rangle$ is a subgroup of G; indeed, it is the image of the group homomorphism discussed in Example. Moreover, this subgroup is the smallest subgroup of G that contains $a_1, ..., a_k$; that is, $\langle a_1, ..., a_k \rangle$ contains the elements $a_1, ..., a_k$, and every subgroup of G that contains these elements must contain everything in $\langle a_1, ..., a_k \rangle$. The subgroup $\langle a_1, ..., a_k \rangle$ it is called the subgroup (of G) generated by $a_1, ..., a_k$.

An abelian group G is called cyclic if $G = \langle a \rangle$ for some $a \in G$, in which case, a is called a generator for G. An abelian group G is called finitely generated if $G = \langle a_1, ..., a_k \rangle$ for some $a_1, ..., a_k \in G$.

Multiplicative notation: if G is written multiplicatively, then $\langle a \rangle := \{a^z : z \in \mathbb{Z}\}$, and $\langle a_1,\dots,a_k \rangle := \left\{a_1^{z_1} \dots a_k^{z_k} : z_1,\dots z_k \in \mathbb{Z}\right\}$; also, for emphasis and clarity, we use the term multiplicative order of a.

Example.

1. For $n > 0$, consider the additive group $\mathbb{Z}_n$. This is a cyclic group, with [1] being a generator:

$$\langle [1] \rangle = \{z[1] : z \in \mathbb{Z}\} = \{[z] : z \in \mathbb{Z}\} = \mathbb{Z}_n.$$

For every $m \in \mathbb{Z}$, we have

$$\langle [m] \rangle = \{z[m] : z \in \mathbb{Z}\} = \{[zm] : z \in \mathbb{Z}\} = \{m[z] : z \in \mathbb{Z}\} = m\mathbb{Z}_n.$$

By Example, the subgroup $m\mathbb{Z}_n$ has order $n/\gcd(m, n)$. Thus, $[m]$ has order $n/\gcd(m, n)$; in particular, $[m]$ generates $\mathbb{Z}_n$ if and only if m is relatively prime to n, and hence, the number of generators of $\mathbb{Z}_n$ is $\varphi(n)$.

2. Consider the additive group $\mathbb{Z}$. This is a cyclic group, with 1 being a generator:

$$\langle 1 \rangle = \{z.1 : z \in \mathbb{Z}\} = \{z : z \in \mathbb{Z}\} = \mathbb{Z}.$$

For every $m \in \mathbb{Z}$, we have

$$\langle m \rangle = \{zm : z \in \mathbb{Z}\} = \{mz : z \in \mathbb{Z}\} = m\mathbb{Z}.$$

It follows that the only elements of $\mathbb{Z}$ that generate $\mathbb{Z}$ are 1 and –1: every other element generates a subgroup that is strictly contained in $\mathbb{Z}$.

Theorem: Let $\rho: G \to G'$ be a group isomorphism.

(*i*) For all $a \in G$, we have $\rho(\langle a \rangle) = \langle \rho(a) \rangle$.

(*ii*) For all $a \in G$, a and $\rho(a)$ have the same order.

(*iii*) G is cyclic if and only if G' is cyclic.

Proof: For all $a \in G$, we have

$$\rho(\langle a \rangle) = \{\rho(za) : z \in \mathbb{Z}\} = \{z\rho(a) : z \in \mathbb{Z}\} = \langle \rho(a) \rangle.$$

That proves (i).

(*ii*) follows from (i) and the fact that ρ is injective.

(*iii*) follows from (i), as follows. If G is cyclic, then $G = \langle a \rangle$, and since ρ is surjective, we have $G' = \rho(G) = \langle p(a) \rangle$. The converse follows by applying the same argument to the inverse isomorphism $\rho^{-1}: G' \to G$.

Theorem: Let G be a cyclic group generated by a. Then for every $m \in \mathbb{Z}$, we have

$$\langle ma \rangle = mG.$$

Proof: We have

$$\langle ma \rangle = \{z(ma): z \in \mathbb{Z}\} = \{m(za): z \in \mathbb{Z}\} = m\langle a \rangle = mG.$$

The following two examples present some groups that are not cyclic.

Example.

1. Consider the additive group $G := \mathbb{Z}_{n_1} \times \mathbb{Z}_{n_2}$. Set

$$\alpha_1 := [1]_{n_1} (\in, [0]_{n_2}) \in G \text{ and } \alpha_2 := ([0]_{n_1}, [1]_{n_2}) \in G.$$

It is not hard to see that $G = \langle \alpha_n, \alpha_2 \rangle$, since for all $z_1, z_2 \in \mathbb{Z}$, we have

$$z_1\alpha_1 + z_2\alpha_2 = ([z_1]_{n_1}, [0]_{n_2}) + ([0]_{n_1}, [z_2]_{n_2}) = ([z]_{n_1}, [z_2]_{n_2}).$$

However, G may or may not be cyclic: it depends on $d := \gcd(n_1, n_2)$.

If $d = 1$, then G is cyclic, with $\alpha := ([1]_{n_1}, [1]_{n_2})$ being a generator. One can see this easily using the Chinese remainder theorem for all $z_1, z_2 \in \mathbb{Z}$, there exists $z \in \mathbb{Z}$ such that

$$z \equiv z_1 \pmod{n_1} \text{ and } z \equiv z_2 \pmod{n_2},$$

which implies

$$z\alpha = \left([z]_{n_1}, [z]_{n_2}\right) = \left([z_1]_{n_1}, [z_2]_{n_2}\right).$$

If $d > 1$, then G is not cyclic. To see this, let $\beta = \left([b_1]_{n_1}, [b_2]_{n_2}\right)$ be an arbitrary element of G. We claim that one of $\alpha 1$ or α_2 does not belong to $\langle \beta \rangle$. Suppose to the contrary that both α_1 and α_2 belong to $\langle \beta \rangle$. This would imply that there exist integers z and z' such that

$$zb_1 \equiv 1 \pmod{n_1},\ zb_2 \equiv 0 \pmod{n_2},$$
$$z'b_1 \equiv 0 \pmod{n_1},\ z'b_2 \equiv 1 \pmod{n_2}.$$

All of these congruences hold modulo d as well, and multiplying the upper left congruence by the lower right, and the upper right by the lower left, we obtain

$$1 \equiv zz'b_1b_2 \equiv 0 \pmod{d},$$

which is impossible.

It should be clear that since a group isomorphism preserves all structural properties of groups, it preserves the property of being cyclic. We state this, along with related facts, as a theorem.

2. Consider the additive group $G:= \mathbb{Z} \times \mathbb{Z}$. Set

$$\alpha_1 := (1, 0) \in G \text{ and } \alpha_2 := (0, 1) \in G.$$

It is not hard to see that $G = \langle \alpha_1, \alpha_2 \rangle$, since for all $z_1, z_2 \in \mathbb{Z}$, we have

$$z_1\alpha_1 + z_2\alpha_2 = (z_1, 0) + (0, z_2) = (z_1, z_2).$$

However, G is not cyclic. To see this, let $\beta = (b_1, b_2)$ be an arbitrary element of G. We claim that one of α_1 or α_2 does not belong to $\langle \beta \rangle$. Suppose to the contrary that both α_1 and α_2 belong to $\langle \beta \rangle$. This would imply that there exist integers z and z' such that

$$zb_1 = 1, zb_2 = 0,$$
$$z'b_1 = 0, z'b_2 = 1.$$

Multiplying the upper left equality by the lower right, and the upper right by the lower left, we obtain

$$1 = zz'b_1b_2 = 0,$$

which is impossible.

CLASSIFICATION OF CYCLIC GROUPS

We shall now demonstrate, every cyclic group is isomorphic either to $\mathbb{Z}$ or to $\mathbb{Z}_n$ for some $n > 0$.

Suppose that G is a cyclic group with generator a. Consider the map $\rho: \mathbb{Z} \to G$ that sends $z \in \mathbb{Z}$ to $za \in G$. As discussed in Example, this map is a group homomorphism, and since a is a generator for G, it must be surjective. There are two cases to consider.

Case 1. Ker $\rho = \{0\}$. In this case, is an isomorphism of $\mathbb{Z}$ with G.

Case 2. Ker $\rho \neq \{0\}$. In this case, since Ker ρ is a subgroup of $\mathbb{Z}$ different from $\{0\}$, by previous theorem, it must be of the form $n\mathbb{Z}$ for some $n > 0$. Hence, by Theorem the map $\bar{\rho}: \mathbb{Z}_n \to G$ that sends $[z]n$ to za is an isomorphism of $\mathbb{Z}_n$ with G.

Theorem: Let G be a finite abelian group of order n. If p is a prime dividing n, then G contains an element of order p.

Proof: We can prove this by induction on n.

If $n = 1$, then the theorem is vacuously true.

Now assume $n > 1$ and that the theorem holds for all groups of order strictly less than n. Let a be any non-zero element of G, and let m be the order of a. Since a is non-zero, we must have $m > 1$. If $p \mid m$, then $(m/p)a$ is an element of order p, and we are done. So assume that $p \mid m$ and consider the quotient group G/H, where H is the subgroup of G generated by a.

Since H has order m, G/H has order n/m, which is strictly less than n, and since $p \mid m$, we must have $p \mid (n/m)$.

So we can apply the induction hypothesis to the group G/H and the prime p, which says that there is an element $b \in G$ such that the coset $[b]_H \in G/H$ has order p. If ℓ is the order of b, then $\ell b = 0_G$, and so $\ell b \equiv 0_G \pmod{H}$, which implies that the order of $[b]_H$ divides ℓ. Thus, $p \mid \ell$, and so $(\ell/p)b$ is an element of G of order p.

Theorem: Let G ban abelian group and let $a \in G$. If there exists a positive integer m such that $ma = 0_G$, then the least such positive integer n is the order of a; in this case, we have:

- For every integer z, $za = 0_G$ if and only if n divides z, and more generally, for all integers z_1, z_2, we have $z_1a = z_2a$ if and only if $z_1 \equiv z_2 \pmod{n}$; the subgroup $\langle a \rangle$ consists of the n distinct elements

  $0. a, 1\ a, \ldots, (n-1). a.$

Otherwise, a has infinite order, and every element of $\langle a \rangle$ can be expressed as za for some unique integer z.

Theorem: Let G be a finite abelian group. Then the primes dividing the exponent of G are the same as the primes dividing its order.

Proof: Since the exponent divides the order, every prime dividing the exponent must divide the order. Conversely, if a prime p divides the order, then since there is an element of order p in the group, the exponent must be divisible by p.

Theorem: Let G be a finite abelian group and let $a \in G$. Then $|G|a = 0_G$ and the order of a divides $|G|$.

Example.

1. As we saw in previous example, all elements of $\mathbb{Z}_{15}^*$ have multiplicative order dividing 4, and since $\mathbb{Z}_{15}^*$ has order 8, we conclude that $\mathbb{Z}_{15}^*$ is not cyclic.

2. The group $\mathbb{Z}_5^*$ is cyclic, with [2] being a generator:

   $$[2]^2 = [4] = [-1], [2]^3 = [-2], [2]^4 = [1].$$

3. Consider again the additive group $G := \mathbb{Z}_{n_1} \times \mathbb{Z}_{n_2}$, discussed in Example. If $d := \gcd(n_1, n_2) > 1$, then one can also see that G is not cyclic as follows: for every $\beta \in G$, we have $(n_1n_2/d)\beta = 0_G$, and hence by Theorem, the order of β divides n_1n_2/d.

The following two theorems completely characterize the subgroup structure of cyclic groups. Actually, we have already proven most of the results in these two theorems, but nevertheless, they deserve special emphasis.

Theorem: Let G be a cyclic group of infinite order.

(*i*) G is isomorphic to $\mathbb{Z}$.

(*ii*) There is a one-to-one correspondence between the non-negative integers and the subgroups of G, where each such integer m corresponds to the cyclic group mG.

(*iii*) For every two non-negative integers m, m', mG Í $m'G$ if and only if $m' \mid m$.

Proof: That $G \cong \mathbb{Z}$ was established in our classification of cyclic groups, and so it suffices to prove the other statements of the theorem for $G = \mathbb{Z}$. As we saw in Example, for every integer m, the subgroup $m\mathbb{Z}$ is cyclic, as it is generated by m. This fact, together with Theorem, establishes all the other statements.

Theorem: Let G be a cyclic group. Then for every subgroup H of G, both H and G/H are cyclic.

Proof: The fact that H is cyclic follows from part (ii) of previous theorem in the case where G is infinite, and part (ii) of previous theorem in the case where G is finite. If G is generated by a, then it is easy to see that G/H is generated by $[a]_H$.

The next three theorems are often useful in calculating the order of a group element. The first generalizes theorem.

Theorem: Let G be a cyclic group of finite order n.

(*i*) G is isomorphic to $\mathbb{Z}_n$.

(*ii*) There is a one-to-one correspondence between the positive divisors of n and the subgroups of G, where each such divisor d corresponds to the subgroup dG; moreover, dG is a cyclic group of order n/d.

(*iii*) For each positive divisor d of n, we have $dG = G\{n/d\}$; that is, the kernel of the (n/d)-multiplication map is equal to the image of the d-multiplication map; in particular, $G\{n/d\}$ has order n/d.

(*iv*) For every two positive divisors d, d' of n, we have $dG \subseteq d'G$ if and only if $d' \mid d$.

(*v*) For every positive divisor d of n, the number of elements of order d in G is $\varphi(d)$.

(*vi*) For every integer m, we have $mG = dG$ and $G\{m\} = G\{d\}$, where $d := \gcd(m, n)$.

Theorem: If G_1 and G_2 are finite cyclic groups of relatively prime order, then $G_1 \times G_2$ is also cyclic. In particular, if G_1 is generated by a_1 and G_2 is generated by a_2, then $G_1 \times G_2$ is generated by (a_1, a_2).

Proof: We give a direct proof, based on Theorem. Let $n_1 := |G_1|$ and $n_2 := |G_2|$, where $\gcd(n_1, n_2) = 1$. Also, let $a_1 \in G_1$ have order n_1 and $a_2 \in G_2$ have order n_2. We want to show that (a_1, a_2) has order $n_1 n_2$. Applying Theorem to (a_1, a_2), we see that the order of (a_1, a_2) is the smallest positive integer k such that $k(a_1, a_2) = (0_{G_1}, 0_{G_2})$. Now, for every integer k, we have $k(a_1, a_2) = (ka_1, ka_2)$, and

$$(ka_1, ka_2) = (0_{G_1}, 0_{G_2}) \Leftrightarrow n_1 \mid k \text{ and } n_2 \mid k$$
$$\Leftrightarrow n_1 n_2 \mid k \text{ (since } \gcd(n_1, n_2) = 1).$$

Theorem: If G is an abelian group of prime order, then G is cyclic.

Proof: Let $|G| = p$, which, by hypothesis, is prime. Let $a \in G$ with $a \neq 0_G$, and let k be the order of a. As the order of an element divides the order of the group, we have $k \mid p$, and so $k = 1$ or $k = p$. Since $a \neq 0_G$, we must have $k \neq 1$, and so $k = p$, which implies that a generates G.

Theorem: Let G be an abelian group, let $a \in G$ be of finite order n, and let m be an arbitrary integer. Then the order of ma is $n/\gcd(m, n)$.

Theorem: Let G be an abelian group of exponent m.

(*i*) For every integer k such that $kG = \{0_G\}$, we have $m \mid k$.

(*ii*) If G has finite order, then m divides $|G|$.

(*iii*) If $m \neq 0$, then for every $a \in G$, the order of a is finite, and the order of a divides m.

(*iv*) If G is cyclic, then the exponent of G is 0 if G is infinite, and is $|G|$ if G is finite.

Theorem: Suppose that a is an element of an abelian group, and for some prime p and integer $e \geq 1$, we have $p^e a = 0_G$ and $p^{e-1} a \neq 0_G$. Then a has order p^e.

Proof: If m is the order of a, then since $p^e a = 0_G$, we have $m \mid p^e$. So $m = p^f$ for some $f = 0,\ldots, e$. If $f < e$, then $p^{e-1} a = 0_G$, contradicting the assumption that $p^{e-1} a \neq 0_G$.

Theorem: If an abelian group G has non-zero exponent m, then G contains an element of order m. In particular, a finite abelian group is cyclic if and only if its order equals its exponent.

Proof: The second statement follows immediately from the first. For the first statement, let $m = \prod_{i=1}^{r} p_i^{e_i}$ be the prime factorization of m.

First, we claim that for each $i = 1,\ldots, r$, there exists a_i Ξ G such that $(m/p_i)a_i$'H 0_G. Suppose the claim were false: then for some i, $(m/p_i)a = 0_G$ for all a Ξ G; however, this contradicts the minimality property in the definition of the exponent m. That proves the claim.

Let $a_1,\ldots, a_r$ be as in the above claim. Then by Theorem, $(m/p_i^{e_i})a_i$ has order $p_i^{e_i}$ for each $i = 1,\ldots, r$. Finally, by Theorem, the group element

$$(m/p_1^{e_1})a_1 + \ldots + (m/p_r^{e_r})a_r$$

has order m.

Theorem: Suppose G is an abelian group with $a_1, a_2 \in G$ such that a_1 is of finite order n_1, a_2 is of finite order n_2, and $\gcd(n_1, n_2) = 1$. Then the order of $a_1 + a_2$ is $n_1 n_2$.

Proof: Let $H_1 := \langle a_1 \rangle$ and $H_2 := \langle a_2 \rangle$ so that $|H_1| = n_1$ and $|H_2| = n_2$.

First, we claim that $H_1 \cap H_2 = \{0_G\}$. To see this, observe that $H_1 \cap H_2$ is a subgroup of H_1, and so $|H_1 \cap H_2|$ divides n_1; similarly, $|H_1 \cap H_2|$ divides n_2. Since $\gcd(n_1, n_2) = 1$, we must have $|H_1 \cap H_2| = 1$, and that proves the claim.

Using the claim, we can apply Theorem, obtaining a group isomorphism between $H_1 + H_2$ and $H_1 \times H_2$. Under this isomorphism, the group element $a_1 + a_2 \in H_1 + H_2$ corresponds to $(a_1, a_2) \in H_1 \times H_2$, which by Theorem (again using the fact that $\gcd(n_1, n_2) = 1$) has order $n_1 n_2$.

For an abelian group G, we say that an integer k kills G if $kG = \{0_G\}$. Consider the set K_G of integers that kill G. Evidently, K_G is a subgroup of $\mathbb{Z}$, and hence of the form $m\mathbb{Z}$ for a uniquely determined non-negative integer m. This integer m is called the exponent of G. If $m \neq 0$, then we see that m is the least positive integer that kills G.

We first state some basic properties.

EXERCISE

1. Show that for abelian groups G_1, G_2 whose exponents are m_1 and m_2, the exponent of $G_1 \times G_2$ is $\operatorname{lcm}(m_1, m_2)$.
2. Show that $\mathbb{Q}$ is not finitely generated.
3. Let G be an abelian group of exponent $m_1 m_2$, where m_1 and m_2 are relatively prime. Show that G is the internal direct product of $m_1 G$ and $m_2 G$.
4. Find $\alpha_1, \alpha_2 \in \mathbb{Z}_{15}^*$ such that $\mathbb{Z}_{15}^* = \langle \alpha_1, \alpha_2 \rangle$.
5. Let G be an abelian group, $a \in G$, and $m \in Z$, such that $m > 0$ and $ma = 0_G$. Let $m = p_1^{e_1} \ldots p_r^{e_r}$ be the prime factorization of m. For $i = 1, \ldots, r$, let f_i be the largest non-negative integer such that $f_i \leq e_i$ and $m/p_i^{f_i}.a = 0_G$. Show that the order of a is equal to $p_1^{e_1 - f_1} \ldots p_r^{e_r - f_r}$.
6. Let G be an abelian group of order n, and let m be an integer. Show that $mG = G$ if and only if $\gcd(m, n) = 1$.
7. Consider the quotient group $G := \mathbb{Q}/\mathbb{Z}$. Show that:

   (*a*) G has exponent 0;

   (*b*) All elements of G have finite order;

(*c*) For all positive integers m, we have $mG = G$ and $G\{m\} \cong \mathbb{Z}_m$;

(*d*) All finite subgroups of G are cyclic.

8. Suppose that G is an abelian group that satisfies the following properties:

   (*a*) For all m Î$\mathbb{Z}$, *Gfmg* is either equal to G or is of finite order;

   (*b*) For some m Î$\mathbb{Z}$, $\{0_G\}$ $(G\{m\}$ Í G.

   Show that $G\{m\}$ is finite for all non-zero $m \in Z$.

9. Let H be a subgroup of an abelian group G. Show that:

   (*a*) If H and G / H are both finitely generated, then so is G;

   (*b*) If G is finite, $\gcd(|H|, |G/H|) = 1$, and H and G / H are both cyclic, then G is cyclic.

HOMOMORPHISM GROUPS

In this section, we study maps that relate the structure of one group to another. Such maps are often very useful, as they may allow us to transfer hard-won knowledge about one group to another, perhaps more mysterious, group.

Definition: A group homomorphism is a function ρ from an abelian group G to an abelian group G' such that $\rho(a + b) = \rho(a) + \rho(b)$ for all $a, b \in G$.

Note that in the equality $\rho(a + b) = \rho(a) + \rho(b)$ in the above definition, the addition on the left-hand side is taking place in the group G while the addition on the right-hand side is taking place in the group G'.

Two sets play a critical role in the study of a group homomorphism $\rho: G \to G'$. The first set is the image of ρ, that is, the set $\rho(G) = \{\rho(a): a \in G\}$. The second set is the kernel of ρ, defined as the set of all elements of G that are mapped to $0_{G'}$ by ρ, that is, the set $\rho^{-1}(\{0_{G'}\}) = \{a \in G: \rho(a) = 0_{G'}\}$. We introduce the following notation for these sets: Im ρ denotes the image of ρ, and Ker ρ denotes the kernel of ρ.

Example.

1. As a special case of the previous example, let n be a positive integer and let be an element of $\mathbb{Z}_n^*$. Let $\rho: \mathbb{Z} \to \mathbb{Z}_n^*$ be the group homomorphism that sends $z \in \mathbb{Z}$ to $\alpha^z \in \mathbb{Z}_n^*$. That is a group homomorphism means that $\alpha^{z+z'} = \alpha^z\alpha^{z'}$ for all $z, z' \in \mathbb{Z}$ (note that the group operation is addition in $\mathbb{Z}$ and multiplication in $\mathbb{Z}_n^*$). If the multiplicative order of α is equal to k, the image of α consists of the k distinct group elements $\alpha^0, \alpha^1,..., \alpha^{k-1}$. The kernel of ρ consists of those integers z such that $\alpha^z = 1$. The kernel of ρ is equal to the subgroup $k\mathbb{Z}$.

2. The reader may verify that if $a_1,\ldots, a_k$ are fixed elements of an abelian group G, then the map

$$\rho: \mathbb{Z}^{\times k} \to G$$
$$(z_1,\ldots, z_k) \mapsto z_1a_1 + \ldots + z_ka_k$$

is a group homomorphism.

$$a \mapsto [a]_H.$$

It is not hard to see that this is a group homomorphism. Indeed, this follows almost immediately from the way we defined addition in the quotient group G/H:

$$\rho(a + b) = [a + b]_H = [a]_H + [b]_H = \rho(a) + \rho(b).$$

It is clear that ρ is surjective. It is also not hard to see that Ker $\rho = H$; indeed, H is the identity element in G/H, and $[a]_H = H$ if and only if $a \in H$. The map ρ is called the natural map from G to G/H.

3. For a given positive integer n, the natural map from $\mathbb{Z}$ to $\mathbb{Z}_n$ sends $a \in \mathbb{Z}$ to the residue class $[a]_n$. This map is a surjective group homomorphism with kernel $n\mathbb{Z}$.
4. Suppose G is an abelian group and a is an element of G. It is easy to see that the map

$$\rho: \mathbb{Z} \to G$$
$$z \mapsto za$$

is a group homomorphism, since

$$\rho(z + z') = (z + z')a = za + z'a = \rho(z) + \rho(z').$$

5. Suppose G is an abelian group and m is an integer. The map

$$\rho: G \to G$$
$$a \mapsto ma$$

is a group homomorphism, since

$$\rho(a + b) = m(a + b) = ma + mb = \rho(a) + \rho(b).$$

The image of this homomorphism is the subgroup mG and the kernel is the subgroup $G\{m\}$. We call this map the m-multiplication map on G. If G is written multiplicatively, then this map, which sends $a \in G$ to $am \in G$, is called the m-power map on G, and its image is G^m.

6. Suppose that $H_1,\ldots, H_k$ are subgroups of an abelian group G. The reader may easily verify that the map

$$\rho: H_1 \times \ldots \times H_k \to G$$
$$(a_1,\ldots, a_k) \mapsto a_1 + \ldots + a_k$$

is a group homomorphism whose image is the subgroup $H_1 + \ldots + H_k$.

7. If H is a subgroup of an abelian group G, then the inclusion map $i: H \to G$ is obviously a group homomorphism.
8. Suppose H is a subgroup of an abelian group G. We define the map

$$\rho: G \to G/H$$

Properties of Homomorphisms Groups

Theorem: Let be a group homomorphism from G to G'.

(*i*) If H is a subgroup of G, then $\rho(H)$ is a subgroup of G'; in particular (setting $H := G$), Im ρ is a subgroup of G'.

(*ii*) If H' is a subgroup of G', then $\rho^{-1}(H')$ is a subgroup of G; in particular (setting $H' := \{0_{G'}\}$), Ker is a subgroup of G.

(*iii*) $\rho(0_G) = 0_{G'}$.

(*iv*) $\rho(-a) = -\rho(a)$ for all $a \in G$.

(*v*) ρ is injective if and only if Ker $\rho = \{0_G\}$.

(*vi*) For all $a, b \in G$, $(a) = (b)$ if and only if $a \equiv b$ (mod Ker ρ).

(*vii*) $\rho(na) = n\rho(a)$ for all $n \in \mathbb{Z}$ and $a \in G$.

Proof: These are all straightforward calculations.

(*i*) For all $a, b \in H$, we have $a + b \in H$ and $-a \in H$; hence, $\rho(H)$ contains $\rho(a + b) = \rho(a) + (b)$ and $\rho(-a) = -(a)$.

(*ii*) $\rho^{-1}(H')$ is non-empty, since $\rho(0_G) = \rho_G' \in H'$. If $\rho(a) \in H'$ and $\rho(b) \in H'$, then $\rho(a + b) = \rho(a) + \rho(b) \in H'$, and $\rho(-a) = -\rho(a) \in H'$.

(*iii*) We have

$$0_{G'} + \rho(0_G) = \rho(0_G) = \rho(0_G + 0_G) = \rho(0_G) + \rho(0_G).$$

Now cancel $\rho(0_G)$ from both sides.

(*iv*) We have

$$0_{G'} = \rho(0_G) = \rho(a + (-a)) = \rho(a) + \rho(-a),$$

and hence $\rho(-a)$ is the inverse of $\rho(a)$.

(*v*) If ρ is injective, then in particular, $\rho^{-1}(\{0_{G'}\})$ cannot contain any other element besides 0_G. If ρ is not injective, then there exist two distinct elements $a, b \in G$ with $\rho(a) = \rho(b)$, and by part (vi), Ker ρ contains the element $a - b$, which is non-zero.

(*vi*) We have

$$\rho(a) = \rho(b) \Leftrightarrow \rho(a) - \rho(b) = 0_{G'} \Leftrightarrow (a - b) = 0_{G'}$$

$$\Leftrightarrow a - b \in \text{Ker } \rho \Leftrightarrow a \equiv b \text{ (mod Ker } \rho).$$

(*vii*) For $n = 0$, this follows from part (i). For $n > 0$, this follows from the definitions by induction on n. For $n < 0$.

The above theorem is particularly useful: to check that a group homomorphism is injective, it suffices to determine if Ker $\rho = \{0_G\}$.

Thus, the injectivity and surjectivity of a given group homomorphism $\rho: G \to G'$ may be characterized in terms of its kernel and image:

- ρ is injective if and only if its kernel is trivial (i.e. Ker $\rho = \{0_G\}$);
- ρ is surjective if and only if Im $\rho = G'$.

We next present two very easy theorems that allow us to compose group homomorphisms in simple ways.

Theorem: Let $\rho_i: G \to G_i'$, for $i = 1,\ldots, k$, be group homomorphisms. Then the map

$$\begin{aligned} \rho: G &\to G_1' \times \ldots \times G_k' \\ a &\mapsto (\rho_1(a),\ldots, \rho_k(a)) \end{aligned}$$

is a group homomorphism.

Proof: For all $a, b \in G$, we have

$$\begin{aligned} \rho(a + b) &= (\rho_1(a + b),\ldots, \rho_k(a + b)) = (\rho_1(a) + \rho_1(b),\ldots, \rho_k(a) + \rho_k(b)) \\ &= \rho(a) + \rho(b). \end{aligned}$$

Consider a group homomorphism $\rho: G \to G'$. If is bijective, then ρ is called a group isomorphism of G with G'. If such a group isomorphism ρ exists, we say that G is isomorphic to G', and write $G \cong G'$. Moreover, if $G = G'$, then ρ is called a group automorphism on G.

Theorem: If $\rho: G \to G'$ and $\rho': G' \to G''$ are group homomorphisms, then so is their composition $\rho': G \to G''$.

Proof: For all $a, b \in G$, we have

$$\rho'(\rho(a + b)) = \rho'(\rho(a) + \rho(b)) = \rho'(\rho(a)) + \rho'((b)).$$

ISOMORPHISM

Theorem: If ρ is a group isomorphism of G with G', then the inverse function ρ^{-1} is a group isomorphism of G' with G.

Proof: For all $a', b' \in G'$, we have

$$\rho(\rho^{-1}(a') + \rho^{-1}(b')) = \rho(\rho^{-1}(a')) + \rho(\rho^{-1}(b')) = a' + b',$$

and hence $\rho^{-1}(a') + \rho^{-1}(b') = \rho^{-1}(a' + b')$.

Because of this theorem, if G is isomorphic to G', we may simply say that "G and G' are isomorphic."

We stress that a group isomorphism $\rho: G \to G'$ is essentially just a "renaming" of the group elements. This can be visualized as follows.

Imagine the addition table for G written out with rows and columns labeled by elements of G, with the entry in row a and column b being $a + b$. Now suppose we use the function ρ to consistently rename all the elements of G appearing in this table: the label on row a is replaced by $\rho(a)$, the label on column b by $\rho(b)$, and the entry in row a and column b by $\rho(a + b)$.

Because ρ is bijective, every element of G' appears exactly once as a label on a row and as a label on a column; moreover, because $\rho(a + b) = \rho(a) + \rho(b)$, what we end up with is an addition table for G'. It follows that all structural properties of the group are preserved, even though the two groups might look quite different syntactically.

Example. If $\gcd(m, n) = 1$, then the m-multiplication map on $\mathbb{Z}_n$ is a group automorphism.

First Isomorphism Theorem

Theorem: Let $\rho: G \to G'$ be a group homomorphism with kernel K and image H'. Then we have a group isomorphism

$$G/K \cong H'.$$

Specifically, the map

$$\bar{\rho} : G/K \to G'$$

$$[a]_K \mapsto \rho(a)$$

is an injective group homomorphism whose image is H'.

Proof: Using part (vi) of previous theorem, we see that for all $a, b \in G$, we have

$$[a]_K = [b]_K \Leftrightarrow a \equiv b \ (\mathrm{mod}\ K) \Leftrightarrow \rho(a) = \rho(b).$$

This immediately implies that the definition of $\bar{\rho}$ is unambiguous ($[a]_K = [b]_K$ implies $\rho(a) = \rho(b)$), and that $\bar{\rho}$ is injective ($\rho(a) = \rho(b)$ implies $[a]_K = [b]_K$). It is clear that $\bar{\rho}$ maps onto H', since every element of H' is of the form (a) for some $a \in G$, and the map $\bar{\rho}$ sends $[a]_K$ to $\rho(a)$. Finally, to see that is a group homomorphism, note that

$$\bar{\rho}([a]_K + [b]_K) = \bar{\rho}([a + b]_K) = \rho(a + b) = \rho(a) + \rho(b) = \bar{\rho}([a]_K) + \bar{\rho}([b]_K).$$

We can generalize the previous theorem, as follows:

Theorem: Let $\rho: G \to G'$ be a group homomorphism. Then for every subgroup H of G with $H \subseteq \mathrm{Ker}\ \rho$, we may define a group homomorphism

$$\bar{\rho} : G/H \to G'$$

$$[a]_H \mapsto \rho(a).$$

Moreover, $\mathrm{Im}\ \bar{\rho} = \mathrm{Im}\ \rho$, and $\bar{\rho}$ is injective if and only if $H = \mathrm{Ker}\ \rho$.

Proof: Using the assumption that $H \subseteq \text{Ker } \rho$, we see that $\overline{\rho}$ is unambiguously defined, since for all $a, b \in G$, we have

$$[a]_H = [b]_H \Rightarrow a \equiv b \ (\text{mod } H) \Rightarrow a \equiv b \ (\text{mod Ker } \rho) \Rightarrow \rho(a) = \rho(b).$$

That $\overline{\rho}$ is a group homomorphism, with $\text{Im}\,\overline{\rho} = \text{Im } \rho$, follows as in the proof of Theorem. If $H = \text{Ker } \rho$, then by Theorem, $\overline{\rho}$ is injective, and if H Ĩ Ker ρ, then $\overline{\rho}$ is not injective, since if we choose $a \in \text{Ker } \rho \backslash H$, we see that $\overline{\rho}([a]_H) = 0_{G'}$, and hence $\text{Ker}\,\overline{\rho}$ is non-trivial.

The next theorem gives us another important construction of a group isomorphism.

Internal Direct Product

Theorem: Let G be an abelian group with subgroups H_1, H_2, where $H_1 \cap H_2 = \{0_G\}$. Then we have a group isomorphism

$$H_1 \times H_2 \cong H_1 + H_2$$

given by the map

$$\rho: H_1 \times H_2 \rightarrow H_1 + H_2$$
$$(a_1, a_2) \mapsto a_1 + a_2.$$

Proof: We have already seen that ρ is a surjective group homomorphism. To see that ρ-is injective, it suffices to show that Ker ρ is trivial; that is, it suffices to show that for all $a_1 \in H_1$ and $a_2 \in H_2$, if $a_1 + a_2 = 0_G$, then $a_1 = a_2 = 0_G$. But $a_1 + a_2 = 0_G$ implies $a_1 = -a_2 \in H_2$, and hence $a_1 \in H_1 \setminus H_2 = \{0_G\}$, and so $a_1 = 0_G$. Similarly, one shows that $a_2 = 0_G$, and that finishes the proof.

If H_1, H_2 are as in the above theorem, then $H_1 + H_2$ is sometimes called the internal direct product of H_1 and H_2.

Example.

1. Let n_1, n_2 be positive integers with $n_1 \mid n_2$. Consider the natural map r: $\mathbb{Z}$ ® $\mathbb{Z}_{n_1}$. This is a surjective group homomorphism with Ker r $= n_1\mathbb{Z}$. Since $H := n_2\mathbb{Z}$ N $n_1\mathbb{Z}$, we may apply Theorem with the subgroup H, obtaining the surjective group homomorphism

$$\overline{\rho}: \mathbb{Z}_{n_2} \rightarrow \mathbb{Z}_{n_1}$$
$$[a]_{n_2} \mapsto [a]_{n_1}.$$

2. We can use the general theory developed so far to get a quick and dirty proof of the Chinese remainder theorem. Let $\{n_i\}_{i=1}^{k}$ be a pairwise relatively prime family of positive integers, and let $n := \prod_{i=1}^{k} n_i$. Consider the map

$$\rho: \mathbb{Z} \to \mathbb{Z}_{n_1} \times \ldots \times \mathbb{Z}_{n_k}$$
$$a \mapsto ([a]_{n_1}, \ldots, [a]_{n_k}).$$

It is easy to see that this map is a group homomorphism; indeed, it is the map constructed in Theorem applied with the natural maps $\rho_i : \mathbb{Z} \to \mathbb{Z}_{n_i}$, for $i = 1, \ldots, k$. Evidently, $a \in \text{Ker}\ \rho$ if and only if $n_i \mid a$ for $i = 1, \ldots, k$, and since $[n_i]_{i=1}^{k}$ is pairwise relatively prime, it follows that $a \in \text{Ker}\ \rho$ if and only if $n \mid a$; that is, $\text{Ker}\ \rho = n\mathbb{Z}$. Theorem then gives us an injective group homomorphism

$$\bar{\rho}: \mathbb{Z}_n \to \mathbb{Z}_{n_1} \times \ldots \times \mathbb{Z}_{n_k}$$
$$[a]_n \mapsto ([a]_{n_1}, \ldots, [a]_{n_k}).$$

But since the sets n and $\mathbb{Z}_{n_1} \times \ldots \times \mathbb{Z}_{n_k}$ have the same size, injectivity implies surjectivity. From this, Theorem is immediate.

The map $\bar{\rho}$ is a group isomorphism

$$\mathbb{Z}_n \cong \mathbb{Z}_{n_1} \times \ldots \times \mathbb{Z}_{n_k}.$$

3. Let n be a positive integer, and let m be any integer. Let $\rho_1 : \mathbb{Z} \to \mathbb{Z}_n$ be the natural map, and let $\rho_2 : \mathbb{Z}_n \to \mathbb{Z}_n$ be the m-multiplication map. The composed map $\rho := \rho_2 \circ \rho_1$ from $\mathbb{Z}$ to $\mathbb{Z}_n$ is also a group homomorphism. For each $z \in \mathbb{Z}$, we have $\rho(z) = m[z]_n = [mz]_n$. The kernel of ρ consists of those integers z such that $mz \equiv 0 \pmod{n}$, and so part (ii) of Theorem implies that $\text{Ker}\ \rho = (n/d)\mathbb{Z}$, where $d := \gcd(m, n)$. The image of ρ is $m\mathbb{Z}_n$. Theorem therefore implies that the map

   $$\bar{\rho}: \mathbb{Z}_{n/d} \to m\mathbb{Z}_n$$
   $$[z]_{n/d} \mapsto m[z]_n$$

   is a group isomorphism.

4. Let $\mathbb{Q}^*$ be the multiplicative group of non-zero rational numbers. Let H_1 be the subgroup $\{\pm 1\}$, and let H_2 be the subgroup of positive rationals. It is easy to see that $\mathbb{Q}^* = H_1 . H_2$ and that $H_1 \cap H_2 = \{1\}$. Thus, $\mathbb{Q}^*$ is the internal direct product of H_1 and H_2, and Theorem gives us a group isomorphism $\mathbb{Q}^* \cong H_1 \text{ X } H_2$.

Let G and G' be abelian groups. Recall that Map(G, G') is the group of all functions $\sigma: G \to G'$, where the group operation is defined point-wise using the group operation of G':

$$(\sigma + \tau)(a) = \sigma(a) + \tau(a) \text{ and } (-\sigma)(a) = -\sigma(a)$$

for all $\sigma, \tau \in$ Map(G, G') and all $a \in G$. The following theorem isolates an important subgroup of this group.

5. Consider the group $\mathbb{Z}^*_p$ where p is an odd prime, and let $\rho: \mathbb{Z}^*_p$! $\mathbb{Z}^*_p$ be the squaring map. By definition, Im $\rho = (\mathbb{Z}^*_p)^2$, and we proved in Theorem that Ker $\rho = \{[\pm 1]\}$. Theorem says that for all $\gamma, \beta \in \mathbb{Z}^*_p, \gamma^2 = \beta^2$ if and only if $\gamma = \pm\beta$. This fact can also be seen to be a special case of part (vi) of Theorem. Theorem says that $\mathbb{Z}^*_p$/Ker $\rho \cong$ Im ρ, and since $|\mathbb{Z}^*_p/\text{Ker } \rho| = |\mathbb{Z}^*_p|/|\text{Ker } \rho| = (p-1)/2$, we see that theorem, which says that $|(\mathbb{Z}^*_p)^2| = (p-1)/2$, follows from this.

Let $H := (\mathbb{Z}^*_p)^2$, and consider the quotient group $\mathbb{Z}^*_p/H$. Since $|H| = (p-1)/2$, we know that $|\mathbb{Z}^*_p / H| = |\mathbb{Z}^*_p| / |H| = 2$, and hence $\mathbb{Z}^*_p / H$ consists of the two cosets H and $\bar{H} := \mathbb{Z}^*_p \setminus H$.

Let α be an arbitrary, fixed element of $\bar{H}$, and consider the map

$$\begin{aligned} t: \mathbb{Z} &\to \mathbb{Z}^*_p / H \\ z &\mapsto [\alpha^z]_H. \end{aligned}$$

It is easy to see that τ is a group homomorphism; indeed, it is the composition of the homomorphism discussed in previsous example and the natural map from $\mathbb{Z}^*_p$ to $\mathbb{Z}^*_p / H$. Moreover, it is easy to see that

$$\alpha^z \in H \Leftrightarrow z \text{ is even.}$$

From this, it follows that Ker $\tau = 2\mathbb{Z}$; also, since $\mathbb{Z}^*_p / H$ consists of just the two cosets H and $\bar{H}$, it follows that "is surjective. Therefore, Theorem. 23 says that the map

$$\begin{aligned} \bar{\tau}: \mathbb{Z}_2 &\to \mathbb{Z}^*_p / H \\ [z]_2 &\mapsto [\alpha^z]_H \end{aligned}$$

is a group isomorphism, under which $[0]_2$ corresponds to H, and $[1]_2$ corresponds to $\bar{H}$.

This isomorphism gives another way to derive Theorem, which says that in $\mathbb{Z}^*_p$, the product of two non-squares is a square; indeed, the statement "non-zero plus non-zero equals zero in $\mathbb{Z}_2$" translates via the isomorphism $\mp$ to the statement "non-square times non-square equals square in $\mathbb{Z}^*_p$."

Theorem: Let G and G' be abelian groups, and consider the group of functions Map(G, G'). Then

Hom(G, G'):= {$\sigma \in$ Map(G, G'): σ is a group homomorphismg}

is a subgroup of Map(G, G').

Proof: First, observe that Hom(G, G') is non-empty, as it contains the map that sends everything in G to $0_{G'}$ (this is the identity element of Map(G, G')).

Next, we have to show that if σ and τ are homomorphisms from G to G', then so are $\sigma + \tau$ and $-\sigma$. But $\sigma + \tau = \rho_2 \circ \rho_1$, where ρ_1: $G' X G' \to G'$ is the map constructed in Theorem, applied with σ and τ, and ρ_2: G'X $G' \to G'$. Also, $-\sigma = \rho_{-1} \circ \sigma$, where ρ_{-1} is the (–1)-multiplication map.

EXERCISE

1. Let ρ: $G \to G'$ be a group homomorphism. Let H be a subgroup of G, and let τ: $H \to G'$ be the restriction of ρ to H. Show that τ is a group homomorphism and that Ker τ = Ker $\rho \cap H$.
2. Verify that the "is isomorphic to" relation on abelian groups is an equivalence relation; that is, for all abelian groups G_1, G_2, G_3, we have:

   (*a*) $G_1 \cong G_2$ and $G_2 \cong G_3$ implies $G_1 \cong G_3$.

   (*b*) $G_1 \cong G_1$;

   (*c*) $G_1 \cong G_2$ implies $G_2 \cong G_1$;
3. Suppose G_1,..., G_k are abelian groups. Show that for each i = 1,..., k, the projection map π_i: G_1 X... X $G_k \to G_i$ that sends (a_1,..., a_k) to a_i is a surjective group homomorphism.
4. Show that if $G = G_1$ X G_2 for abelian groups G_1 and G_2, and H_1 is a subgroup of G_1 and H_2 is a subgroup of G_2, then we have a group isomorphism $G/(H_1$ X $H_2) \cong G_1/H_1$ X G_2/H_2.
5. Let $\mathbb{Z}^+$ denote the set of positive integers, and let $\mathbb{Q}$ be the multiplicative group of non-zero rational numbers. Consider the abelian groups Map# ($\mathbb{Z}^+$, $\mathbb{Z}$) and Map# ($\mathbb{Z}^+$, $\mathbb{Z}_2$). Show that we have group isomorphisms

   (*a*) $\mathbb{Q}^*/(\mathbb{Q}^*)^2 \cong$ Map# ($\mathbb{Z}^+$, $\mathbb{Z}_2$).

   (*b*) $\mathbb{Q}^*$ '= $\mathbb{Z}_2 \times$ Map# ($\mathbb{Z}^+$, $\mathbb{Z}$), and
6. Let G be an abelian group with subgroups H and K.

(*a*) Show that if H and K are finite, then $|H+K| = |H||K|/|H \cap K|$.

(*b*) Show that we have a group isomorphism $(H+K)/K \cong H/(H \cap K)$.

7. Suppose that G, G_1, and G_2 are abelian groups, and that $\rho: G_1 \times G_2 \to G$ is a group isomorphism. Let $H_1 := \rho(G_1 \times \{0_{G_2}\})$ and $H_2 := \rho(\{0_{G_1}\} \times G_2)$. Show that G is the internal direct product of H_1 and H_2.

8. Let G be an abelian group with subgroups H, K, and A, where $K \subseteq H$. Show that $(H \cap A)/(K \cap A)$ is isomorphic to a subgroup of H/K.

9. Let $\rho: G \to G'$ be a surjective group homomorphism. Let S be the set of all subgroups of G that contain Ker ρ, and let S' be the set of all subgroups of G'. Show that the sets S and S′ are in one-to-one correspondence, via the map that sends $H \in S$ to $\rho(H) \in S'$. Moreover, show that this correspondence preserves inclusions, that is, for all $H_1, H_2 \in S$, we have $H_1 \subseteq H_2 \Leftrightarrow (H_1) \subseteq (H_2)$.

10. Let n be an odd, positive integer whose factorization into primes is $n = p_1^{e_1} \ldots p_r^{e_r}$. Show that:

    (a) We have a group isomorphism $\mathbb{Z}_n^*/(\mathbb{Z}_n^*)^2 \cong \mathbb{Z}_2^{\times r}$;

    (b) If p_i 'I 3 (mod 4) for each $i = 1, \ldots, r$, then the squaring map on $(\mathbb{Z}_n^*)^2$ is a group automorphism.

11. Which of the following pairs of groups are isomorphic? Why or why not?

    (a) $\mathbb{Z}_2 \times \mathbb{Z}_2$ and $\mathbb{Z}_4$, (b) $\mathbb{Z}_{12}^*$ and $\mathbb{Z}_8^*$, (c) $\mathbb{Z}_5^*$ and $\mathbb{Z}_4$, (d) $\mathbb{Z}_2 X \mathbb{Z}$ and $\mathbb{Z}$, (e) $\mathbb{Q}$ and $\mathbb{Z}$, (f) $\mathbb{Z}$ X $\mathbb{Z}$ and $\mathbb{Z}$.

12. Let $\rho_i: G_i \to G_i'$, for $i = 1, \ldots, k$, be group homomorphisms. Show that the map

    $\rho: G_1 \times \ldots \times G_k \to G_1' \times \ldots \times G_k'$

    $(a_1, \ldots, a_k) \mapsto (\rho_1(a_1), \ldots, \rho_k(a_k))$

    is a group homomorphism. Also show that if each ρ_i is an isomorphism, then so is ρ.

13. Let $\rho: G \to G'$ be a group homomorphism. Let H, K be subgroups of G and let m be a positive integer. Show that $\rho(H+K) = \rho(H) + \rho(K)$ and $\rho(mH) = m\rho(H)$.

14. Let $\rho: G \to G'$ be a group homomorphism with kernel K. Let S be a subgroup of G. Show that we have a group isomorphism $G/(H+K) \cong \rho(G)/\rho(H)$.

QUOTIENT GROUPS

We now generalize the notion of a congruence relation.

Let G be an abelian group, and let H be a subgroup of G. For $a, b \in G$, we write $a \equiv b \pmod{H}$ if $a - b \in H$. In other words, $a \equiv b \pmod{H}$ if and only if $a = b + h$ for some $h \in H$.

Analogously to Theorem, if we view the subgroup H as fixed, then the following theorem says that the binary relation "$\cdot \equiv \cdot \pmod{H}$" is an equivalence relation on the set G:

Theorem: Let G be an abelian group and H a subgroup of G. For all $a, b, c \in G$, we have:

(*i*) $a \equiv a \pmod{H}$;

(*ii*) $a \equiv b \pmod{H}$ implies $b \equiv a \pmod{H}$;

(*iii*) $a \equiv b \pmod{H}$ and $b \equiv c \pmod{H}$ implies $a \equiv c \pmod{H}$.

Proof: For (i), observe that H contains $0_G = a - a$. For (ii), observe that if H contains $a - b$, then it also contains $-(a - b) = b - a$. For (iii), observe that if H contains $a - b$ and $b - c$, then it also contains $(a - b) + (b - c) = a - c$.

Since the binary relation "$\cdot \equiv \cdot \pmod{H}$" is an equivalence relation, it partitions G into equivalence classes. For $a \in G$, we denote the equivalence class containing a by $[a]_H$. By definition, we have

$$x \in [a]_H \Leftrightarrow x \equiv a \pmod{H} \Leftrightarrow x = a + h \text{ for some } h \in H,$$

and hence

$$[a]_H = a + H := \{a + h : h \in H\}.$$

It is also clear that $[0_G]_H = H$.

Historically, these equivalence classes are called cosets of H in G, and we shall adopt this terminology here as well. Any member of a coset is called a representative of the coset.

Multiplicative notation: if G is written multiplicatively, then $a \equiv b \pmod{H}$ means $ab^{-1} \in H$, and $[a]_H = aH := \{ah : h \in H\}$.

Example.

1. Let $G := \mathbb{Z}_6$, which consists of the residue classes [0], [1], [2], [3], [4], [5]. Let H be the subgroup $3G = \{[0], [3]\}$ of G. The coset of H containing the residue class [1] is $[1] + H = \{[1], [4]\}$, and the coset of H containing the residue class [2] is $[2] + H = \{[2], [5]\}$. The cosets $\{[0], [3]\}$, $\{[1], [4]\}$, and $\{[2], [5]\}$ are the only cosets of H in G, and they clearly partition the set $\mathbb{Z}_6$. Note that each coset of H in G contains two elements, each of which is itself a coset of $6\mathbb{Z}$ in $\mathbb{Z}$ (i.e., a residue classes modulo 6).
2. Let $G := \mathbb{Z}$ and $H := n\mathbb{Z}$ for some positive integer n. Then $a \equiv b \pmod{H}$ if and only if $a \equiv b \pmod{n}$. The coset $[a]_H$ is exactly the same thing as the residue class $[a]_n \in \mathbb{Z}_n$.

In the previous example, we saw that each coset contained the same number of elements. As the next theorem shows, this was no accident.

Theorem: Let G be an abelian group and H a subgroup of G. For all $a, b \in G$, the function

$$f: G \to G$$
$$x \mapsto b - a + x$$

is a bijection, which when restricted to the coset $[a]_H$, yields a bijection from $[a]_H$ to the coset $[b]_H$. In particular, every two cosets of H in G have the same cardinality.

Proof: First, we claim that f is a bijection. Indeed, if $f(x) = f(x')$, then $b - a + x = b - a + x'$, and subtracting b and adding a to both sides of this equation yields $x = x'$. That proves that f is injective. To prove that f is surjective, observe that for any given $x' \in G$, we have $f(a - b + x') = x'$.

Second, we claim that for all $x \in G$, we have $x \in [a]_H$ if and only if $f(x) \in [b]_H$. On the one hand, suppose that $x \in [a]_H$, which means that $x = a + h$ for some $h \in H$. Subtracting a and adding b to both sides of this equation yields $b - a + x = b + h$, which means $f(x) \in [b]_H$. Conversely, suppose that $f(x) \in [b]_H$, which means that $b - a + x = b + h$ for some $h \in H$. Subtracting b and adding a to both sides of this equation yields $x = a + h$, which means that $x \in [a]_H$.

The theorem is now immediate from these two claims.

Lagrange's Theorem

Theorem: If G is a finite abelian group, and H is a subgroup of G, then the order of H divides the order of G.

Proof: This is an immediate consequence of the previous theorem, and the fact that the cosets of H in G partition G.

Theorem: Suppose G is an abelian group and H is a subgroup of G. For all $a, a', b, b' \in G$, if $a \equiv a' \pmod{H}$ and $b \equiv b' \pmod{H}$, then we have $a + b \equiv a' + b' \pmod{H}$.

Proof: Now, $a \equiv a' \pmod{H}$ and $b \equiv b' \pmod{H}$ means that $a = a' + x$ and $b = b' + y$ for some $x, y \in H$. Therefore, $a + b = (a' + x) + (b' + y) = (a' + b') + (x + y)$, and since $x + y \in H$, this means that $a + b \equiv a' + b' \pmod{H}$.

Let G be an abelian group and H a subgroup. Let G/H denote the set of all cosets of H in G. Theorem allows us to define a binary operation on G/H in the following natural way: for $a, b \in G$, define

$$[a]_H + [b]_H := [a + b]_H.$$

That this definition is unambiguous follows immediately from theorem if $[a]_H = [a']_H$ and $[b]_H = [b']_H$, then $[a + b]_H = [a' + b']_H$.

We can easily verify that this operation makes G/H into an abelian group.

Properties of Lagrange's Theorem

(*i*) *Associativity*:

$$[a]_H + ([b]_H + [c]_H) = [a]_H + [b + c]_H = [a + (b + c)]_H$$
$$= [(a + b) + c]_H = [a + b]_H + [c]_H$$
$$= ([a]_H + [b]_H) + [c]_H.$$

Here, we have used the definition of addition of cosets, and the corresponding associativity property for G.

(*ii*) Identity element: the coset $[0_G]_H = H$ acts as the identity element, since

$$[a]_H + [0_G]_H = [a + 0_G]_H = [a]_H = [0_G + a]_H = [0_G]_H + [a]_H.$$

(*iii*) *Inverses*: the inverse of the coset $[a]_H$ is $[-a]_H$, since

$$[a]_H + [-a]_H = [a + (-a)]_H = [0_G]_H = [(-a) + a]_H = [-a]_H + [a]_H.$$

(*iv*) *Commutativity*:

$$[a]_H + [b]_H = [a + b]_H = [b + a]_H = [b]_H + [a]_H.$$

The group G/H is called the quotient group of G modulo H. The order of the group G/H is sometimes denoted $[G: H]$ and is called the index of H in G. Note that if $H = G$, then the quotient group G/H is the trivial group, and so $[G: H] = 1$.

Multiplicative notation: if G is written multiplicatively, then the definition of the group operation of G/H is expressed $[a]_H \cdot [b]_H := [a \cdot b]_H$; the identity element of G/H is $[1_G]_H = H$, and the inverse of $[a]_H$ is $[a^{-1}]_H$.

Theorem: Suppose G is a finite abelian group and H is a subgroup of G. Then $[G: H] = | G | = | H |$. Moreover, if K is a subgroup of H, then

$$[G: K] = [G: H][H: K].$$

Proof: The fact that $[G: H] = | G | / | H |$ follows directly from Theorem. The fact that $[G: K] = [G: H][H: K]$ follows from a simple calculation:

$$[G:H] = \frac{|G|}{|H|} = \frac{|G|/|K|}{|H|/|K|} = \frac{[G:K]}{[H:K]}.$$

Example.

1. For each $n \geq 1$, the group $\mathbb{Z}_n$ is precisely the quotient group $\mathbb{Z}/n\mathbb{Z}$.
2. The multiplicative group $\mathbb{Z}_{15}^*$, as we saw, is of order 8. The subgroup $(\mathbb{Z}_{15}^*)^2$ of $\mathbb{Z}_{15}^*$ has order 2. Therefore, the quotient group $\mathbb{Z}_{15}^*/(\mathbb{Z}_{15}^*)^2$ has order 4. Indeed, the cosets are

$$\alpha_{00} := (\mathbb{Z}_{15}^*)^2 = \{[1], [4]\}, \alpha_{01} := [-1](\mathbb{Z}_{15}^*)^2 = \{[-1], [-4]\},$$

$$\alpha_{10} := [2](\mathbb{Z}_{15}^*)^2 = \{[2], [-7]\}, \alpha_{11} := [-2](\mathbb{Z}_{15}^*)^2 = \{[-2], [7]\}.$$

We can write down the multiplication table for the quotient group:

$\cdot$	α_{00}	α_{01}	α_{10}	α_{11}
α_{00}	α_{00}	α_{01}	α_{10}	α_{11}
α_{01}	α_{01}	α_{00}	α_{11}	α_{10}
α_{10}	α_{10}	α_{11}	α_{00}	α_{01}
α_{11}	α_{11}	α_{10}	α_{01}	α_{00}

Note that this group is essentially just a "renaming" of the additive group $\mathbb{Z}_2 \times \mathbb{Z}_2$.

3. As we saw in Example, $(\mathbb{Z}_{15}^*)^2 = \{[1]\}$. Therefore, the quotient group $\mathbb{Z}_{15}^* / (\mathbb{Z}_{15}^*)^2$ has order 2. The cosets of $(\mathbb{Z}_{15}^*)^2$ in $\mathbb{Z}_{15}^*$ are $0 := \{[\pm 1]\}$ and $a_1 := \{[\pm 2]\}$, and the multiplication table looks like this:

$\cdot$	α_0	α_1
α_0	α_0	α_1
α_1	α_1	α_0

We see that the quotient group is essentially just a "renaming" of $\mathbb{Z}_2$.

EXERCISE

1. Let H be a subgroup of an abelian group G, and let $a, b \in G$ with $a \equiv b \pmod{H}$. Show that $ka \equiv kb \pmod{H}$ for all $k \in \mathbb{Z}$.

2. Let G be an abelian group and H a subgroup with $[G: H] = 2$. Show that if $a, b \in G \setminus H$, then $a + b \in H$.

3. Let G be an abelian group, and let $\sim$ be an equivalence relation on G. Further, suppose that for all $a, a', b \in G$, if $a \sim a'$, then $a + b \sim a' + b$. Let $H := \{a \in G: a \sim 0_G\}$. Show that H is a subgroup of G, and that for all $a, b \in G$, we have $a \sim b$ if and only if $a \equiv b \pmod{H}$.

4. Let n be an odd, positive integer whose factorization into primes is $n = p_1^{e_1} \ldots p_r^{e_r}$.

   Show that $[\mathbb{Z}_n^* : (\mathbb{Z}_n^*)^2] = 2^r$.

5. Let H be a subgroup of an abelian group G, and let $a, b \in G$. Show that $[a + b]_H = \{x + y: x \in [a]_H, y \in [b]_H\}$.

6. Write down the cosets of $(\mathbb{Z}_{35}^*)^2$ in $\mathbb{Z}_{35}^*$, along with the multiplication table for the

   quotient group $\mathbb{Z}_{35}^*/(\mathbb{Z}_{35}^*)^2$.

7. Let n be a positive integer, and let m be any integer. Show that $[\mathbb{Z}_n: m\mathbb{Z}_n] = n/\gcd(m, n)$.

THE STRUCTURE OF FINITE ABELIAN GROUPS

We next state a theorem that classifies all finite abelian groups up to isomorphism. A finite abelian group (with more than one element) is isomorphic to a direct product of cyclic groups

$$\mathbb{Z}_{p_1^{e_1}} \times \ldots \times \mathbb{Z}_{p_r^{e_r}},$$

where the p_i's are primes (not necessarily distinct) and the e_i's are positive integers. This direct product of cyclic groups is unique up to the order of the factors.

Theorem: A finite abelian group (with more than one element) is isomorphic to a direct product of cyclic groups

$$\mathbb{Z}_{m_1} \times \ldots \times \mathbb{Z}_{m_t},$$

where each $mi > 1$, and where for $i = 1, \ldots, t - 1$, we have $m_i \mid m_{i+1}$. Moreover, the integers $m_1, \ldots, m_t$ are uniquely determined, and m_t is the exponent of the group.

The statements of these theorems are much more important than their proofs. Even if the reader does not study the proofs, he is urged to understand what the theorems actually say.

Lemma: A finite abelian group (with more than one element) is isomorphic to a direct product of cyclic groups

$$\mathbb{Z}_{m_1} \times \ldots \times \mathbb{Z}_{m_t},$$

where each $m_i > 1$, and where for $i = 1, \ldots, t - 1$, we have $m_i \mid m_{i+1}$; moreover, m_t is the exponent of the group.

Proof: Let G be a finite abelian group with more than one element, and let m be the exponent of G. By Theorem, there exists an element $a \in G$ of order m. Let $A = \langle a \rangle$. Then $A \cong \mathbb{Z}_m$. Now, if $A = G$, the lemma is proved. So assume that $A \subseteq G$.

We will show that there exists a subgroup B of G such that $G = A + B$ and $A \cap B = \{0_G\}$. From this, Theorem gives us an isomorphism of G with A X B. Moreover, the exponent of B is clearly a divisor of m, and so the lemma will follow by induction (on the order of the group).

So it suffices to show the existence of a subgroup B as above. We prove this by contradiction. Suppose that there is no such subgroup, and among all subgroups B such that $A \cap B = \{0_G\}$, assume that B is maximal, meaning that there is no subgroup B' of G such that $B \subseteq B'$ and $A \cap B' = \{0_G\}$. By assumption $C := A + B \subseteq G$.

Let d be any element of G that lies outside of C. Consider the quotient group G/C, and let r be the order of $[d]_C \in G/C$. Note that $r > 1$ and $r \mid m$. We shall define a group element d' with slightly nicer properties than d, as follows.

Since $rd \in C$, we have $rd = sa + b$ for some $s \in \mathbb{Z}$ and $b \in B$. We claim that $r \mid s$. To see this, note that $0_G = md = (m/r)rd = (m/r)sa + (m/r)b$, and since $A \cap B = \{0_G\}$, we have $(m/r)sa = 0_G$, which can only happen if $r \mid s$. That proves the claim. This allows us to define $d' := d - (s/r)a$. Since $d \equiv d' \pmod{C}$, we see not only that $[d']_C \in G/C$ has order r, but also that $rd' \in B$.

We next show that $A \cap (B + \langle d' \rangle) = \{0_G\}$, which will yield the contradiction we seek, and thus prove the lemma. Because $A \cap B = \{0_G\}$, it will suffice to show that $A \cap (B + \langle d' \rangle) \subseteq B$. Now, suppose we have a group element $b' + xd' \in A$, with $b' \in B$ and $x \in \mathbb{Z}$. Then in particular, $xd' \in C$, and so $r \mid x$, since $[d']_C \in G/C$ has order r. Further, since $rd' \in B$, we have $xd' \in B$, whence $b' + xd' \in B$.

Lemma: Suppose that $G := \mathbb{Z}_{m_1} \times \ldots \times \mathbb{Z}_{m_t}$ and $H := \mathbb{Z}_{n_1} \times \ldots \times \mathbb{Z}_{n_t}$ are isomorphic, where the m_i's and n_i's are positive integers (possibly 1) such that $m_i \mid m_{i+1}$ and $n_i \mid n_{i+1}$ for $i = 1,\ldots, t - 1$. Then $m_i = n_i$ for $i = 1,\ldots, t$.

Proof: Clearly, $\prod_i m_i = |G| = |H| = \prod_i n_i$. We prove the lemma by induction on the order of the group. If the group order is 1, then clearly all the m_i's and n_i's must be 1, and we are done. Otherwise, let p be a prime dividing the group order.

Now, suppose that p divides $m_r,\ldots, m_t$ but not $m_1,\ldots, m_{r-1}$, and that p divides $n_s,\ldots, n_t$ but not $n_1,\ldots, n_{s-1}$, where $r \leq t$ and $s \leq t$. Evidently, the groups pG and pH are isomorphic. Moreover,

$$pG \cong \mathbb{Z}_{m_1} \times \ldots \times \mathbb{Z}_{m_{r-1}} \times \mathbb{Z}_{m_r/p} \times \ldots \times \mathbb{Z}_{m_t/p},$$

and

$$pH \cong \mathbb{Z}_{n_1} \times \ldots \times \mathbb{Z}_{m_{s-1}} \times \mathbb{Z}_{n_s/p} \times \ldots \times \mathbb{Z}_{n_t/p}.$$

Thus, we see that $|pG| = |G| / p^{t-r+1}$ and $|pH| = |H| / p^{t-s+1}$, from which it follows that $r = s$, and the lemma then follows by induction.

EXERCISE

1. Suppose $G \cong \mathbb{Z}_{m_1} \times \ldots \times \mathbb{Z}_{m_t}$ with $m_i \mid m_{i+1}$ for $i = 1,\ldots, t - 1$, and that H is a subgroup of G. Show that $H \cong \mathbb{Z}_{n_1} \times \ldots \times \mathbb{Z}_{n_t}$, where $n_i \mid n_{i+1}$ for $i = 1,\ldots, t - 1$ and $n_i \mid m_i$ for $i = 1,\ldots, t$.

2. In our proof of Euler's criterion, we really only used the fact that $\mathbb{Z}_p^*$ has a unique element of multiplicative order 2. This exercise develops a proof of a generalization of Euler's criterion, based on the fundamental theorem of finite abelian groups. Suppose G is an abelian group of even order n that contains a unique element of order 2.

(a) Show that $G \cong \mathbb{Z}_{2^e} \times \mathbb{Z}_{m_1} \times \ldots \times \mathbb{Z}_{m_k}$, where $e > 0$ and the m_i's are odd integers.

(b) Using part (a), show that $2G = G\{n/2\}$.

3. Suppose that G is an abelian group such that for all $m > 0$, we have $mG = G$ and $|G\{m\}| = m^2$ (note that G is not finite). Show that $G\{m\} \cong \mathbb{Z}_m \text{ X } \mathbb{Z}_m$ for all $m > 0$.

4. Suppose $G \cong \mathbb{Z}_{m_1} \times \ldots \times \mathbb{Z}_{m_t}$. Let p be a prime, and let s be the number of m_i's divisible by p. Show that $G\{p\} \cong \mathbb{Z}_p^{\times s}$.

SUBGROUPS

Definition: Let G be an abelian group, and let H be a non-empty subset of G such that

- $a + b \in H$ for all $a, b \in H$, and
- $-a \in H$ for all $a \in H$.

Then H is called a subgroup of G.

In words: H is a subgroup of G if it is closed under the group operation and taking inverses.

Multiplicative notation: if the abelian group G in the above definition is written using multiplicative notation, then H is a subgroup if $ab \in H$ and $a^{-1} \in H$ for all $a, b \in H$.

Theorem: Let G be an abelian group, and let m be an integer. Then

$$mG := \{ma : a \in G\}$$

is a subgroup of G.

Proof: m'/G is non-empty, since $0_G = m0_G \in mG$. For $ma, mb \in mG$, we have $ma + mb = m(a + b) \in mG$, and $-(ma) = m(-a) \in mG$.

Theorem: If G is an abelian group, and H is a subgroup of G, then H contains 0_G; moreover, the binary operation of G, when restricted to H, yields a binary operation that makes H into an abelian group whose identity is 0_G.

Proof: First, to see that $0_G \in H$, just pick any $a \in H$, and using both properties of the definition of a subgroup, we see that $0_G = a + (-a) \in H$.

H is closed under addition, which means that the restriction of the binary operation "+" on G to H induces a well defined binary operation on H.

So now it suffices to show that H, together with this operation, satisfies the defining properties of an abelian group. Associativity and commutativity follow directly from the corresponding properties for G.

Clearly, for an abelian group G, the subsets G and $\{0_G\}$ are subgroups. These are not very interesting subgroups. Other, more interesting subgroups may sometimes be found by using the following two theorems.

Theorem: Let G be an abelian group, and let m be an integer. Then

$$G\{m\} := \{a \in G: ma = 0_G\}$$

is a subgroup of G.

Proof: $G\{m\}$ is non-empty, since $m0_G = 0_G$, and so $G\{m\}$ contains 0_G. If $ma = 0_G$ and $mb = 0_G$, then $m(a + b) = ma + mb = 0_G + 0_G = 0_G$ and $m(-a) = -(ma) = -0_G = 0_G$.

Multiplicative notation: if the abelian group G in the above two theorems is written using multiplicative notation, then we write the subgroup of the first theorem as $G^m := \{a^m: a \in G\}$. The subgroup in the second theorem is denoted in the same way: $G\{m\} := \{a \in G: a^m = 1_G\}$.

Example.

1. For every integer m, the set $m\mathbb{Z}$ is the subgroup of the additive group $\mathbb{Z}$ consisting of all multiples of m. This is the same as the ideal of $\mathbb{Z}$ generated by m, which we already studied in some detail in §1.2. Two such subgroups $m\mathbb{Z}$ and $m'\mathbb{Z}$ are equal if and only if $m = \pm m'$. The subgroup $\mathbb{Z}\{m\}$ is equal to $\mathbb{Z}$ if $m = 0$, and is equal to $\{0\}$ otherwise.
2. Let n be a positive integer, let $m \in \mathbb{Z}$, and consider the subgroup $m\mathbb{Z}_n$ of the additive group $\mathbb{Z}_n$. Now, for every residue class $[z] \in \mathbb{Z}_n$, we have $m[z] = [mz]$. Therefore, $[b] \in m\mathbb{Z}_n$ if and only if there exists $z \in \mathbb{Z}$ such that $mz \equiv b \pmod{n}$. By part (i) of Theorem such a z exists if and only if $d \mid b$, where $d := \gcd(m, n)$. Thus, $m\mathbb{Z}_n$ consists precisely of the n/d distinct residue classes

   $$[i.\ d]\ (i = 0, \ldots, n/d - 1),$$

   and in particular, $m\mathbb{Z}_n = d\mathbb{Z}_n$.

Now consider the subgroup $\mathbb{Z}_n\{m\}$ of $\mathbb{Z}_n$. The residue class $[z]$ is in $\mathbb{Z}_n\{m\}$ if and only if $mz \equiv 0 \pmod{n}$. By part (ii) of Theorem, this happens if and only if $z \equiv 0 \pmod{n/d}$, where $d := \gcd(m, n)$ as above. Thus, $\mathbb{Z}_n\{m\}$ consists precisely of the d residue classes

$$[i.\ n/d]\ (i = 0, \ldots, d - 1),$$

and in particular, $\mathbb{Z}_n\{m\} = \mathbb{Z}_n\{d\} = (n/d)\mathbb{Z}_n$.

3. We already proved that $(\mathbb{Z}_n^*)^m$ is a subgroup of $\mathbb{Z}_n^*$ in Theorem. Also, the proof of previous theorem clearly works for an arbitrary abelian group G: for each $a \in G$, and all $\ell, m \in \mathbb{Z}$ with $\gcd(\ell, m) = 1$, if $\ell a \in mG$, then $a \in mG$.

Theorem: If G is a subgroup of $\mathbb{Z}_n$, then there exists a unique positive integer d dividing n such that $G = d\mathbb{Z}_n$. Also, for all positive divisors d_1, d_2 of n, we have $d_1\mathbb{Z}_n \subseteq d_2\mathbb{Z}_n$ if and only if $d_2 \mid d_1$.

Proof: Note that the second statement implies the uniqueness part of the first statement, so it suffices to prove just the existence part of the first statement and the second statement.

Let G be an arbitrary subgroup of $\mathbb{Z}_n$, and let $H := \{z \in \mathbb{Z} : [z] \in G\}$. We claim that H is a subgroup of $\mathbb{Z}$. To see this, observe that if $a, b \in H$, then $[a]$ and $[b]$ belong to G, and hence so do $[a + b] = [a] + [b]$ and $[-a] = -[a]$, and thus $a + b$ and $-a$ belong to H. That proves the claim, and Theorem implies that $H = d\mathbb{Z}$ for some non-negative integer d. It follows that

$$G = \{[y]: y \in H\} = \{[dz]: z \in \mathbb{Z}\} = d\mathbb{Z}_n.$$

Evidently, $n \in H = d\mathbb{Z}$, and hence $d \mid n$. That proves the existence part of the first statement of the theorem.

To prove the second statement of the theorem, observe that if d_1 and d_2 are arbitrary integers, then

$$d_1\mathbb{Z}_n \subseteq d_2\mathbb{Z}_n \Leftrightarrow d_2 z \equiv d_1 \pmod{n} \text{ for some } z \in \mathbb{Z}$$

$$\Leftrightarrow \gcd(d_2, n) \mid d_1 \text{ (by part (i) of previous theorem).}$$

In particular, if d_2 is a positive divisor of n, then $\gcd(d_2, n) = d_2$, which proves the second statement.

Of course, not all abelian groups have such a simple subgroup structure.

Theorem: If G is a subgroup of $\mathbb{Z}$, then there exists a unique non-negative integer m such that $G = m\mathbb{Z}$. Moreover, for two non-negative integers m_1 and m_2, we have $m_1\mathbb{Z} \subseteq m_2\mathbb{Z}$ if and only if $m_2 \mid m_1$.

Proof: Actually, we have already proven this. One only needs to observe that a subset G of $\mathbb{Z}$ is a subgroup if and only if it is an ideal of $\mathbb{Z}$. The first statement of the theorem then follows from Theorem.

Example.

1. Consider the group $\mathbb{Z}_{15}^*$. We can enumerate its elements as

$$[\pm 1], [\pm 2], [\pm 4], [\pm 7].$$

Therefore, the elements of $(\mathbb{Z}_{15}^*)^2$ are

$$[1]^2 = [1], [2]^2 = [4], [4]^2 = [16] = [1], [7]^2 = [49] = [4];$$

thus, $(\mathbb{Z}_{15}^*)^2$ has order 2, consisting as it does of the two distinct elements [1] and [4].

Going further, one sees that $(\mathbb{Z}_{15}^*)^4 = \{[1]\}$. Thus, $\alpha^4 = [1]$ for all $\alpha \in \mathbb{Z}_{15}^*$.

By direct calculation, one can determine that $(\mathbb{Z}_{15}^*)^3 = \mathbb{Z}_{15}^*$; that is, cubing simply permutes $\mathbb{Z}_{15}^*$.

For any given integer m, write $m = 4q + r$, where $0 \leq r < 4$. Then for every $\alpha \in \mathbb{Z}_{15}^*$, we have $\alpha^m = \alpha^{4q+r} = \alpha^{4q}\alpha^r = \alpha^r$. Thus, $(\mathbb{Z}_{15}^*)^m$ is either $\mathbb{Z}_{15}^*$, $(\mathbb{Z}_{15}^*)^2$, or $\{[1]\}$.

However, there are certainly other subgroups of $\mathbb{Z}_{15}^*$ – for example, the subgroup $\{[\pm 1]\}$.

2. Consider the group $\mathbb{Z}_{15}^* = \{[\pm 1], [\pm 2]\}$. The elements of $(\mathbb{Z}_{15}^*)^2$ are

$$[1]^2 = [1],\ [2]^2 = [4] = [-1];$$

thus, $(\mathbb{Z}_{15}^*)^2 = \{[\pm 1]\}$ and has order 2.

There are in fact no other subgroups of $\mathbb{Z}_{15}^*$ besides $\mathbb{Z}_{15}^*$, $\{[\pm 1]\}$, and $\{[1]\}$. Indeed, if H is a subgroup containing [2], then we must have $H = \mathbb{Z}_{15}^*$: $[2] \in H$ implies $[2]^2 = [4] = [-1] \in H$, which implies $[-2] \in H$ as well. The same holds if H is a subgroup containing [–2].

3. Consider the group $G = \mathbb{Z} \in \times \mathbb{Z}_2$. For every non-zero $\alpha \in G$, $\alpha + \alpha = 0_G$. From this, it is clear that the set $H = \{0_G, \alpha\}$ is a subgroup of G. However, for every integer m, $mG = G$ if m is odd, and $mG = \{0_G\}$ if m is even. Thus, the subgroup H is not of the form mG for any m.

Theorem: If H_1 and H_2 are subgroups of an abelian group G, then so is $H_1 \cap H_2$.

Proof: It is evident that $H_1 \cap H_2$ is non-empty, as both H_1 and H_2 contain 0_G, and hence so does their intersection. If $a \in H_1 \cap H_2$ and $b \cap H_1 \cap H_2$, then since $a, b \in H_1$, we have $a + b \in H_1$, and since $a, b \in H_2$, we have $a + b \in H_2$; therefore, $a + b \in H_1 \cap H_2$. Similarly, $-a \in H_1$ and $-a \in H_2$, and therefore, $-a \in H_1 \cap H_2$.

Let G be an abelian group and H_1, H_2, H_3 subgroups of G. The reader may verify that $H_1 + H_2 = H_2 + H_1$ and $(H_1 + H_2) + H_3 = H_1 + (H_2 + H_3)$. It follows that if $H_1, \ldots, H_k$ are subgroups of G, then we can write $H_1 + \ldots + H_k$ without any parentheses, and there can be no ambiguity; moreover, the order of the H_i's does not matter. The same holds with "+" replaced by "$\cap$".

Theorem: If H_1 and H_2 are subgroups of an abelian group G, then so is

$$H_1 + H_2 := \{a_1 + a_2 : a_1 \in H_1, a_2 \neq H_2\}.$$

Proof: It is evident that $H_1 + H_2$ is non-empty, as it contains $0_G + 0_G = 0_G$. Consider two elements in $H_1 + H_2$, which we can write as $a_1 + a_2$ and $b_1 + b_2$, where $a_1, b_1 \in H_1$ and $a_2, b_2 \in H_2$. Then by the closure properties of subgroups, $a_1 + b_1 \in H_1$ and $a_2 + b_2 \in H_2$, and hence $(a_1 + a_2) + (b_1 + b_2) = (a_1 + b_1) + (a_2 + b_2) \in H_1 + H_2$. Similarly, $-(a_1 + a_2) = (-a_1) + (-a_2) \in H_1 + H_2$.

Theorem: if the abelian group G in the above theorem is written multiplicatively, then the subgroup defined in the theorem is written $H_1H_2 := \{a_1a_2 : a_1 \in H_1, a_2 \in H_2\}$.

Theorem: If H is a subgroup of an abelian group G, then in general, we have $H + H \neq 2H$. For example, $\mathbb{Z} + \mathbb{Z} = \mathbb{Z}$, while $\mathbb{Z} \in \neq \mathbb{Z}$.

EXERCISE

1. Let G be an abelian group.
   (*a*) Show that if H is a subgroup of G and $a, b \in H$, then $a - b \in H$.
   (*b*) Suppose that H is a non-empty subset of G such that $a - b \in H$ for all $a, b \in H$. Show that H is a subgroup of G.
2. Let G be an abelian group.
   (*a*) Show that if H is a subgroup of G, $h \in H$, and $g \in G \setminus H$, then $h + g \in G \setminus H$.
   (*b*) Suppose that H is a non-empty subset of G such that for all $h, g \in G$: (i) $h \in H$ implies $-h \in H$, and (ii) $h \in H$ and $g \in G \setminus H$ implies $h + g \in G \setminus H$. Show that H is a subgroup of G.
3. Show that if H is a subgroup of an abelian group G, then a set $K \subseteq H$ is a subgroup of G if and only if K is a subgroup of H.
4. Let G be an abelian group with subgroups H_1 and H_2. Show that every subgroup H of G that contains $H_1 \cup H2$ must contain all of $H_1 + H_2$, and that $H_1 \subseteq H_2$ if and only if $H_1 + H_2 = H_2$.
5. Let H_1 be a subgroup of an abelian group G_1 and H_2 a subgroup of an abelian group G_2. Show that $H_1 \times H_2$ is a subgroup of G_1 X G_2.
6. Show that if G_1 and G_2 are abelian groups, and m is an integer, then $m(G_1 \times G_2) = mG_1 \times mG_2$.
7. Let G_1 and G_2 be abelian groups, and let H be a subgroup of G_1 X G_2. Define
   $$H_1 := \{a_1 \in G_1 : (a_1, a_2) \in H \text{ for some } a_2 \in G_2\}.$$
   Show that H_1 is a subgroup of G_1.
8. Let I be a set and G be an abelian group, and consider the group Map(I, G) of functions $f: I \rightarrow G$. Let Map$^\#$ (I, G) be the set of functions $f \in$ Map(I, G) such that $f(i) \neq 0_G$ for at most finitely many $i \in I$. Show that Map$^\#$ (I, G) is a subgroup of Map(I, G).

Chapter 4

Ring Theory

In mathematics, ring theory is the study of rings, algebraic structures in which addition and multiplication are defined and have similar properties to those familiar from the integers. Ring theory studies the structure of rings, their representations, or, in different language, modules, special classes of rings (group rings, division rings, universal enveloping algebras), as well as an array of properties that proved to be of interest both within the theory itself and for its applications, such as homological properties and polynomial identities.

Commutative rings are much better understood than noncommutative ones. Due to its intimate connections with algebraic geometry and algebraic number theory, which provide many natural examples of commutative rings, their theory, which is considered to be part of commutative algebra and field theory rather than of general ring theory, is quite different in flavour from the theory of their noncommutative counterparts.

A fairly recent trend, started in the 1980s with the development of noncommutative geometry and with the discovery of quantum groups, attempts to turn the situation around and build the theory of certain classes of noncommutative rings in a geometric fashion as if they were rings of functions on (non-existent) 'noncommutative spaces'.

ELEMENTARY INTRODUCTION

Definition: Formally, a ring is an Abelian group $(R, +)$, together with a second binary operation * such that for all a, b and c in R,

$$a * (b * c) = (a * b) * c$$

$$a * (b + c) = (a * b) + (a * c)$$

$$(a + b) * c = (a * c) + (b * c)$$

also, if there exists a *multiplicative identity* in the ring, that is, an element e such that for all a in R,

$$a * e = e * a = a$$

Then it is said to be a *ring with unity*. The number 1 is a common example of a unity.

It is simple to show that any ring in which 1 = 0 must have just one element; any such ring is called a zero ring.

Rings that sit inside other rings are called subrings. Maps between rings which respect the ring operations are called ring homomorphisms. Rings, together with ring homomorphisms, form a category (the category of rings).

Closely related is the notion of ideals, certain subsets of rings which arise as kernels of homomorphisms and can serve to define factor rings. Basic facts about ideals, homomorphisms and factor rings are recorded in the isomorphism theorems and in the Chinese remainder theorem.

A ring is called *commutative* if its multiplication is commutative. Commutative rings resemble familiar number systems, and various definitions for commutative rings are designed to recover properties known from the integers.

Commutative rings are also important in algebraic geometry. In commutative ring theory, numbers are often replaced by ideals, and the definition of prime ideal tries to capture the essence of prime numbers. Integral domains, non-trivial commutative rings where no two non-zero elements multiply to give zero, generalize another property of the integers and serve as the proper realm to study divisibility.

Principal ideal domains are integral domains in which every ideal can be generated by a single element, another property shared by the integers. Euclidean domains are integral domains in which the Euclidean algorithm can be carried out. Important examples of commutative rings can be constructed as rings of polynomials and their factor rings.

Summary: Euclidean domain => principal ideal domain ⇒ unique factorization domain ⇒ integral domain ⇒ Commutative ring.

Non-commutative rings resemble rings of matrices in many respects. Following the model of algebraic geometry, attempts have been made recently at defining non-commutative geometry based on non-commutative rings.

Non-commutative rings and associative algebras (rings that are also vector spaces) are often studied via their categories of modules. A module over a ring is an Abelian group that the ring acts on as a ring of endomorphisms, very much akin to the way fields (integral domains in which every non-zero element is invertible) act on vector spaces. Examples of non-commutative rings are given by rings of square matrices or more generally by rings of endomorphisms of Abelian groups or modules, and by monoid rings.

PROPERTIES OF RING THEORY

Definition: A commutative ring with unity is a set R together with addition and multiplication operations on R, such that:

(*i*) The set R under addition forms an abelian group, and we denote the additive identity by 0_R;

(*ii*) Multiplication is associative; that is, for all $a, b, c \in R$, we have $a(bc) = (ab)c$;

(*iii*) Multiplication distributes over addition; that is, for all $a, b, c \in R$, we have $a(b + c) = ab + ac$ and $(b + c)a = ba + ca$;

(*iv*) There exists a multiplicative identity; that is, there exists an element $1_R \in R$, such that $1_R . a = a = a . 1_R$ for all $a \in R$;

(*v*) Multiplication is commutative; that is, for all $a, b \in R$, we have $ab = ba$.

There are other, more general (and less convenient) types of rings – one can drop properties (iv) and (v), and still have what is called a ring. We shall not, however, be working with such general rings in this text. Therefore, to simplify terminology, from now on, by a "ring," we shall always mean a commutative ring with unity.

Let R be a ring. Notice that because of the distributive law, for any fixed $a \in R$, the map from R to R that sends $b \in R$ to $ab \in R$ is a group homomorphism with respect to the underlying additive group of R. We call this the a-multiplication map.

Theorem: Let R be a ring. Then:

(i) The multiplicative identity 1_R is unique;

(ii) $0_R . a = 0_R$ for all $a \in R$;

(iii) $(-a)b = a(-b) = -(ab)$ for all $a, b \in R$;

(iv) $(-a)(-b) = ab$ for all $a, b \in R$;

(v) $(ka)b = a(kb) = k(ab)$ for all $k \in \mathbb{Z}$ and $a, b \in R$.

Proof: Part (i) may be proved using the same argument as was used to prove part (i) of Previous theorem. Parts (ii), (iii), and (v) follow directly from parts (i), (ii), and (iii) of previous theorem, using appropriate multiplication maps, discussed above. Part (iv) follows from part (iii), along with part (iv) of Theorem:

$$(-a)(-b) = -(a(-b)) = -(-(ab)) = ab.$$

Example.

1. The set $\mathbb{Q}$ of rational numbers under the usual rules of multiplication and addition forms a ring.
2. Generalizing Example, if $R_1,..., R_k$ are rings, then we can form the direct product $S := R_1 \text{ X } ... \text{ X } R_k$, which consists of all k-tuples $(a_1,..., a_k)$ with $a_1 \in R_1,..., a_k \in R_k$. We can view S in a natural way as a ring, with addition and multiplication defined component-wise. The additive identity is $(0_{R_1},...,0_{R_k})$ and the multiplicative identity is $(1_{R_1},...,1_{R_k})$. When $R = R_1 =... = R_k$, the k-wise direct product of R is denoted R^{Xk}.

3. The set $\mathbb{R}$ of real numbers under the usual rules of multiplication and addition forms a ring.
4. The set $\mathbb{Z}$ under the usual rules of multiplication and addition forms a ring.
5. The ring $\mathbb{Z}$ has characteristic zero, $\mathbb{Z}_n$ has characteristic *n*, and $\mathbb{Z}_{n_1} \times \mathbb{Z}_{n_2}$ has characteristic $\text{lcm}(n_1, n_2)$.
6. The set $\mathbb{C}$ of complex numbers under the usual rules of multiplication and addition forms a ring. Every $\alpha \in \mathbb{C}$ can be written (uniquely) as $\alpha = a + bi$, where $a, b \in \mathbb{R}$ and $i = \sqrt{-1}$. If $0 = a' + b'i$ is another complex number, with $a', b' \in \mathbb{R}$, then

$$\alpha + \alpha' = (a + a') + (b + b')i \text{ and } \alpha\alpha' = (aa' - bb') + (ab' + a'b)i.$$

The fact that $\mathbb{C}$ is a ring can be verified by direct calculation; however, we shall see later that this follows easily from more general considerations.

Recall the complex conjugation operation, which sends α to $\bar{\alpha} := a - bi$. One can verify by direct calculation that complex conjugation is both additive and multiplicative; that is, $\overline{\alpha + \alpha'} = \bar{\alpha} + \overline{\alpha'}$ and $\bar{\alpha}.\overline{\alpha'}$.

The norm of α is $N(\alpha) := \alpha\bar{\alpha} = a^2 + b^2$. So we see that $N(\alpha)$ is a non-negative real number, and is zero if and only if $\alpha = 0$. Moreover, from the multiplicativity of complex conjugation, it is easy to see that the norm is multiplicative as well:

$$N(\alpha\alpha') = \alpha\alpha'\overline{\alpha\alpha'} = \alpha\alpha'\overline{\alpha\alpha'} = \alpha\bar{\alpha}\alpha'\overline{\alpha'} = N(\alpha)N(\alpha').$$

7. Consider the set *F* of all arithmetic functions, that is, functions mapping positive integers to reals. Let us define addition of arithmetic functions point-wise (i.e., $(f + g)(n) = f(n) + g(n)$ for all positive integers *n*) and multiplication using the Dirichlet product. The reader should verify that with addition and multiplication so defined, F forms a ring, where the all-zero function is the additive identity, and the special function.
8. If *I* is an arbitrary set and *R* is a ring, then Map(*I*, *R*), which is the set of all functions $f : I \to R$, may be naturally viewed as a ring, with addition and multiplication defined point-wise: for $f, g \in$ Map(*I*, *R*), we define

$$(f + g)(i) := f(i) + g(i) \text{ and } (f \,.\, g)(i) := f(i).\, g(i) \text{ for all } i \in I.$$

We leave it to the reader to verify that Map(*I*, *R*) is indeed a ring, where the additive identity is the all-zero function, and the multiplicative identity is the all-one function.

Note that in a ring *R*, if $1_R = 0_R$, then for all $a \in R$, we have $a = 1_R \,.\, a = 0_R \,.\, a = 0_R$, and hence the ring *R* is trivial, in the sense that it consists of the single element 0_R, with $0_R + 0_R = 0_R$ and $0_R \,.\, 0_R = 0_R$. If $1_R \neq 0_R$, we say that *R* is non-trivial. Trivial rings are not very interesting, but they naturally arise in certain constructions.

For $a_1, \ldots, a_k \in R$, the product $a_1 \ldots a_k$ needs no parentheses, because multiplication is associative; moreover, we can reorder the a_i's without changing the value of the product, since multiplication is commutative. We can also write this product as $\prod_{i=1}^{k} a_i$. By convention, such a product is defined to be 1_R when $k = 0$. When $a = a_1 = \ldots = a_k$, we can write this product as a^k. The reader may verify the usual power laws: for all $a, b \in R$, and all non-negative integers k and ℓ, we have

$$(a^{\ell})^k = a^{k\ell} = (a^k)^{\ell},\ a^{k+\ell} = a^k a^{\ell},\ (ab)^k = a^k b^k.$$

For all $a_1, \ldots, a_k, b_1, \ldots, b_{\ell} \in R$, the distributive law implies

$$(a_1 + \ldots + a_k)(b_1 + \ldots + b_{\ell}) = \sum_{\substack{1 \le i \le k \\ 1 \le j \le \ell}} a_i b_j .$$

A ring R is in particular an abelian group with respect to addition. We shall call a subgroup of the additive group of R an additive subgroup of R. The characteristic of R is defined as the exponent of this group. Note that for all $m \in \mathbb{Z}$ and $a \in R$, we have

$$ma = m(1_R \cdot a) = (m \cdot 1_R)a,$$

so that if $m \cdot 1_R = 0_R$, then $ma = 129129129\ 0_R$ for all $a \in R$. Thus, if the additive order of 1_R is infinite, the characteristic of R is zero, and otherwise, the characteristic of R is equal to the additive order of 1_R.

EXERCISE

1. Let R be a ring. For additive subgroups A and B of R, we define their ring-theoretic product AB as the set of all elements of R that can be expressed as

   $$a_1 b_1 + \ldots + a_k b_k$$

   for some $a_1, \ldots, a_k \in A$ and $b_1, \ldots, b_k \in B$; by definition, this set includes the "empty sum" 0_R. Show that for all additive subgroups A, B, and C of R:

   (*a*) AB is also an additive subgroup of R;

   (*b*) $AB = BA$;

   (*c*) $A(BC) = (AB)C$;

   (*d*) $A(B + C) = AB + AC$.

2. Show that the familiar binomial theorem holds in an arbitrary ring R; that is, for all $a, b \in R$ and every positive integer n, we have

   $$(a+b)^n = \sum_{k=0}^{n} \binom{n}{k} a^{n-k} b^k .$$

ZERO DIVISORS AND INTEGRAL DOMAINS

Let R be a ring. If a and b are non-zero elements of R such that $ab = 0$, then a and b are both called zero divisors. If R is non-trivial and has no zero divisors, then it is called an integral domain. Note that if a is a unit in R, it cannot be a zero divisor (if $ab = 0$, then multiplying both sides of this equation by a^{-1} yields $b = 0$). In particular, it follows that every field is an integral domain.

Example.

1. For two non-trivial rings R_1, R_2, an element $(a_1, a_2) \in R_1 \times R_2$ is a zero divisor if and only if a_1 is a zero divisor, a_2 is a zero divisor, or exactly one of a_1 or a_2 is zero. In particular, $R_1 \times R_2$ is not an integral domain.
2. For $n > 1$, $\mathbb{Z}_n$ is an integral domain if and only if n is prime. In particular, if n is composite, so $n = ab$ with $1 < a < n$ and $1 < b < n$, then $[a]_n$ and $[b]_n$ are zero divisors: $[a]_n[b]_n = [0]_n$, but $[a]_n \neq [0]_n$ and $[b]_n \neq [0]_n$.
3. $\mathbb{Z}$ is an integral domain.
4. $\mathbb{Q}, \mathbb{R}$ and $\mathbb{C}$ are fields, and hence are also integral domains.

The next two theorems establish certain results that are analogous to familiar facts about integer divisibility. These results hold in a general ring, provided one avoids zero divisors. The first is a cancellation law:

Theorem: Every finite field F must be of cardinality p^w, where p is prime, w is a positive integer, and p is the characteristic of F.

Proof: The characteristic of F is either zero or a prime, and since F is finite, it must be prime. Let p denote the characteristic. By definition, p is the exponent of the additive group of F, and by Theorem, the primes dividing the exponent are the same as the primes dividing the order, and hence F must have cardinality p^w for some positive integer w.

Theorem: If R is a ring, and $a, b, c \in R$ such that $a \neq 0$ and a is not a zero divisor, then $ab = ac$ implies $b = c$.

Proof: $ab = bc$ implies $a(b - c) = 0$. The fact that $a \neq 0$ and a is not a zero divisor implies that we must have $b - c = 0$, and so $b = c$.

Theorem: The characteristic of an integral domain is either zero or a prime.

Proof: By way of contradiction, suppose that D is an integral domain with characteristic m that is neither zero nor prime.

Since, by definition, D is not a trivial ring, we cannot have $m = 1$, and so m must be composite. Say $m = st$, where $1 < s < m$ and $1 < t < m$. Since m is the additive order of 1_D, it follows that $(s \cdot 1_D) \neq 0_D$ and $(t \cdot 1_D) \neq 0_D$; moreover, since D is an integral domain, it follows that $(s \cdot 1_D)(t \cdot 1_D) \neq 0_D$. So we have

$$0_D = m \,.\, 1_D = (st).\, 1_D = (s \,.\, 1_D)(t \,.\, 1D) \neq 0_D,$$

a contradiction.

Theorem: Every finite integral domain is a field.

Proof: Let D be a finite integral domain, and let a be any non-zero element of D. Consider the a-multiplication map that sends $b \in D$ to ab, which is a group homomorphism on the additive group of D. Since a is not a zero-divisor, it follows that the kernel of the a-multiplication map is $\{0_D\}$, hence the map is injective, and by finiteness, it must be surjective as well. In particular, there must be an element $b \in D$ such that $ab = 1_D$.

Theorem: Let R be a ring.

(*i*) Suppose $a, b \in R$, and that either a or b is not a zero divisor. Then $a \mid b$ and $b \mid a$ if and only if $ar = b$ for some $r \in R^*$.

(*ii*) Suppose $a, b \in R$, $a \mid b$, $a \neq 0$, and a is not a zero divisor. Then there exists a unique $r \in R$ such that $ar = b$, which we denote by b/a.

Proof: For the first statement, if $ar = b$ for some $r \in R^*$, then we also have $br^{-1} = a$; thus, $a \mid b$ and $b \mid a$. For the converse, suppose that $a \mid b$ and $b \mid a$. We may assume that b is not a zero divisor (otherwise, exchange the roles of a and b). We may also assume that b is non-zero (otherwise, $b \mid a$ implies $a = 0$, and so the conclusion holds with any r). Now, $a \mid b$ implies $ar = b$ for some $r \in R$, and $b \mid a$ implies $br' = a$ for some $r' \in R$, and hence $b = ar = br'r$. Canceling b from both sides of the equation $b = br'r$, we obtain $1 = r'r$, and so r is a unit.

For the second statement, $a \mid b$ means $ar = b$ for some $r \in R$. Moreover, this value of r is unique: if $ar = b = ar'$, then we may cancel a, obtaining $r = r'$.

Of course, in the previous two theorems, if the ring is an integral domain, then there are no zero divisors, and so the hypotheses may be simplified in this case, dropping the explicit requirement that certain elements are not zero divisors. In particular, if a, b, and c are elements of an integral domain, such that $ab = ac$ and $a \neq 0$, then we can cancel a, obtaining $b = c$.

EXERCISE

1. Show that for all $n \geq 1$, and for all $a, b \in \mathbb{Z}_n$, if $a \mid b$ and $b \mid a$, then $ar = b$ for some $r \in \mathbb{Z}_n^*$.
2. Let R be a ring, and let $a, b \in R$ such that $ab \neq 0$. Show that ab is a zero divisor if and only if a is a zero divisor or b is a zero divisor.
3. Let R be a ring of characteristic $m > 0$, and let n be an integer.

   Show that:

(*a*) If $\gcd(n, m) = 1$, then $n \,.\, 1_R$ is a unit;

(*b*) If $1 < \gcd(n, m) < m$, then $n \,.\, 1_R$ is a zero divisor;

(*c*) Otherwise, $n \,.\, 1_R = 0$.

4. Let D be an integral domain, $m \in \mathbb{Z}$ and $a \in D$. Show that $ma = 0$ if and only if m is a multiple of the characteristic of D or $a = 0$.
5. Suppose that R is a non-trivial ring in which the cancellation law holds in general: for all $a, b, c \in R$, if $a \neq 0$ and $ab = ac$, then $b = c$. Show that R is an integral domain.

DIVISIBILITY

For elements a, b in a ring R, we say that a divides b if $ar = b$ for some $r \in R$. If a divides b, we write $a \mid b$, and we may say that a is a divisor of b, or that b is a multiple of a, or that b is divisible by a. If a does not divide b, then we write $a \mid b$.

We call $a \in R$ a unit if $a \mid 1_R$, that is, if $ar = 1_R$ for some $r \in R$. Using the same argument it is easy to see that r is uniquely determined; it is called the multiplicative inverse of a, and we denote it by a^{-1}.

Also, for $b \in R$, we may write b/a to denote ba^{-1}. Evidently, if a is a unit, then $a \mid b$ for every $b \in R$.

We denote the set of units by R^*. It is easy to see that $1_R \in R$. Moreover, R^* is closed under multiplication; indeed, if a and b are elements of R^*, then $(ab)^{-1} = a^{-1} b^{-1}$. It follows that with respect to the multiplication operation of the ring, R^* is an abelian group, called the multiplicative group of units of R.

If $a \in R^*$ and k is a positive integer, then $a^k \in R^*$; indeed, the multiplicative inverse of a^k is $(a^{-1})^k$, which we may also write as a^{-k} (which is consistent with our notation for abelian groups). For all $a, b \in R^*$, the identities hold for all integers k and ℓ.

If R is non-trivial and every non-zero element of R has a multiplicative inverse, then R is called a field.

Example.

1. Every non-zero element of $\mathbb{Q}$ is a unit. Hence, $\mathbb{Q}$ is a field.
2. Every non-zero element of $\mathbb{R}$ is a unit. Hence, $\mathbb{R}$ is a field.
3. The only units in the ring $\mathbb{Z}$ are ± 1. Hence, $\mathbb{Z}$ is not a field.
4. For non-zero $\alpha = a + bi \in \mathbb{C}$, with $a, b \in \mathbb{R}$, we have $c := N(\alpha) = a^2 + b^2 > 0$. It follows that the complex number $\bar{\alpha} c^{-1} = (ac^{-1}) + (-bc^{-1})i$ is the multiplicative inverse of α, since $\bar{\alpha} . \alpha\, c^{-1} = (\alpha \bar{\alpha})c^{-1} = 1$. Hence, every non-zero element of $\mathbb{C}$ is a unit, and so $\mathbb{C}$ is a field.

5. If I is a set and R is a ring, then the units in Map(I, R) are those functions $f: I \to R$ such that $f(i) \in R^*$ for all $i \in I$.
6. For rings $R_1,..., R_k$, it is easy to see that the multiplicative group of units of the direct product $R_1 \times ... \times R_k$ is equal to $R_1^* \times ... \times R_k^*$. Indeed, by definition, $(a_1,..., a_k)$ has a multiplicative inverse if and only if each individual a_i does.
7. Let n be a positive integer. The units in $\mathbb{Z}_n$ are the residue classes $[a]_n$ with gcd(a, n) = 1. In particular, if n is prime, all non-zero residue classes are units, and if n is composite, some non-zero residue classes are not units. Hence, $\mathbb{Z}_n$ is a field if and only if n is prime. The notation $\mathbb{Z}_n^*$ introduced in this section for the group of units of the ring $\mathbb{Z}_n$ is consistent with the notation.

SUBRINGS

Definition: A subset S of a ring R is called a subring if

(*i*) S is an additive subgroup of R,

(*ii*) S is closed under multiplication, and

(*iii*) $1_R \in S$.

It is clear that the operations of addition and multiplication on a ring R make a subring S of R into a ring, where 0_R is the additive identity of S and 1_R is the multiplicative identity of S. One may also call R an extension ring of S.

Some texts do not require that 1_R belongs to a subring S, and instead require only that S contains a multiplicative identity, which may be different than that of R. This is perfectly reasonable, but for simplicity, we restrict ourselves to the case where $1_R \in S$.

Expanding the above definition, we see that a subset S of R is a subring if and only if $1_R \in S$ and for all $a, b \in S$, we have

$$a + b \in S, -a \in S, \text{ and } ab \in S.$$

In fact, to verify that S is a subring, it suffices to show that $-1_R \in S$ and that S is closed under addition and multiplication; indeed, if $-1_R \in S$ and S is closed under multiplication, then S is closed under negation, and further, $1_R = -(-1_R) \in S$.

Example.

1. $\mathbb{Q}$ is a subring of $\mathbb{R}$.
2. $\mathbb{Z}$ is a subring of $\mathbb{Q}$.
3. Suppose R is a non-trivial ring. Then the set $\{0_R\}$ is not a subring of R: although it satisfies the first two requirements of the definition of a subring, it does not satisfy the third.

4. $\mathbb{R}$ is a subring of $\mathbb{C}$. Note that for all $\alpha := a + bi \in \mathbb{C}$, with $a, b \in \mathbb{R}$, we have $\overline{\alpha} = \alpha \Leftrightarrow a + bi = a - bi \Leftrightarrow b = 0$. That is, $\overline{\alpha} = \alpha \Leftrightarrow \in \mathbb{R}$.

5. Let m be a positive integer, and let $\mathbb{Q}^{(m)}$ be the set of rational numbers which can be written as a/b, where a and b are integers, and b is relatively prime to m. Then $\mathbb{Q}^{(m)}$ is a subring of $\mathbb{Q}$, since for all $a, b, c, d \in \mathbb{Z}$ with $\gcd(b, m) = 1$ and $\gcd(d, m) = 1$, we have

$$\frac{a}{b}+\frac{c}{d}=\frac{ad+bc}{bd} \text{ and } \frac{a}{b}.\frac{c}{d}=\frac{ac}{bd},$$

and since $\gcd(bd, m) = 1$, it follows that the sum and product of any two elements of $\mathbb{Q}^{(m)}$ are again in $\mathbb{Q}^{(m)}$. Clearly, $\mathbb{Q}^{(m)}$ contains -1, and so it follows that $\mathbb{Q}^{(m)}$ is a subring of $\mathbb{Q}$. The units of $\mathbb{Q}^{(m)}$ are precisely those rational numbers of the form a/b, where $\gcd(a, m) = \gcd(b, m) = 1$.

6. The set $\mathbb{Z}[i]$ of complex numbers of the form $a + bi$, with $a, b \in \mathbb{Z}$, is a subring of $\mathbb{C}$. It is called the ring of Gaussian integers. Since $\mathbb{C}$ is a field, it contains no zero divisors, and hence $\mathbb{Z}[i]$ contains no zero divisors either. Hence, $\mathbb{Z}[i]$ is an integral domain.

Let us determine the units of $\mathbb{Z}[i]$. Suppose $\alpha \in \mathbb{Z}[i]$ is a unit, so that there exists $\alpha' \in \mathbb{Z}[i]$ such that $\alpha\alpha' = 1$. Taking norms, we obtain

$$1 = N(1) = N(\alpha\alpha') = N(\alpha)N(\alpha').$$

Since the norm of any Gaussian integer is itself a non-negative integer, and since $N(\alpha)N(\alpha') = 1$, we must have $N(\alpha) = 1$. Now, if $\alpha = a + bi$, with $a, b \in \mathbb{Z}$, then $1 = N(\alpha) = a^2 + b^2$, which implies that $\alpha = \pm 1$ or $\alpha = \pm i$. Conversely, it is easy to see that ± 1 and $\pm i$ are indeed units, and so these are the only units in $\mathbb{Z}[i]$.

EXERCISE

1. Consider the ring Map(R, R) of functions $f : \mathbb{R} \to \mathbb{R}$, with addition and multiplication defined point-wise.
   (*a*) Show that Map($\mathbb{R}, \mathbb{R}$) is not an integral domain, and that Map($\mathbb{R}, \mathbb{R}$)* consists of those functions that never vanish.
   (*b*) Let $a, b \in$ Map($\mathbb{R}, \mathbb{R}$). Show that if $a \mid b$ and $b \mid a$, then $ar = b$ for some $r \in$ Map($\mathbb{R}, \mathbb{R}$).
   (*c*) Let $\mathbb{C}$ be the subset of Map($\mathbb{R}, \mathbb{R}$) of continuous functions. Show that $\mathbb{C}$ is a subring of Map($\mathbb{R}, \mathbb{R}$), and that all functions in C^* are either everywhere positive or everywhere negative.

(*d*) Find elements $a, b \in C$, such that in the ring C, we have $a \mid b$ and $b \mid a$, yet there is no $r \in C^*$ such that $ar = b$.

2. Show that if S and T are subrings of R, then so is $S \cap T$.
3. Let S_1 be a subring of R_1, and S_2 a subring of R_2. Show that $S_1 \times S_2$ is a subring of $R_1 \times R_2$.
4. Show that the set $\mathbb{Q}[i]$ of complex numbers of the form $a + bi$, with $a, b \in \mathbb{Q}$, is a subfield of $\mathbb{C}$.
5. Suppose that S and T are subrings of a ring R. Show that their ring-theoretic product ST is a subring of R that contains $S \cup T$, and is the smallest such subring.

THE STRUCTURE OF $\mathbb{Z}_n^*$

We are now in a position to precisely characterize the structure of the group $\mathbb{Z}_n^*$, for an arbitrary integer $n > 1$. This characterization will prove to be very useful in a number of applications.

Suppose $n = p_1^{e_1} \ldots p_r^{e_r}$ is the factorization of n into primes. By the Chinese remainder theorem, we have the ring isomorphism

$$\theta : \mathbb{Z}_n \to \mathbb{Z}_{p_1^{e_1}} \times \ldots \times \mathbb{Z}_{p_r^{e_r}}$$

$$[a]_n \mapsto ([a]_{p_1^{e_1}}, \ldots, [a]_{p_r^{e_r}})$$

and restricting θ to $\mathbb{Z}_n^*$ yields a group isomorphism

$$\mathbb{Z}_n^* \cong \mathbb{Z}_{p_1^{e_1}}^* \times \ldots \times \mathbb{Z}_{p_r^{e_r}}^* .$$

Thus, to determine the structure of the group $\mathbb{Z}_n^*$ for general n, it suffices to determine the structure for $n = p^e$, where p is prime. By Theorem, we already know the order of the group $\mathbb{Z}_{p^e}^*$, namely, $\varphi(p^e) = p^{e-1}(p - 1)$, where φ is Euler's phi function.

Theorem: Let D be an integral domain and G a subgroup of D^* of finite order. Then G is cyclic.

Proof: Suppose G is not cyclic. Then by theorem, we know that the exponent m of G is strictly less than $\mid G \mid$. It follows that $a^m = 1$ for all $a \in G$. That is, all the elements of G are roots of the polynomial $X^m - 1 \in D[X]$. But by Theorem, a polynomial of degree m over an integral domain has at most m roots, and this contradicts the fact that $m < |G|$.

This theorem immediately implies that $\mathbb{Z}_p^*$ is cyclic for every prime p, since $\mathbb{Z}_p$ is a field; however, we cannot directly use this theorem to prove that $\mathbb{Z}_{p^e}^*$ is cyclic for $e > 1$ (and p odd), because $\mathbb{Z}_{p^e}$ is not a field. To deal with the case $e > 1$, we need a few simple facts.

Theorem: If p is an odd prime, then for every positive integer e, the group $\mathbb{Z}_{p^e}^*$ is cyclic. The group $\mathbb{Z}_{2^e}^*$ is cyclic for $e = 1$ or 2, but not for $e \geq 3$. For $e \geq 3$, $\mathbb{Z}_{2^e}^*$ is isomorphic to the additive group $\mathbb{Z}_2 \text{ X } \mathbb{Z}_{2^{e-2}}$.

In the case where $e = 1$, this theorem is a special case of the following, more general, theorem:

Lemma: Let p be a prime. For every positive integer e, if $a \equiv b \pmod{p^e}$, then $ap \equiv bp \pmod{p^{e+1}}$.

Proof: We have $a = b + cp^e$ for some $c \in \mathbb{Z}$. Thus, $a^p = b^p + pb^{p-1}cp^e + dp^{2e}$ for an integer d. It follows that $a^p \equiv b^p \pmod{p^{e+1}}$.

Lemma: Let p be a prime, and let e be a positive integer such that $p^e > 2$. If $a \equiv 1 + p^e \pmod{p^{e+1}}$, then $a^p \equiv 1 + p^{e+1} \pmod{p^{e+2}}$.

Proof: By Lemma 7.30, $a^p \equiv (1 + p^e)^p \pmod{p^{e+2}}$. Expanding $(1 + p^e)^p$, we have

$$(1 + p^e)^p = 1 + p \,.\, p^e + \sum_{k=2}^{p-1} \binom{p}{k} p^{ek} + p^{ep}.$$

All of the terms in the sum on k are divisible by p^{1+2e}, and $1 + 2e \geq e + \in$ for all $e \geq 1$. For the term p^{ep}, the assumption that $p^e > 2$ means that either $p \geq 3$ or $e \geq 2$, which implies $ep \geq e + 2$.

Now consider Theorem in the case where p is odd. As we already know that $\mathbb{Z}_p^*$ is cyclic, assume $e > 1$. Let $x \in \mathbb{Z}$ be chosen so that $[x]_p$ generates $\mathbb{Z}_p^*$. Suppose the multiplicative order of $[x]_{p^e} \in \mathbb{Z}_p^*$ is m. Then as $x^m \equiv 1 \pmod{p^e}$ implies $x^m \equiv 1 \pmod{p}$, it must be the case that $p - 1$ divides m, and so $[x^{m/(p-1)}]_{pe}$ has multiplicative order exactly $p - 1$. By Theorem, if we find an integer y such that $[y]_{pe}$ has multiplicative order p^{e-1}, then $[x^{m/(p-1)}y]_{pe}$ has multiplicative order $(p-1)p^{e-1}$, and we are done. We claim that $y :=$ $1+p$ does the job. Any integer between 0 and $p^e - 1$ can be expressed as an e-digit number in base p; for example, $y = (0 \ldots. 0\ 1\ 1)_p$. If we compute successive pth powers of y modulo p^e, then by Lemma we have

$$
\begin{aligned}
&y \bmod p^e = (0 \ldots 011)_p,\\
&y^p \bmod p^e = (* \ldots * 101)_p,\\
&y^{p^2} \bmod p^e = (* \ldots *1001)_p,\\
&\quad\vdots\\
&y^{p^{e-2}} \bmod p^e = (10 \ldots 01)_p,\\
&y^{p^{e-1}} \bmod p^e = (0 \ldots 01)_p.
\end{aligned}
$$

Here, "*" indicates an arbitrary digit. From this table of values, it is clear that $[y]_{p^e}$ has multiplicative order p^{e-1}.

We now prove Theorem in the case $p = 2$. For $e = 1$ and $e = 2$, the theorem is easily verified. Suppose $e \geq 3$. Consider the subgroup $G \subseteq \mathbb{Z}^*_{2^e}$ generated by $[5]_{2^e}$. Expressing integers between 0 and $2^e - 1$ as e-digit binary numbers, we have

$$
\begin{aligned}
&5 \bmod 2^e = (0 \ldots 0101)_2,\\
&5^2 \bmod 2^e = (* \ldots 1001)_2,\\
&\quad\vdots\\
&5^{2^{e-3}} \bmod 2^e = (10 \ldots 01)_2,\\
&5^{2^{e-2}} \bmod 2^e = (0 \ldots 01)_2.
\end{aligned}
$$

So it is clear that [5]2*e* has multiplicative order 2^{e-2}. We claim that $[-1]_{2^e} \notin G$. If it were, then since it has multiplicative order 2, and since every cyclic group of even order has precisely one element of order $\in$, it must be equal to $\left[5^{2^{e-3}}\right]_{2^e}$; however, it is clear from the above calculation that $5^{2^{e-3}} \not\equiv -1 \pmod{2^e}$. Let $H \subseteq \mathbb{Z}^*_{2^e}$ be the subgroup generated by $[-1]_{2^e}$. Then from the above, $G \cap H = \left\{[1]_{2^e}\right\}$, and hence by Theorem, $G\,X$

H is isomorphic to the subgroup $G\,.\,H$ of $\mathbb{Z}^*_{2^e}$. But since the orders of $G \times H$ and $\mathbb{Z}^*_{2^e}$ are equal, we must have $G\,.\,H = \mathbb{Z}^*_{2^e}$. That proves the theorem.

Example. Let p be an odd prime, and let d be a positive integer dividing $p - 1$. Since $\mathbb{Z}^*_p$ is a cyclic group of order $p - 1$, Theorem, implies that $(\mathbb{Z}^*_p)^d$ is the unique subgroup of $\mathbb{Z}^*_p$ of order $(p - 1)/d$, and moreover, $(\mathbb{Z}^*_p)^d = \mathbb{Z}^*_p\{(p - 1)/d\}$; that is, for all $\alpha \in \mathbb{Z}^*_p$, we have

$$\alpha = \beta^d \text{ for some } \beta \in \mathbb{Z}^*_p \Leftrightarrow \alpha^{(p-1)/d} = 1.$$

Setting $d = 2$, we arrive again at Euler's criterion, but by a very different, and perhaps more elegant, route than that taken in our original proof of that theorem.

EXERCISE

1. Let $n = pq$, where p and q are distinct primes such that $p = 2p' + 1$ and $q = 2q$Ά $+ 1$, where pΆ and qΆ are themselves prime. Show that the subgroup $(\mathbb{Z}^*_p)^2$ of squares is a cyclic group of order pΆqΆ.
2. Show that if n is a positive integer, the group $\mathbb{Z}^*_p$ is cyclic if and only if

   $n = 1, 2, 4, p^e$, or $2p^e$,

   where p is an odd prime and e is a positive integer.
3. Let $n = pq$, where p and q are distinct primes such that $p \mid (q - 1)$ and $q \mid (p - 1)$.

   (*a*) Show that the map that sends $([k]_n, [a]_n) \in \mathbb{Z}_n \times \mathbb{Z}^*_n$ to $[(1 + kn)a^n]_{n^2} \in \mathbb{Z}^*_{n^2}$ is a group isomorphism.

   (*b*) Show that the map that sends $[a]n \in \mathbb{Z}^*_n$ to $[a^n]_{n^2} \in (\mathbb{Z}^*_{n^2})^n$ is a group isomorphism (in particular, you need to show that this map is unambiguously defined).

   (*c*) Consider the element $a_k := [1 + n]_{n^2}$ Î Zn2; show that for every non-negative integer k, $a^k = [1 + kn]_{n^2}$; deduce that a has multiplicative order n, and also that the identity $a^k = [1 + kn]_{n^2}$ holds for all integers k.

POLYNOMIAL RINGS

If R is a ring, then we can form the ring of polynomials $R[\mathrm{X}]$, consisting of all polynomials $g = a_0 + a_1\mathrm{X} + ... + a_k\mathrm{X}^k$ in the indeterminate, or "formal" variable, X, with coefficients a_i in R, and with addition and multiplication defined in the usual way.

Example. Let us define a few polynomials over the ring $\mathbb{Z}$:

$$a: = 3 + \mathrm{X}^2,\ b := 1 + 2\mathrm{X} - \mathrm{X}^3,\ c := 5,\ d := 1 + \mathrm{X},\ e := \mathrm{X}, f := 4\mathrm{X}^3.$$

We have

$$a + b = 4 + 2X + \mathrm{X}^2 - \mathrm{X}^3,\ ab = 3 + 6\mathrm{X} + \mathrm{X}^2 - \mathrm{X}^3 - \mathrm{X}^5,\ cd + ef = 5 + 5\mathrm{X} + 4\mathrm{X}^4.$$

As illustrated in the previous example, elements of R are also polynomials. Such polynomials are called constant polynomials; all other polynomials are called non-constant polynomials. The set R of constant polynomials forms a subring of $R[\mathrm{X}]$.

In particular, 0_R is the additive identity in $R[\mathrm{X}]$ and 1_R is the multiplicative identity in $R[\mathrm{X}]$. Note that if R is the trivial ring, then so is $R[\mathrm{X}]$; also, if R is a subring of E, then $R[\mathrm{X}]$ is a subring of $E[\mathrm{X}]$.

So as to keep the distinction between ring elements and indeterminates clear, we shall use the symbol "X" only to denote the latter. Also, for a polynomial $g \in R[\mathrm{X}]$, we shall in general write this simply as "g," and not as "$g(\mathrm{X})$." Of course, the choice of the symbol "X" is arbitrary; occasionally, we may use another symbol, such as "Y," as an alternative.

MULTI-VARIATE POLYNOMIALS

One can naturally generalize the notion of a polynomial in a single variable to that of a polynomial in several variables.

Consider the ring $R[X]$ of polynomials over a ring R. If Y is another indeterminate, we can form the ring $R[X][Y]$ of polynomials in Y whose coeffcients are themselves polynomials in X over the ring R. One may write $R[X, Y]$ instead of $R[X][Y]$. An element of $R[X, Y]$ is called a bivariate polynomial.

Consider a typical element $g \in R[X, Y]$, which may be written

$$g = \sum_{j=0}^{\ell}\left(\sum_{i=0}^{k} a_{ij}X^i\right)Y^j.$$

Rearranging terms, this may also be written as

$$g = \sum_{\substack{0 \le i \le k \\ 0 \le j \le \ell}} a_{ij}X^iY^j,$$

or as

$$g = \sum_{i=0}^{k}\left(\sum_{j=0}^{\ell} a_{ij}Y^{j}\right)X^{j}.$$

If g is written as in, the terms X^iY^j are called monomials. The total degree of such a monomial X^iY^j is defined to be $i + j$, and if g is non-zero, then the total degree of g, denoted Deg(g), is defined to be the maximum total degree among all monomials X^iY^j appearing in with a non-zero coefficient a_{ij}. We define the total degree of the zero polynomial to be $-\infty$.

When g is written as in, one sees that we can naturally view g as an element of $R[Y][X]$, that is, as a polynomial in X whose coefficients are polynomials in Y. From a strict, syntactic point of view, the rings $R[Y][X]$ and $R[X][Y]$ are not the same, but there is no harm done in blurring this distinction when convenient. We denote by $\deg_X(g)$ the degree of g, viewed as a polynomial in X, and by $\deg_Y(g)$ the degree of g, viewed as a polynomial in Y.

Example1. Let us illustrate, with a particular example, the three different forms of expressing a bivariate polynomial. In the ring $\mathbb{Z}[X, Y]$ we have

$$\begin{aligned} g &= (5X^2 - 3X + 4)Y + (2X^2 + 1) \\ &= 5X^2Y + 2X^2 - 3XY + 4Y + 1 \\ &= (5Y + 2)X^2 + (-3Y)X + (4Y + 1). \end{aligned}$$

We have Deg $(g) = 3$, $\deg_X(g) = 2$, and $\deg_Y(g) = 1$.

More generally, we can form the ring $R[X_1,..., X_n]$ of multi-variate polynomials over R in the variables $X_1,..., X_n$. Formally, we can define this ring recursively as $R[X_1,..., X_{n-1}]$ $[X_n]$, that is, the ring of polynomials in the variable X_n, with coefficients in $R[X_1,..., X_{n-1}]$. A monomial is a term of the form $X_1^{e_1}...X_n^{e_n}$, and the total degree of such a monomial is $e_1 + ... + e_n$. Every non-zero multivariate polynomial g can be expressed uniquely (up to a re-ordering of terms) as $a_1\mu_1 + ... + a_k\mu_k$, where each a_i is a non-zero element of R, and each i is a monomial; we define the total degree of g, denoted Deg (g), to be the maximum of the total degrees of the μ_i's. As usual, the zero polynomial is defined to have total degree $-\infty$.

Just as for bivariate polynomials, the order of the indeterminates is not important, and for every $i = 1,..., n$, one can naturally view any $g \in R[X_1,..., X_n]$ as a polynomial in X_i over the ring $R[X_1,..., X_{i-1}, X_{i+1},..., X_n]$, and define $\deg X_i(g)$ to be the degree of g when viewed in this way.

Just as polynomials in a single variable define polynomial functions, so do polynomials in several variables. If R is a subring of E, $g \in R[X_1,..., X_n]$, and $\alpha_1,..., \alpha_n \in E$, we define $g(\alpha_1,..., \alpha_n)$ to be the element of E obtained by evaluating the expression obtained by substituting α_i for X_i in g.

EXERCISE

1. Let D be an integral domain, and let $g \in D[X_1,..., X_n]$, with $\text{Deg}(g) = k \geq 0$. Let S be a finite, non empty subset of D. Show that the number of elements $(x_1,..., x_n) \in S^{Xn}$ such that $g(x_1,..., x_n) = 0$ is at most $k \mid S \mid^{n-1}$.
2. Let R be a ring, and consider the ring of multi-variate polynomials $R[X_1,..., X_n]$. For $m \geq 0$, define H_m to be the subset of polynomials that can be expressed as $a_1\mu_1 +... + a_k\mu_k$, where each ai belongs to R and each μ_i is a monomial of total degree m (by definition, Hm includes the zero polynomial, and $H_0 = R$). Polynomials that belong to H_m for some m are called homogeneous polynomials. Show that:
   (*a*) If $g, h \in H_m$, then $g + h \in H_m$;
   (*b*) If $g \in H_\ell$ and $h \in H_m$, then $gh \in H_{\ell+m}$;
   (*c*) Every non-zero polynomial g can be expressed uniquely as $g_0 +... + g_d$, where $g_i \in H_i$ for $i = 0,..., d$, $g_d \neq 0$, and $d = \text{Deg}(g)$;
   (*d*) For all polynomials g, h, we have $\text{Deg}(gh) \leq \text{Deg}(g) + \text{Deg}(h)$, and if R is an integral domain, then $\text{Deg}(gh) = \text{Deg}(g) + \text{Deg}(h)$.
3. Suppose that D is an integral domain, and g, h are non-zero, multi-variate polynomials over D such that gh is homogeneous. Show that g and h are also homogeneous.
4. Let R be a ring, and let $x_1,..., x_n$ be elements of R. Show that every polynomial $g \in R[X_1,..., X_n]$ can be expressed as
   $$g = (X_1 - x_1)q_1 +... + (X_n - x_n)q_n + g(x_1,..., x_n),$$
   where $q_1,..., q_n \in R[X_1,..., X_n]$.
5. Show that if $R = \mathbb{Z}$ and $I = (d)$, where $d = p_1^{e_1} ....p_r^{e_r}$ is the prime factorization of d, then $\text{Rad}(I) = (p_1 \cdots p_r)$.

FORMALITIES

For completeness, we present a more formal definition of the ring $R[X]$. The reader should bear in mind that this formalism is rather tedious, and may be more distracting than it is enlightening. Formally, a polynomial $g \in R[X]$ is an infinite sequence $\{a_i\}_{i=0}^{\infty}$, where each $a_i \in R$, but only finitely many of the a_i's are nonzero (intuitively, a_i represents the coefficient of X^i). For each non-negative integer j, it will be convenient to define the function $\varepsilon_j : R \rightarrow R[X]$ that maps $c \in R$ to the sequence $\{c_i\}_{i=0}^{\infty} \in R[X]$, where $c_j := c$ and $c_i := 0_R$ for $i \neq j$ (intuitively, $\varepsilon_j(c)$ represents the polynomial cX^j).

For

$$g = \{a_i\}_{i=0}^{\infty} \in R[X] \text{ and } h = \{b_i\}_{i=0}^{\infty} \in R[X],$$

we define

$$g + h := \{s_i\}_{i=0}^{\infty} \text{ and } gh := \{p_i\}_{i=0}^{\infty},$$

where for $i = 0, 1, 2,...,$

$$s_i := a_i + b_i$$

and

$$p_i := \sum_{i=j+k} a_j b_k,$$

the sum being over all pairs (j, k) of non-negative integers such that $i = j + k$ (which is a finite sum). We leave it to the reader to verify that $g + h$ and gh are polynomials (i.e., only finitely many of the s_i's and p_i's are non-zero). The additive identity is the all-zero sequence $\varepsilon_0(0_R)$, and the multiplicative identity is $\varepsilon_0(1_R)$.

One can easily verify that for all $c, d \in R$, we have

$$\varepsilon_0(c + d) = \varepsilon_0(c) + \varepsilon_0(d) \text{ and } \varepsilon_0(cd) = \varepsilon_0(c)\, \varepsilon_0(d).$$

We shall identify $c \in R$ with $\varepsilon_0(c) \in R[X]$, viewing the ring element c as simply "shorthand" for the polynomial $\varepsilon_0(c)$ in contexts where a polynomial is expected. Note that while c and $\varepsilon_0(c)$ are not the same mathematical object, there will be no confusion in treating them as such.

Thus, from a narrow, legalistic point of view, R is not a subring of $R[X]$, but we shall not let such annoying details prevent us from continuing to speak of it as such. Indeed, by appropriately renaming elements, we can make R a subring of $R[X]$ in the literal sense of the term.

We also define $X := \varepsilon_1(1_R)$. One can verify that $X^i = \varepsilon_i(1_R)$ for all $i \geq 0$. More generally, for any polynomial $g = \{a_i\}_{i=0}^{\infty}$, if $a_i = 0_R$ for all i exceeding some value k, then we have $g = \sum_{i=0}^{k} \varepsilon_0(a_i) X^i$. Writing a_i in place of $\varepsilon_0(a_i)$, we have $g = \sum_{i=0}^{k} a_i X^i$, and so we can return to the standard practice of writing polynomials as we did in Example, without any loss of precision.

POLYNOMIAL EVALUATION[H]

A polynomial $g = \sum_{i=0}^{k} a_i X^i \in R[X]$ naturally defines a polynomial function on R

that sends $x \in R$ to $\sum_{i=0}^{k} a_i x^i \in R$, and we denote the value of this function as $g(x)$ (note that "X" denotes an indeterminate, while "x" denotes an element of R). It is important to regard polynomials over R as formal expressions, and not to identify them with their corresponding functions. In particular, two polynomials are equal if and only if their coefficients are equal, while two functions are equal if and only if their values agree at all points in R.

This distinction is important, since there are rings R over which two different polynomials define the same function. One can of course define the ring of polynomial functions on R, but in general, that ring has a different structure from the ring of polynomials over R.

Example. In the ring $\mathbb{Z}_p$, for prime p, by Fermat's little theorem, we have $x^p = x$ for all $x \in \mathbb{Z}_p$. However, the polynomials X^p and X are not the same polynomials (in particular, the former has degree p, while the latter has degree 1).

More generally, suppose R is a subring of a ring E. Then every polynomial $g = \sum_{i=0}^{k} a_i x^i \in R[X]$ defines a polynomial function from E to E that sends $\alpha \in E$ to $\sum_{i=0}^{k} a_i \alpha^i \in E$, and, again, the value of this function is denoted $g(\alpha)$. We say that α is a root of g if $g(\alpha) = 0$.

Theorem: Let D be an integral domain, and suppose that $g \in D[X]$, with $\deg(g) = k \geq 0$. Then g has at most k distinct roots.

Proof: If g had $k + 1$ distinct roots $x_1, \ldots, x_{k+1}$, then by the previous theorem, the polynomial $\prod_{i=1}^{k+1}(X - x_i)$, which has degree $k + 1$, would divide g, which has degree k – an impossibility.

Theorem: Let R be a subring of a ring E. For all $g, h \in R[X]$ and $\alpha \in E$, if $s := g + h \in R[X]$ and $p := gh \in R[X]$, then we have

$$s(\alpha) = g(\alpha) + h(\alpha) \text{ and } p(\alpha) = g(\alpha)h(\alpha).$$

Also, if $c \in R$ is a constant polynomial, then $c(\alpha) = c$ for all $\alpha \in E$.

Proof: The statement about evaluating a constant polynomial is clear from the definitions. The proof of the statements about evaluating the sum or product of polynomials is really just symbol pushing. Indeed, suppose $g = \sum_i a_i X^i$ and $h = \sum_i b_i X^i$. Then $s = \sum_i (a_i + b_i) X^i$, and so

$$s(\alpha) = \sum_i (a_i + b_i)\alpha^i = \sum_i a_i\alpha^i + \sum_i b_i\alpha^i = g(\alpha) + h(\alpha).$$

Also, we have

$$p = \left(\sum_i a_i X^i\right)\left(\sum_j b_j X^j\right) = \sum_{ij} a_i b_j X^{i+j},$$

and employing the result for evaluating sums of polynomials, we have

$$p(\alpha) = \sum_{i,j} a_i b_j \alpha^{i+j} = \left(\sum_i a_i\alpha^i\right)\left(\sum_j b_j\alpha^j\right) = g(\alpha)\, h(\alpha).$$

Example. If $E = R[X]$, then evaluating a polynomial $g \in R[X]$ at a point $\alpha \in E$ amounts to polynomial composition. For example, if $g := X^2 + X$ then

$$g(X+1) = (X+1)^2 + (X+1) = X^2 + 3X + 2.$$

The reader is perhaps familiar with the fact that over the real or the complex numbers, every polynomial of degree k has at most k distinct roots, and the fact that every set of k points can be interpolated by a unique polynomial of degree less than k. As we will now see, these results extend to much more general, though not completely arbitrary, coefficient rings.

Example. Consider the polynomial $g := 2X^3 - 2X^2 + X - 1 \in \mathbb{Z}[X]$. We can write $g = (2X^2 + 1)(X - 1)$. For any element α of $\mathbb{Z}$, or an extension ring of $\mathbb{Z}$, we have $g(\alpha) = (2\alpha^2 + 1)(\alpha - 1)$. From this, it is clear that in $\mathbb{Z}$, g has a root only at 1; moreover, it has no other roots in $\mathbb{R}$, but in $\mathbb{C}$, it also has roots $\pm i/\sqrt{2}$.

Theorem: Let D be an integral domain, and let $x_1,\ldots, x_k$ be distinct elements of D. Then for every polynomial $g \in D[X]$, the elements $x_1,\ldots, x_k$ are roots of g if and only if the polynomial $\prod_{i=1}^{k}(X - x_i)$ divides g.

Proof: One direction is trivial: if $\prod_{i=1}^{k}(X - x_i)$ divides g, then it is clear that each x_i is a root of g. We prove the converse by induction on k. The base case $k = 1$ is just Theorem. So assume $k > 1$, and that the statement holds for $k - 1$. Let $g \in D[X]$ and let $x_1,\ldots, x_k$ be distinct roots of g. Since xk is a root of g, then by Theorem, there exists $q \in D[X]$ such that $g = (X - x_k)q$. Moreover, for each $i = 1,\ldots, k - 1$, we have

$$0 = g(x_i) = (x_i - x_k)q(x_i),$$

and since $x_i - x_k \neq 0$ and D is an integral domain, we must have $q(x_i) = 0$. Thus, q has roots $x_1,\ldots, x_{k-1}$, and by induction $\prod_{i=1}^{k-1}(X - x_i)$ divides q, from which it then follows that $\prod_{i=1}^{k}(X - x_i)$ divides g.

Note that in this theorem, we can slightly weaken the hypothesis: we do not need to assume that the coefficient ring is an integral domain; rather, all we really need is that for all $i \neq j$, the difference $x_i - x_j$ is not a zero divisor.

Lagrange Interpolation

Theorem: Let F be a field, let $x_1,..., x_k$ be distinct elements of F, and let $y_1,..., y_k$ be arbitrary elements of F. Then there exists a unique polynomial $g \in F[X]$ with deg $(g) < k$ such that $g(x_i) = y_i$ for $i = 1,..., k$, namely

$$g := \sum_{i=1}^{k} y_i \frac{\prod_{j \neq i}(X - x_j)}{\prod_{j \neq i}(x_i - x_j)}.$$

Proof: For the existence part of the theorem, one just has to verify that $g(x_i) = y_i$ for the given g, which clearly has degree less than k. This is easy to see: for $i = 1,..., k$, evaluating the ith term in the sum defining g at x_i yields y_i, while evaluating any other term at x_i yields 0. The uniqueness part of the theorem follows almost immediately from Theorem: if g and h are polynomials of degree less than k such that $g(x_i) = y_i = h(x_i)$ for $i = 1,..., k$, then $g - h$ is a polynomial of degree less than k with k distinct roots, which, by the previous theorem, is impossible.

Again, we can slightly weaken the hypothesis of this theorem: we do not need to assume that the coefficient ring is a field; rather, all we really need is that for all $i \neq j$, the difference $x_i - x_j$ is a unit.

Theorem: Let R be a ring, $g \in R[X]$, and $x \in R$. Then there exists a unique polynomial $q \in R[X]$ such that $g = (X - x)q + g(x)$. In particular, x is a root of g if and only if $(X - x)$ divides g.

Proof: If R is the trivial ring, there is nothing to prove, so assume that R is nontrivial. Using the division with remainder property for polynomials, there exist unique $q, r \in R[X]$ such that $g = (X - x)q + r$, with $q, r \in R[X]$ and $\deg(r) < 1$, which means that $r \in R$. Evaluating at x, we see that $g(x) = (x - x)q(x) + r = r$. That proves the first statement. The second follows immediately from the first.

Note that the above theorem says that $X - x$ divides $g - g(x)$, and the polynomial q in the theorem may be expressed as

$$q = \frac{g - g(x)}{X - x}.$$

EXERCISE

1. Let D be an integral domain, let $g, h \in D[X]$, and let $x \in D$. Show that $m_x(gh) = m_x(g) + m_x(h)$.

2. Let D be an infinite integral domain; and let $g, h \in D[X]$. Show that if $g(x) = h(x)$ for all $x \in D$, then $g = h$. Thus, for an infinite integral domain D, there is a one-to-one correspondence between polynomials over D and polynomial functions on D.

4. Let D be an integral domain, and suppose that $g \in D[X]$, with deg $(g) = k \geq 0$. Show that

$$\sum_{x \in D} m_x(g) \leq k.$$

3. Let F be a field.

   (a) Show that for all $b \in F$, we have $b^2 = 1$ if and only if $b = \pm 1$.

   (b) Show that for all $a, b \in F$, we have $a^2 = b^2$ if and only if $a = \pm b$.

   (c) Show that the familiar quadratic formula holds for F, assuming F has characteristic other than 2, so that $2_F \neq 0_F$. That is, for all $a, b, c \in F$ with $a \neq 0$, the polynomial $g := aX^2 + bX + c \in F[X]$ has a root in F if and only if there exists $e \in F$ such that $e^2 = d$, where d is the discriminant of g, defined as $d := b^2 - 4ac$, and in this case the roots of g are $(-b \pm e)/2a$.

4. Let R be a ring, let $g \in R[X]$, with $\deg(g) = k \geq 0$, and let x be an element of R. Show that:

   (a) There exist an integer m, with $0 \leq m \leq k$, and a polynomial $q \in R[X]$, such that $g = (X - x)^m q$ and $q(x) \neq 0$, and moreover, the values of m and q are uniquely determined;

   (b) If we evaluate g at $X + x$, we have

$$g(X + x) = \sum_{i=0}^{k} b_i X^i,$$

   where $b_0 = ... = b_{m-1} = 0$ and $b_m = q(x) \neq 0$.

Let $m_x(g)$ denote the value m in the previous exercise; for completeness, one can define $m_x(g) := 1$ if g is the zero polynomial. If $m_x(g) > 0$, then x is called a root of g of multiplicity $m_x(g)$; if $m_x(g) = 1$, then x is called a simple root of g, and if $m_x(g) > 1$, then x is called a multiple root of g.

BASIC PROPERTIES OF POLYNOMIAL RINGS

Let R be a ring. For non-zero $g \in R[X]$, if $g = \sum_{i=0}^{k} a_i X^i$ with $ak \neq 0$, then we call k the degree of g, denoted $\deg(g)$, we call a_k the leading.coefficient of g, denoted $\operatorname{lc}(g)$, and we call a_0 the constant term of g. If $\operatorname{lc}(g) = 1$, then g is called monic.

Suppose $g = \sum_{i=0}^{k} a_i X^i$ and $h = \sum_{i=0}^{\ell} b_i X^i$ are polynomials such that $a_k \neq 0$ and $b_\ell \neq 0$, so that deg(g) = k and lc(g) = ak, and deg(h) = ℓ and lc(h) = b_ℓ. When we multiply these two polynomials, we get

$$gh = a_0b_0 + (a_0b_1 + a_1b_0)X + ... + a_kb_\ell X^{k+\ell}.$$

In particular, deg(gh) $\leq$ deg(g) + deg(h). If either of a_k or b_ℓ are not zero divisors, then a_kb_ℓ is not zero, and hence deg(gh) = deg(g) + deg(h). However, if both a_k and b_ℓ are zero divisors, then we may have $a_kb_\ell = 0$, in which case, the product gh may be zero, or perhaps $gh \neq 0$ but deg(gh) < deg(g) + deg(h).

For the zero polynomial, we establish the following conventions: its leading coefficient and constant term are defined to be 0_R, and its degree is defined to be $-\infty$. With these conventions, we may succinctly state that

for all $g, h \in R[X]$, we have deg($g\,h$) $\leq$ deg(g) + deg(h), with equality guaranteed to hold unless the leading coefficients of both g and h are zero divisors.

In particular, if the leading coefficient of a polynomial is not a zero divisor, then the polynomial is not a zero divisor. In the case where the ring of coefficients is an integral domain, we can be more precise:

Theorem: Let D be an integral domain. Then:

(*i*) For all $g, h \in D[X]$, we have deg(gh) = deg(g) + deg(h);

(*ii*) $D[X]$ is an integral domain;

(*iii*) $(D[X])^* = D^*$.

Division with Remainder Property

Theorem: Let R be a ring. For all $g, h \in R[X]$ with $h \neq 0$ and lc(h) $\in R^*$, there exist unique $q, r \in R[X]$ such that $g = hq + r$ and deg(r) < deg(h).

Proof: Consider the set $S := \{g - ht : t \in R[X]\}$. Let $r = g - hq$ be an element of S of minimum degree. We must have deg(r) < deg(h), since otherwise, we could subtract an appropriate multiple of h from r so as to eliminate the leading coefficient of r, obtaining

$$r' := r - h\,.\,(\text{lc}(r)\,\text{lc}(h)^{-1}\,X^{\deg(r)-\deg(h)}) \in S,$$

where deg(r') < deg(r), contradicting the minimality of deg(r).

That proves the existence of r and q. For uniqueness, suppose that $g = hq + r$ and $g = hq' + r'$, where deg(r) < deg(h) and deg(r') < deg(h). This implies $r' - r = h\,.\,(q - q')$. However, if $q \neq q'$, then

$$\deg(h) > \deg(r' - r) = \deg(h\,.\,(q - q')) = \deg(h) + \deg(q - q') \geq \deg(h),$$

which is impossible. Therefore, we must have $q = q'$, and hence $r = r'$.

If $g = hq + r$ as in the above theorem, we define $g \bmod h := r$. Clearly, $h \mid g$ if and only if $g \bmod h = 0$. Moreover, note that if $\deg(g) < \deg(h)$, then $q = 0$ and $r = g$; otherwise, if $\deg(g) \geq \deg(h)$, then $q \neq 0$ and $\deg(g) = \deg(h) + \deg(q)$.

HOMOMORPHISMS AND ISOMORPHISMS OF RINGS

Definition: A function ρ from a ring R to a ring R' is called a ring homomorphism if

- ρ is a group homomorphism with respect to the underlying additive groups of R and R',
- $\rho(ab) = \rho(a)\,\rho(b)$ for all $a, b \in R$, and
- $\rho(1_R) = 1_{R'}$.

Expanding the definition, the requirements that ρ must satisfy in order to be a ring homomorphism are that for all $a, b \in R$, we have $\rho(a + b) = \rho(a) + \rho(b)$ and $\rho(ab) = \rho(a)\,\rho(b)$, and that $\rho(1_R) = 1_{R'}$.

Note that some texts do not require that a ring homomorphism satisfies part (iii) of our definition. Since a ring homomorphism is also an additive group homomorphism, we use the same notation and terminology for image and kernel.

Example.

1. For a given positive integer n, the natural map from $\mathbb{Z}$ to $\mathbb{Z}_n$ sends $a \in \mathbb{Z}$ to the residue class $[a]_n$. This is a surjective ring homomorphism, whose kernel is $n\mathbb{Z}$.
2. If S is a subring of a ring R, then the inclusion map $i : S \to R$ is obviously a ring homomorphism.
3. We can generalize the previous example to multi-variate polynomials. If R is a subring of a ring E and $\alpha_1,\ldots, \alpha_n \in E$, then the map

   $$\rho : R[X_1,\ldots, X_n] \to E$$
   $$g \mapsto g(\alpha_1,\ldots, \alpha_n)$$

   is a ring homomorphism. Its image consists of all polynomial expressions in $\alpha_1,\ldots, \alpha_n$ with coefficients in R, and is denoted $R[\alpha_1,\ldots, \alpha_n]$. Moreover, this image is a subring of E containing $\alpha_1,\ldots, \alpha_n$ and all of R, and is the smallest such subring of E. Note that $R[\alpha_1,\ldots, \alpha_n] = R[\alpha_1,\ldots, \alpha_{n-1}][\alpha_n]$.
4. Let $\rho : R \to R'$ be a ring homomorphism. We can extend the domain of definition of ρ from R to $R[X]$ by defining $\rho(\sum_i a_i X^i) := \sum_i \rho(a_i) X^i$. This yields a ring homomorphism from $R[X]$ into $R'[X]$. To verify this, suppose $g = \sum_i a_i X^i$ and $h = \sum_i b_i X^i$ are polynomials in $R[X]$. Let $s := g + h \in R[X]$ and $p := gh \in R[X]$, and write $s = \sum_i s_i X^i$ and $p = \sum_i p_i X^i$, so that

$$s_i = a_i + b_i \text{ and } p_i = \sum_{i=j+k} a_j b_k.$$

Then we have

$$\rho(s_i) = \rho(a_i + b_i) = \rho(a_i) + \rho(b_i),$$

which is the coefficient of X^i in $\rho(g) + \rho(h)$, and

$$\rho(p_i) = \rho\left(\sum_{i=j+k} a_j b_k\right) = \sum_{i=j+k} \rho(a_j b_k) = \sum_{i=j+k} \rho(a_j)\rho(b_k),$$

which is the coefficient of X^i in $\rho(g)\rho(h)$.

Sometimes a more compact notation is convenient: we may prefer to write $\bar{a}$ for the image of $a \in R$ under ρ, and if we do this, then for $g = \sum_i a_i X^i \in R[X]$, we write g for the image $\sum_i \bar{a}_i X^i$ of g under the extension of ρ to $R[X]$.

5. Consider the natural map that sends $a \in \mathbb{Z}$ to $\bar{a} := [a]_n \in \mathbb{Z}_n$. As in the previous example, we may extend this to a ring homomorphism from $\mathbb{Z}[X]$ to $\mathbb{Z}_n[X]$ that sends $g = \sum_i a_i X^i \in \mathbb{Z}[X]$ to $\bar{g} = \sum_i \bar{a}_i X^i \in \mathbb{Z}_n[X]$.

   This homomorphism is clearly surjective. Let us determine its kernel. Observe that if $g = \sum_i a_i X^i$, then $\bar{g} = 0$ if and only if $n \mid a_i$ for each i; therefore, the kernel is the ideal $n\mathbb{Z}[X]$ of $\mathbb{Z}[X]$.

6. Suppose I is an ideal of a ring R. Analogous to Example, we may define the natural map from the ring R to the quotient ring R/I as follows:

$$\rho : R \to R/I$$
$$a \mapsto [a]_I.$$

Not only is this a surjective homomorphism of additive groups, with kernel I, it is a ring homomorphism. Indeed, we have

$$\rho(ab) = [ab]_I = [a]_I \,.\, [b]_I = \rho(a).\, \rho(b),$$

and $\rho(1_R) = [1_R]_I$, which is the multiplicative identity in R/I.

7. Let R be a subring of a ring E, and fix $\alpha \in E$. The polynomial evaluation map

$$\rho : R[X] \to E$$
$$g \mapsto g(\alpha)$$

is a ring homomorphism. The image of consists of all polynomial expressions in with coefficients in R, and is denoted $R[\alpha]$. As the reader may verify, $R[\alpha]$ is a subring of E containing α and all of R, and is the smallest such subring of E.

8. Let R be a ring of prime characteristic p. For all $a, b \in R$, we have

$$(a+b)^p = \sum_{k=0}^{p} \binom{p}{k} a^{p-k} b^k.$$

However, by Exercise 1.14, all of the binomial coefficients are multiples of p, except for $k = 0$ and $k = p$, and hence in the ring R, all of these terms vanish, leaving us with

$$(a+b)^p = a^p + b^p.$$

This result is often jokingly referred to as the "freshman's dream," for somewhat obvious reasons.

Of course, as always, we have

$$(ab)^p = a^p b^p \text{ and } 1_R^p = 1R,$$

and so it follows that the map that sends $a \in R$ to $a^p \in R$ is a ring homomorphism from R into R.

Theorem: Let $\rho_i : R \to R_i'$, for $i = 1,..., k$, be ring homomorphisms. Then the map

$$\rho : R \to R_1' \times ... \times R_k'$$

$$a \mapsto (\rho_1(a),..., \rho_k(a))$$

is a ring homomorphism.

If a ring homomorphism $\rho : R \to R'$ is a bijection, then it is called a ring isomorphism of R with R'. If such a ring isomorphism exists, we say that R is isomorphic to R', and write $R \cong R'$. Moreover, if $R = R'$, then is called a ring automorphism on R.

Theorem: Let $\rho : R \to R'$ be a ring homomorphism.

(*i*) If S is a subring of R, then $\rho(S)$ is a subring of R'; in particular (setting $S := R$), Im ρ is a subring of R'.

(*ii*) If S' is a subring of R', then $\rho^{-1}(S')$ is a subring of R.

(*iii*) If I is an ideal of R, then $\rho(I)$ is an ideal of Im ρ.

(*iv*) If I' is an ideal of Im ρ, then $\rho^{-1}(I')$ is an ideal of R; in particular (setting $I' := \{0_{R'}\}$), Ker ρ is an ideal of R.

Proof: In each part, we already know that the relevant object is an additive subgroup, and so it suffices to show that the appropriate additional properties are satisfied.

(*i*) For all $a, b \in S$, we have $ab \in S$, and hence $\rho(S)$ contains $\rho(ab) = \rho(a)\rho(b)$. Also, $1_R \in S$, and hence $\rho(S)$ contains $\rho(1_R) = 1_{R'}$.

(*ii*) If $\rho(a) \in S'$ and $\rho(b) \in S'$, then $\rho(ab) = \rho(a)\rho(b) \in S'$. Moreover, $\rho(1_R) = 1_{R'} \in S'$.

(*iii*) For all $a \in I$ and $r \in R$, we have $ar \in I$, and hence $\rho(I)$ contains $\rho(ar) = \rho(a)\rho(r)$.

(*iv*) For all $a \in \rho^{-1}(I')$ and $r \in R$, we have $\rho(ar) = \rho(a)\rho(r)$, and since $\rho(a)$ belongs to the ideal I', so does $\rho(a)\rho(r)$, and hence $\rho^{-1}(I')$ contains ar.

Theorem: If is a ring isomorphism of R with R', then the inverse function ρ^{-1} is a ring isomorphism of R' with R.

Proof: Exercise.

Because of this theorem, if R is isomorphic to R', we may simply say that "R and R' are isomorphic." We stress that a ring isomorphism is essentially just a "renaming" of elements; in particular.

Theorem: If $\rho : R \to R'$ and $\rho' : R' \to R''$ are ring homomorphisms, then so is their composition $\rho' \circ \rho : R \to R''$.

First Isomorphism Theorem

Theorem: Let $\rho : R \to R'$ be a ring homomorphism with kernel K and image S'. Then we have a ring isomorphism

$$R/K \cong S'.$$

Specifically, the map

$$\bar{\rho} : R/K \to R'$$

$$[a]_K \mapsto \rho(a)$$

is an injective ring homomorphism whose image is S'.

Theorem: Let $\rho : R \to R'$ be a ring homomorphism. Then for every ideal I of R with I Ker , we may define a ring homomorphism

$$\bar{\rho} : R/I \to R'$$

$$[a]_I \mapsto \rho(a).$$

Moreover, Im $\bar{\rho}$ = Im ρ, and $\bar{\rho}$ is injective if and only if I = Ker ρ.

Theorem: Let $\rho : R \to R'$ be a ring isomorphism.

(*i*) For all $a \in R$, a is a zero divisor if and only if $\rho(a)$ is a zero divisor.

(*ii*) For all $a \in R$, a is a unit if and only if $\rho(a)$ is a unit.

(*iii*) The restriction of R to R^* is a group isomorphism of R^* with $(R')^*$.

Proof: Exercise.

An injective ring homomorphism $\rho : R \to E$ is called an embedding of R in E. In this case, Im ρ is a subring of E and $R \cong$ Im ρ. If the embedding is a natural one that is clear from context, we may simply identify elements of R with their images in E under the embedding; that is, for $a \in R$, we may simply write "a," and it is understood that this really means "$\rho(a)$" if the context demands an element of E.

As a slight abuse of terminology, we shall say that R is a subring of E. Indeed, by appropriately renaming elements, we can always make R a subring of E in the literal sense of the term.

This practice of identifying elements of a ring with their images in another ring under a natural embedding is very common. We have already seen an example of this, namely, when we formally defined the ring of polynomials $R[X]$ over R, we defined the map $\varepsilon_0 : R \to R[X]$ that sends $c \in R$ to the polynomial whose constant term is c, and all other coefficients zero.

This map ε_0 is an embedding, and it was via this embedding that we identified elements of R with elements of $R[X]$, and so viewed R as a subring of $R[X]$. We shall see more examples of this later.

Example.

1. Extending Example, if n_1 and n_2 are positive integers with $n_1 \mid n_2$, then the map

   $$\bar{\rho} : \mathbb{Z}_{n_2} \to \mathbb{Z}_{n_1}$$

   $$[a]_{n_2} \mapsto [a]_{n_1}$$

   is a surjective ring homomorphism.

2. We can generalize Example by evaluating polynomials at several points. This is most fruitful when the underlying coefficient ring is a field, and the evaluation points belong to the same field. So let F be a field, and let $x_1, \ldots, x_k$ be distinct elements of F. Define the map

   $$\rho : F[X] \to F^{\times k}$$

   $$g \mapsto (g(x_1), \ldots, g(x_k)).$$

   This is a ring homomorphism. By Theorem, Ker $\rho = (f)$, where $f := \prod_{i=1}^{k}(X - x_i)$. By Theorem, ρ is surjective. Therefore, by Theorem, we get a ring isomorphism

   $$\bar{\rho} : F[X]/(f) \to F^{\times k}$$

   $$[g]_f \mapsto (g(x_1), \ldots, g(x_k)).$$

3. For a ring R, consider the map $\rho : \mathbb{Z} \to R$ that sends $m \in \mathbb{Z}$ to $m \cdot 1_R$ in R. It is easily verified that ρ is a ring homomorphism. Since Ker ρ is an ideal of $\mathbb{Z}$, it is

either $\{0\}$ or of the form $n\mathbb{Z}$ for some $n > 0$. In the first case, if Ker $\rho = \{0\}$, then Im $\rho \cong \mathbb{Z}$, and so the ring $\mathbb{Z}$ is embedded in R, and R has characteristic zero. In the second case, if Ker $\rho = n\mathbb{Z}$ for some $n > 0$, then by Theorem, Im $\rho \cong \mathbb{Z}n$, and so the ring $\mathbb{Z}n$ is embedded in R, and R has characteristic n.

4. Let f be a polynomial over a ring R with $\deg(f) = \ell$ and $\mathrm{lc}(f) \in R$, but now assume that $\ell > 0$. Consider the natural map ρ from $R[X]$ to the quotient ring $E := R[X]/(f)$ that sends $g \in R[X]$ to $[g]_f$. Let τ be the restriction of ρ to the subring R of $R[X]$. Evidently, τ is a ring homomorphism from R into E. Moreover, since distinct polynomials of degree less than ℓ belong to distinct residue classes modulo f, we see that τ is injective. Thus, τ is an embedding of R into E. As τ is a very natural embedding, we can identify elements of R with their images in E under τ, and regard R as a subring of E. Taking this point of view, we see that if $g = \Sigma_i a_i X^i$, then

$$[g]_f = \left[\sum_i a_i X^i\right]_f = \sum_i [a_i]_f \left([X]_f\right)^i = \sum_i a_i \xi^i = g(\xi),$$

where $\zeta := [X]_f \in E$. Therefore, the natural map ρ may be viewed as the polynomial evaluation map that sends $g \in R[X]$ to $g(\zeta) \in E$.

Note that we have $E = R[\zeta]$; moreover, every element of E can be expressed uniquely as $g(\xi)$ for some $g \in R[X]$ of degree less than ℓ, and more generally, for arbitrary $g, h \in R[X]$, we have $g(\xi) = h(\xi)$ if and only if $g \; h \pmod f$. Finally, note that $f(\xi) = [f]_f = [0]_f$; that is, ξ is a root of f.

5. Consider the polynomial evaluation map

$$\rho : \mathrm{R}[X] \to \mathbb{C} = R[X]/(X^2 + 1)$$
$$g \mapsto g(-i).$$

For every $g \in \mathbb{R}[X]$, we may write $g = (X^2 + 1)q + a + bX$, where $q \in \mathbb{R}[X]$ and $a, b \in \mathbb{R}$. Since $(-i)^2 + 1 = i^2 + 1 = 0$, we have

$$g(-i) = ((-i)^2 + 1)q(-i) + a - bi = a - bi.$$

Clearly, then, ρ is surjective and the kernel of ρ is the ideal of $\mathbb{R}[X]$ generated by the polynomial $X^2 + 1$. By Theorem, we therefore get a ring automorphism $\bar{\rho}$ on $\mathbb{C}$ that sends $a + bi \in \mathbb{C}$ to $a - bi$. In fact, $\bar{\rho}$ is none other than the complex conjugation map. Indeed, this is the "algebraically correct" way of defining complex conjugation in the first place.

6. Let p an odd prime, and let $d \in \mathbb{Z}_p^*$. Let $f := X^2 - d \in \mathbb{Z}_p[X]$, and consider the ring $E := \mathbb{Z}_p[X]/(f) = \mathbb{Z}_p[\xi]$, where $\xi^2 := [X]_f \in E$. We have $E = \{a + b\xi : a, b \in \mathbb{Z}_p\}$ and $|E| = p^2$. Note that $\xi^2 = d$, and the general rules for arithmetic in E look like

this: for all $a, b, a', b' \in \mathbb{Z}_p$, we have

$$(a + b\xi) + (a' + b'\xi) = (a + a') + (b + b')\xi$$

and

$$(a + b\xi).(a' + b'\xi) = (aa' + bb'd) + (ab' + a'b)\xi.$$

Suppose that $d \in (\mathbb{Z}_p^*)^2$, so that $d = c^2$ for some $c \in \mathbb{Z}_p^*$. Then $f = (X - c)(X + c)$, and like in previous example, we have a ring isomorphism $E \cong \mathbb{Z}_p \times \mathbb{Z}_p$ (which maps $a + b\xi \in E$ to $(a + bc, a - bc) \in \mathbb{Z}_p \times \mathbb{Z}_p$); in particular, E is not a field.

Suppose that $d \notin (\mathbb{Z}_p^*)^2$. This implies that for all $a, b \in \mathbb{Z}_p$, not both zero, we have $a^2 - b^2d \neq 0$. Using this, we get the following formula for multiplicative inverses in E:

$$(a + b\xi)^{-1} = \frac{a - b\xi}{a^2 - b^2d}.$$

Therefore, E is a field in this case.

By Theorem, we know that $|(\mathbb{Z}_p^*)^2| = (p - 1)/2$, and hence there exists $d \in \mathbb{Z}_p^* \setminus (\mathbb{Z}_p^*)^2$ for all odd primes p. Thus, we have a general (though not explicit) construction for finite fields of cardinality p^2 for all odd primes p.

7. Let p be a prime, and consider the quotient ring $E := \mathbb{Z}_p[X]/(f)$, where $f := X^2 + 1$. If we set $i := [X]_f \in E$, then $E = \mathbb{Z}_p[i] = \{a + bi : a, b \in \mathbb{Z}_p\}$. In particular, E is a ring of cardinality p^2. Moreover, we have $i^2 = -1$, and the rules for addition and multiplication in E look exactly the same as they do in $\mathbb{C}$: for all $a, b, a', b' \in \mathbb{Z}_p$, we have

   $$(a + bi) + (a' + b'i) = (a + a') + (b + b')i$$

   and $$(a + bi).(a' + b'i) = (aa' - bb') + (ab' + a'b)i.$$

   The ring E may or may not be a field. We now determine for which primes p we get a field.

   If $p = 2$, then $0 = 1 + i^2 = (1 + i)^2$, and so in this case, $1 + i$ is a zero divisor and E is not a field.

   Now suppose p is odd. There are two subcases to consider: $p \equiv 1 \pmod 4$ and $p \equiv 3 \pmod 4$.

   Suppose $p \equiv 1 \pmod 4$. By Theorem, there exists $c \in \mathbb{Z}_p$ such that $c^2 = -1$, and therefore $f = X^2 + 1 = X^2 - c^2 = (X - c)(X + c)$, we have a ring isomorphism $E \cong \mathbb{Z}_p \times \mathbb{Z}_p$ (which maps $a + bi \in E$ to $(a + bc, a - bc) \in \mathbb{Z}_p \times \mathbb{Z}_p$); in particular, E is not a field. Indeed, $c + i$ is a zero divisor, since $(c + i)(c - i) = c^2 - i^2 = c^2 + 1 = 0$.

Suppose $p \equiv 3 \pmod 4$. By Theorem1, there is no $c \in \mathbb{Z}_p$ such that $c^2 = -1$. It follows that for all $a, b \in \mathbb{Z}_p$, not both zero, we must have $a^2 + b^2 \neq 0$; indeed, suppose that $a^2 + b^2 = 0$, and that, say, $b \neq 0$; then we would have $(a/b)^2 = -1$, contradicting the assumption that -1 has no square root in $\mathbb{Z}_p$. Therefore, $a^2 + b^2$ has a multiplicative inverse in $\mathbb{Z}_p$, from which it follows that the formula for multiplicative inverses in C applies equally well in E; that is,

$$(a + bi)^{-1} = \frac{a - bi}{a^2 + b^2}.$$

Therefore, in this case, E is a field.

We saw a finite field of cardinality 4. The previous example provides us with an explicit construction of a finite field of cardinality p^2, for every prime p congruent to 3 modulo 4. As the next example shows, there exist finite fields of cardinality p^2 for all primes p.

8. Let $f := X^2 + 1 \in \mathbb{R}[X]$, and consider the quotient ring $R[X]/(f)$. If we set $i := [X]_f \in \mathbb{R}[X]/(f)$, then every element of $\mathbb{R}[X]/(f)$ can be expressed uniquely as $a + bi$, where $a, b \in \mathbb{R}$. Moreover, we have $i^2 = -1$, and more generally, for all $a, b, a', b' \in \mathbb{R}$, we have

$$(a + b_i) + (a' + b'i) = (a + a') + (b + b')i$$

and

$$(a + bi).(a' + b'i) = (aa' - bb') + (ab' + a'b)i.$$

Thus, the rules for arithmetic in $\mathbb{R}[X]=(f)$ are precisely the familiar rules of complex arithmetic, and so $\mathbb{C}$ and $\mathbb{R}[X]/(f)$ are essentially the same, as rings. Indeed, the "algebraically correct" way of defining the field of complex numbers $\mathbb{C}$ is simply to define it to be the quotient ring $\mathbb{R}[X]/(f)$ in the first place. This will be our point of view from now on.

EXERCISE

1. Verify that the "is isomorphic to" relation on rings is an equivalence relation; that is, for all rings R_1, R_2, R_3, we have:

   (*a*) $R_1 \cong R_1$;

   (*b*) $R_1 \cong R_2$ implies $R_2 \cong R_1$;

   (*c*) $R_1 \cong R_2$ and $R_2 \cong R_3$ implies $R_1 \cong R_3$.

3. Let $\rho_i : R_i \to R_i'$, for $i = 1,..., k$, be ring homomorphisms. Show that the map

$$\rho : R_1 \times ... \times R_k \to R_1' \times ... \times R_k'$$
$$(a_1,..., a_k) \mapsto (\rho_1(a_1),..., \rho_k(a_k))$$

is a ring homomorphism.

2. Let R be a ring, $a \in R^*$, and $b \in R$. Define the map $\rho : R[X] \to R[X]$ that sends $g \in R[X]$ to $g(aX + b)$. Show that ρ is a ring automorphism.

3. Suppose $R_1,..., R_k$ are rings. Show that for each $i = 1,..., k$, the projection map $\pi_i : R_1 \text{ X} ... \text{ X } R_k \to R_i$ that sends $(a_1,..., a_k)$ to a_i is a surjective ring homomorphism.

4. Show that if $\rho : F \to R$ is a ring homomorphism from a field F into a ring R, then either R is trivial or ρ is injective.

5. Show that if $R = R_1 X R_2$ for rings R_1 and R_2, and I_1 is an ideal of R_1 and I_2 is an ideal of R_2, then we have a ring isomorphism $R/(I_1 \text{ X } I_2) \cong R_1/I_1 \text{ X } R_2/I_2$.

6. Let n be a positive integer whose factorization into primes is $n = p_1^{e_1} ... p_r^{e_r}$. What are the prime ideals of $\mathbb{Z}_n$?

7. Let I be an ideal of R, and S a subring of R. $I \cap S$ is an ideal of S, and I is an ideal of the subring $I + S$. Show that we have a ring isomorphism $(I + S)/I \cong S/(I \cap S)$.

8. Let $\rho : R \to R'$ be a surjective ring homomorphism. Let S be the set of all ideals of R that contain Ker ρ, and let S′ be the set of all ideals of R'. Show that the sets S and S′ are in one-to-one correspondence, via the map that sends $I \in$ S to $(I) \in$ S′. Moreover, show that under this correspondence, prime ideals in S correspond to prime ideals in S′, and maximal ideals in S correspond to maximal ideals in S0.

9. Let $\rho : R \to R0$ be a ring homomorphism with kernel K. Let I be an ideal of R. Show that we have a ring isomorphism $R/(I + K) \cong \rho(R)/\rho(I)$.

10. Consider the subring $\mathbb{Z}[1/2]$ of $\mathbb{Q}$. Show that $\mathbb{Z}[1/2] = \{a/2i : a, i \in \mathbb{Z}, i \geq 0\}$, that $(\mathbb{Z}[1/2]) = \{2^i : i \in \mathbb{Z}\}$, and that every non-zero ideal of $\mathbb{Z}[1/2]$ is of the form (m), for some uniquely determined, odd integer m.

11. Generalize the previous exercise, showing that $R/(I_1 .... I_k)$ is isomorphic to R/I_1 X ... X R/I_k, where R is a ring, and $I_1,..., I_k$ are ideals of R, provided $I_i + I_j = R$ for all i, j such that $i \neq j$.

12. Let n be a positive integer, and consider the natural map that sends $a \in \mathbb{Z}$ to $\overline{a} := [a]_n \in \mathbb{Z}_n$, which we may extend coefficient-wise to a ring homomorphism from $\mathbb{Z}[X]$ to $\mathbb{Z}_n[X]$, as in Example. Show that for every $f \in \mathbb{Z}[X]$, we have a ring isomorphism $\mathbb{Z}[X]/(f, n) \cong \mathbb{Z}_n[X]/(\overline{f})$.

13. Let $\mathbb{Q}^{(m)}$ be the subring of $\mathbb{Q}$ defined. Let us define the map $\rho : \mathbb{Q}^{(m)} \to \mathbb{Z}_m$ as follows. For $a/b \in \mathbb{Q}$ with b relatively prime to m, $\rho(a/b) := [a]_m([b]_m)^{-1}$. Show that ρ is unambiguously defined, and is a surjective ring homomorphism. Also, describe the kernel of ρ.

14. Let p be a prime with $p \equiv 1 \pmod 4$. Show that we have a ring isomorphism $\mathbb{Z}[X]/(X^2 + 1, p) \cong \mathbb{Z}_p \times \mathbb{Z}_p$.
15. Let $\rho : R \to S$ be a ring homomorphism. Show that $\rho(R^*) \subseteq S^*$, and that the restriction of ρ to R^* yields a group homomorphism $\rho^*: R \to S^*$.
16. Let $\rho : R \to R'$ be a ring homomorphism, and let $a \in R$. Show that $\rho(aR) = \rho(a)\rho(R)$.
17. Let R be a ring, and let $x_1,..., x_n$ be elements of R. Show that the rings R and $R[X_1,..., X_n]/(X_1 - x_1,..., X_n - x_n)$ are isomorphic.
18. Let n be a positive integer. Show that we have ring isomorphisms $\mathbb{Z}[X]/(n) \cong \mathbb{Z}_n[X]$, $\mathbb{Z}[X]/(X) \cong \mathbb{Z}$, and $\mathbb{Z}[X]/(X, n) \cong \mathbb{Z}_n$.
19. Let $n = pq$, where p and q are distinct primes. Show that we have a ring isomorphism $\mathbb{Z}_n[X] \cong \mathbb{Z}_p[X] \times \mathbb{Z}_q[X]$.
20. This exercise and the next generalize the Chinese remainder theorem to arbitrary rings. Suppose I and J are two ideals of a ring R such that $I + J = R$. Show that the map $\rho : R \to R/I \times R/J$ that sends $a \in R$ to $([a]_I, [a]_J)$ is a surjective ring homomorphism with kernel IJ. Conclude that $R/(IJ)$ is isomorphic to $R/I \times R/J$.

IDEALS RINGS

Definition: Let R be a ring. An ideal of R is an additive subgroup I of R that is closed under multiplication by elements of R, that is, for all $a \in I$ and $r \in R$, we have $ar \in I$.

Expanding the above definition, we see that a non-empty subset I of R is an ideal of R if and only if for all $a, b \in I$ and $r \in R$, we have

$$a + b \in I, -a \in I, \text{ and } ar \in I.$$

Since R is commutative, the condition $ar \in I$ is equivalent to $ra \in I$. The condition $-a \in I$ is redundant, as it is implied by the condition $ar \in I$ with $r := -1_R$. In the case when R is the ring $\mathbb{Z}$, this definition of an ideal is consistent with that given in §1.2.

Clearly, $\{0_R\}$ and R are ideals of R. From the fact that an ideal I is closed under multiplication by elements of R, it is easy to see that $I = R$ if and only if $1_R \in I$.

Example.

1. Let R be a ring and $x \in R$. Consider the set

   $$I := \{g \in R[X]: g(x) = 0\}.$$

   It is not hard to see that I is an ideal of $R[X]$. Indeed, for all $g, h \in I$ and $q \in R[X]$, we have

   $$(g + h)(x) = g(x) + h(x) = 0 + 0 = 0 \text{ and } (gq)(x) = g(x)q(x) = 0.q(x) = 0.$$

   Moreover, we have $I = \{(X - x)q : q \in R[X]\}$.

2. For each $m \in \mathbb{Z}$, the set $m\mathbb{Z}$ is not only an additive subgroup of the ring $\mathbb{Z}$, it is also an ideal of this ring.
3. For each $m \in \mathbb{Z}$, the set $m\mathbb{Z}n$ is not only an additive subgroup of the ring $\mathbb{Z}_n$, it is also an ideal of this ring. 2

Theorem: If I_1 and I_2 are ideals of a ring R, then so are $I_1 + I_2$ and $I_1 \cap I_2$.

Proof: We already know that $I_1 + I_2$ and $I_1 \cap I_2$ are additive subgroups of R, so it suffices to show that they are closed under multiplication by elements of R. The reader may easily verify that this is the case.

Let $a_1,..., a_k$ be elements of a ring R. The ideal $a_1R +... + a_kR$ is called the ideal of R generated by $a_1,..., a_k$. When the ring R is clear from context, one often writes $(a_1,..., a_k)$ to denote this ideal. This ideal is that smallest ideal of R containing $a_1,..., a_k$.

Example. Let n be a positive integer, and let x be any integer. Define $I := \{g \in \mathbb{Z}[X]: g(x) \equiv 0 \pmod{n}\}$. We claim that I is the ideal $(X - x, n)$ of $\mathbb{Z}[X]$. To see this, consider any fixed $g \in \mathbb{Z}[X]$. Using Theorem, we have $g = (X - x)q + g(x)$ for some $q \in \mathbb{Z}[X]$. Using the division with remainder property for integers, we have $g(x) = nq' + r$ for some $r \in \{0,..., n - 1\}$ and $q' \in \mathbb{Z}$. Thus, $g(x) \equiv r \pmod{n}$, and if $g(x) \equiv 0 \pmod{n}$, then we must have $r = 0$, and hence $g = (X - x)q + nq' \in (X - x, n)$. Conversely, if $g \in (X - x, n)$, we can write $g = (X - x)q + nq'$ for some $q, q' \in \mathbb{Z}[X]$, and from this, it is clear that $g(x) = nq'(x) \equiv 0 \pmod{n}$.

Let I be an ideal of a ring R. Since I is an additive subgroup of R, we may adopt the congruence notation, writing $a \equiv b \pmod{I}$ to mean $a - b \in I$, and we can form the additive quotient group R/I of cosets. Recall that for $a \in R$, the coset of I containing a is denoted $[a]I$, and that $[a]I = a + I = \{a + x : x \in I\}$. Also recall that addition in R/I was defined in terms of addition of coset representatives; that is, for $a, b \in I$, we defined

$$[a]_I + [b]_I := [a + b]_I.$$

Our goal now is to make R/I into a ring by similarly defining multiplication in R/I in terms of multiplication of coset representatives. To do this, we need the following multiplicative analog of Theorem, which exploits in an essential way the fact that an ideal is closed under multiplication by elements of R; in fact, this is one of the main motivations for defining the notion of an ideal as we did.

Theorem: Let R be a ring and let $a \in R$. Then $aR := \{ar: r \in R\}$ is an ideal of R.

Proof: This is an easy calculation. For all $ar, ar' \in aR$ and $r'' \in R$, we have

$$ar + ar' = a(r + r') \in aR \text{ and } (ar)r'' = a(rr'') \in aR.$$

The ideal aR in the previous theorem is called the ideal of R generated by a. An ideal of this form is called a principal ideal. Since R is commutative, one could also write this ideal as $Ra := \{ra : r \in R\}$. This ideal is the smallest ideal of R containing a; that is, aR contains a, and every ideal of R that contains a must contain everything in aR.

Theorem: Suppose I is an ideal of a ring R. For all $a, a', b, b' \in R$, if $a \equiv a' \pmod I$ and $b \equiv b' \pmod I$, then $ab \equiv a'b' \pmod I$.

Proof: If $a = a' + x$ for some $x \in I$ and $b = b' + y$ for some $y \in I$, then $ab = a'b' + a'y + b'x + xy$. Since I is closed under multiplication by elements of R, we see that $a'y, b'x, xy \in I$, and since I is closed under addition, $a'y + b'x + xy \in I$.

Hence, $ab - a'b' \in I$.

Using this theorem we can now unambiguously define multiplication on R/I as follows: for $a, b \in R$,

$$[a]_I \cdot [b]_I := [ab]_I \, .$$

Once that is done, it is straightforward to verify that all the properties that make R a ring are inherited by R/I – we leave the details of this to the reader. The multiplicative identity of R/I is the coset $[1_R]_I$.

The ring R/I is called the quotient ring or residue class ring of R modulo I. Elements of R/I may be called residue classes.

Note that if $I = dR$, then $a \equiv b \pmod I$ if and only if $d \mid (a - b)$, and as a matter of notation, one may simply write this congruence as $a \equiv b \pmod d$. We may also write $[a]_d$ instead of $[a]_I$.

Finally, note that if $I = R$, then R/I is the trivial ring.

Example.

1. Let f be a polynomial over a ring R with $\deg(f) = \ell \geq 0$ and $\mathrm{lc}(f) \in R^*$, and consider the quotient ring $E := R[X] / fR[X]$. By the division with remainder property for polynomials, for every $g \in R[X]$, there exists a unique polynomial $h \in R[X]$ such that g'I $h \pmod f$ and $\deg(h) < \ell$. From this, it follows that every element of E can be written uniquely as $[h]_f$, where $h \in R[X]$ is a polynomial of degree less than ℓ Note that in this situation, we will generally prefer the more compact notation $R[X]/(f)$, instead of $R[X]/fR[X]$.
2. ≥ 1, the ring $\mathbb{Z}_n$ is precisely the quotient ring $\mathbb{Z}/n\mathbb{Z}$.
3. Consider the polynomial $f := X^2 + X + 1 \in \mathbb{Z}_2[X]$ and the quotient ring $E := \mathbb{Z}_2[X]/(f)$. Let us name the elements of E as follows:

$$00 := [0]_f,\ 01 := [1]_f,\ 10 := [X]_f,\ 11 := [X + 1]_f \, .$$

With this naming convention, addition of two elements in E corresponds to just computing the bit-wise exclusive-or of their names. More precisely, the addition table for E is the following:

+	00	01	10	11
00	00	01	10	11
01	01	00	11	10
10	10	11	00	01
11	11	10	01	00

Note that 00 acts as the additive identity for E, and that as an additive group, E is isomorphic to the additive group $\mathbb{Z}_2 \times \mathbb{Z}_2$.

As for multiplication in E, one has to compute the product of two polynomials, and then reduce modulo f. For example, to compute 10 . 11, using the identity $X^2 \in X + 1 \pmod{f}$, one sees that

$$X.(X + 1) \equiv X^2 + X \equiv (X + 1) + X \equiv 1 \pmod{f};$$

thus, 10.11 = 01. The reader may verify the following multiplication table for E:

	00	01	10	11
00	00	00	00	00
01	00	01	10	11
10	00	10	11	01
11	00	11	01	10

Observe that 01 acts as the multiplicative identity for E. Notice that every non-zero element of E has a multiplicative inverse, and so E is in fact a field. Observe that E is cyclic: the reader may verify that both 10 and 11 have multiplicative order 3.

This is the first example we have seen of a finite field whose cardinality is not prime.

EXERCISE

1. Let I be an ideal of R, and S a subring of R. Show that $I + S$ is a subring of R, and that I is an ideal of $I + S$.
2. Let F be a field. Show that in the ring $F[X, Y]$, the ideal (X, Y) is not a principal ideal.
3. Let R be a ring. Show that if I is a non-empty subset of $R[X]$ that is closed under addition, multiplication by elements of R, and multiplication by X, then I is an ideal of $R[X]$.
4. Let I_1 be an ideal of R_1, and I_2 an ideal of R_2. Show that $I_1 \times I_2$ is an ideal of $R_1 \times R_2$.
5. Write down the multiplication table for $\mathbb{Z}_2[X]/(X^2 + X)$. Is this a field?
6. Let R be a ring, and let $\{I_i\}_{i=0}^{\infty}$ be a sequence of ideals of R such that $I_i \subseteq I_{i+1}$ for all $i = 0, 1, 2, \ldots$. Show that the union $\bigcup_{i=0}^{\infty} I_i$ is also an ideal of R.

7. Let p be a prime, and consider the ring $\mathbb{Q}^{(p)}$. Show that every non-zero ideal of $\mathbb{Q}^{(p)}$ is of the form (p^i), for some uniquely determined integer $i \geq 0$.
8. Show that if F is a field, then the only ideals of F are $\{0_F\}$ and F.
9. Let I be an ideal of a ring R, and let x and y be elements of R with $x \equiv y \pmod{I}$. Let $g \in R[X]$. Show that $g(x) \equiv g(y) \pmod{I}$.
10. Let R be a ring, and fix $x_1, ..., x_n \in R$. Let
    $I := \{g \in R[X_1, ..., X_n]: g(x_1, ..., x_n) = 0\}$.
    Show that I is an ideal of $R[X_1, ..., X_n]$, and that $I = (X_1 - x_1, ..., X_n - x_n)$.
11. Let a, b be elements of a ring R. Show that
    $a \mid b \Leftrightarrow b \in aR \Leftrightarrow bR \subseteq aR$.
12. Let p be a prime. Show that in the ring $\mathbb{Z}[X]$, the ideal (X, p) is not a principal ideal.
13. Let F be a field, let $f \in F[X, Y]$, and let $E := F[X, Y]/(f)$.
    Define $V(f) := \{(x, y) \in F \times F : f(x, y) = 0\}$.
    (*a*) Every element α of E naturally defines a function from $V(f)$ to F, as follows: if $\alpha = [g]_f$, with $g \in F[X, Y]$, then for $P = (x, y) \in V(f)$, we define $\alpha(P) := g(x, y)$. Show that this definition is unambiguous, that is, $g \equiv h \pmod{f}$ implies $g(x, y) = h(x, y)$.
    (*b*) For $P = (x, y) \in V(f)$, define $M_P := \{\alpha \in E : \alpha(P) = 0\}$. Show that M_P is a maximal ideal of E, and that $M_P = \mu E + \nu E$, where $\mu := [X - x]_f$ and $\nu := [Y - y]_f$.
14. Let R be a ring. An ideal I of R is called prime if $I \subseteq R$ and if for all $a, b \in R$, $ab \in I$ implies $a \in I$ or $b \in I$. An ideal I of R is called maximal if $I \subseteq R$ and there are no ideals J of R such that $I \subseteq J \subseteq R$. Show that:
    (*a*) An ideal I of R is prime if and only if R/I is an integral domain;
    (*b*) An ideal I of R is maximal if and only if R/I is a field;
    (*c*) All maximal ideals of R are also prime ideals.
15. Continuing with the previous exercise, now assume that the characteristic of F is not 2, and that $f = Y^2 - \phi$, where $\phi \in F[X]$ is a non-zero polynomial with no multiple roots in F.
    (a) Show that if $P = (x, y) \in V(f)$, then so is $\overline{P} := (x, -y)$, and that $P = \overline{P} \Leftrightarrow y = 0 \Leftrightarrow \phi(x) = 0$.
    (*b*) Let $P = (x, y) \in V(f)$ and $\mu := [X - x]_f \in E$. Show that $\mu E = M_P M_{\overline{P}}$ (the ring-theoretic product).

16. This exercise explores some examples of prime and maximal ideals. Show that:
    (*a*) In the ring $\mathbb{Z}$, the ideal $\{0\}$ is prime but not maximal, and that the maximal ideals are precisely those of the form $p\mathbb{Z}$, where p is prime;
    (*b*) In an integral domain D, the ideal f0g is prime, and this ideal is maximal if and only if D is a field;
    (*c*) If p is a prime, then in the ring $\mathbb{Z}[X]$, the ideal (X, p) is maximal, while the ideals (X) and (p) are prime, but not maximal;
    (*d*) If F is a field, then in the ring $F[X, Y]$, the ideal (X, Y) is maximal, while the ideals (X) and (Y) are prime, but not maximal.
17. Let R be a ring, and let I and J be ideals of R. With the ringtheoretic product show that:
    (*a*) IJ is an ideal;
    (*b*) If I and J are principal ideals, with $I = aR$ and $J = bR$, then $IJ = abR$, and so is also a principal ideal;
    (*c*) $IJ \subseteq I \cap J$;
    (*d*) If $I + J = R$, then $IJ = I \cap J$.
18. Let R be a subring of E, and I an ideal of R. Show that the ring-theoretic product IE is an ideal of E that contains I, and is the smallest such ideal.
19. Let M be a maximal ideal of a ring R, and let $a, b \in R$. Show that if $ab \in M^2$ and $b \notin M$, then $a \in M^2$. Here, $M^2 := MM$, the ring-theoretic product.
20. Let R be a ring, and I an ideal of R. Define Rad (I) to be the set of all $a \in R$ such that $a^n \in I$ for some positive integer n.
    (*a*) Show that Rad (I) is an ideal of R containing I.
21. Let I be an ideal of R, and S a subring of R. Show that $I \cap S$ is an ideal of S.

ADVANCES IN RING THEORY

THE FIELD OF FRACTIONS OF AN INTEGRAL DOMAIN

Let D be an integral domain. Just as we can construct the field of rational numbers by forming fractions involving integers, we can construct a field consisting of fractions whose numerators and denominators are elements of D. This construction is quite straightforward, though a bit tedious.

To begin with, let S be the set of all pairs of the form (a, b), with $a, b \in D$ and $b \neq 0_D$. Intuitively, such a pair (a, b) is a "formal fraction," with numerator a and denominator b. We define a binary relation $\sim$ on S as follows: for $(a_1, b_1), (a_2, b_2) \in S$, we say $(a_1, b_1) \sim$

(a_2, b_2) if and only if $a_1b_2 = a_2b_1$. Our first task is to show that this is an equivalence relation:

Lemma: Let (a_1, b_1), (a_1', b_1'), (a_2, b_2), $(a_2', b_2') \in S$, where $(a_1, b_1) \sim (a_1', b_1')$ and $(a_2, b_2) \sim (a_2', b_2')$. Then we have

$$(a_1b_2 + a_2b_1,\, b_1b_2) \sim (a_1'b_2' + a_2'b_1',\, b_1'b_2')$$

and

$$(a_1a_2,\, b_1b_2) \sim (a_1'a_2',\, b_1'b_2').$$

Proof: This is a straightforward calculation. Since $a_1b_1' = a_1'b_1$ and $a_2b_2' = a_2'b_2$, we have

$$(a_1b_2 + a_2b_1)b_1'b_2' = a_1b_2b_1'b_2' + a_2b_1b_1'b_2' = a_1'b_2b_1b_2' + a_2'b_1b_1'b_2$$
$$= (a_1'b_2' + a_2'b_1')b_1b_2$$

and

$$a_1a_2b_1'b_2' = a_1'a_2b_1b_2' = a_1'a_2'b_1b_2.$$

In light of this lemma, we may unambiguously define addition and multiplication on K as follows: for $[a_1, b_1], [a_2, b_2] \in K$, we define

$$[a_1, b_1] + [a_2, b_2] := [a_1b_2 + a_2b_1,\, b_1b_2]$$

and

$$[a_1, b_1].\, [a_2, b_2] := [a_1a_2,\, b_1b_2].$$

The next task is to show that K is a ring – we leave the details of this (which are quite straight forward) to the reader.

Lemma: For all (a_1, b_1), (a_2, b_2), $(a_3, b_3) \in S$, we have

(*i*) $(a_1, b_1) \sim (a_1, b_1)$;

(*ii*) $(a_1, b_1) \sim (a_2, b_2)$ implies $(a_2, b_2) \sim (a_1, b_1)$;

(*iii*) $(a1, b_1) \sim (a_2, b_2)$ and $(a_2, b_2) \sim (a_3, b_3)$ implies $(a_1, b_1) \sim (a_3, b_3)$.

Proof: (*i*) and (*ii*) are rather trivial, and we do not comment on these any further. As for (*iii*), assume that $a_1b_2 = a_2b_1$ and $a_2b_3 = a_3b_2$. Multiplying the first equation by b_3, we obtain $a_1b_2b_3 = a_2b_1b_3$ and substituting a_3b_2 for a_2b_3 on the right-hand side of this last equation, we obtain $a_1b_2b_3 = a_3b_2b_1$. Now, using the fact that b_2 is non-zero and that D is an integral domain, we may cancel b_2 from both sides, obtaining $a_1b_3 = a_3b_1$.

Since $\sim$ is an equivalence relation, it partitions S into equivalence classes, and for $(a, b) \in S$, we denote by $[a, b]$ the equivalence class containing (a, b), and we denote by K

the set of all such equivalence classes. Our next task is to define addition and multiplication operations on equivalence classes, mimicking the usual rules of arithmetic with fractions. We want to define the sum of $[a_1, b_1]$ and $[a_2, b_2]$ to be $[a_1b_2 + a_2b_1, b_1b_2]$, and the product of $[a_1, b_1]$ and $[a_2, b_2]$ to be $[a_1a_2, b_1b_2]$.

Note that since D is an integral domain, if b_1 and b_2 are non-zero, then so is the product b_1b_2, and therefore $[a_1b_2 + a_2b_1, b_1b_2]$ and $[a_1a_2, b_1b_2]$ are indeed equivalence classes. However, to ensure that this definition is unambiguous, and does not depend on the particular choice of representatives of the equivalence classes $[a_1, b_1]$ and $[a_2, b_2]$, we need the following lemma.

Lemma: With addition and multiplication as defined above, K is a ring, with additive identity $[0D, 1D]$ and multiplicative identity $[1_D, 1_D]$.

Finally, we observe that K is in fact a field: it is clear that $[a, b]$ is a non-zero element of K if and only if $a \neq 0_D$, and hence any non-zero element $[a, b]$ of K has a multiplicative inverse, namely, $[b, a]$.

The field K is called the field of fractions of D. Consider the map $\tau : D \to K$ that sends $a \in D$ to $[a, 1_D] \in K$. It is easy to see that this map is a ring homomorphism, and one can also easily verify that it is injective. So, starting from D, we can synthesize "out of thin air" its field of fractions K, which essentially contains D as a subring, via the natural embedding $\tau : D \to K$.

Now suppose that we are given a field L that contains D as a subring. Consider the set K' consisting of all elements of L of the form ab^{-1}, where $a, b \in D$ and $b \neq 0_D$ – note that here, the arithmetic operations are performed using the rules for arithmetic in L. One may easily verify that K' is a subfield of L that contains D, and it is easy to see that this is the smallest subfield of L that contains D. The subfield K' of L may be referred to as the field of fractions of D within L. One may easily verify that the map $\rho : K \to L$ that sends $[a, b] \in K$ to $ab^{-1} \in L$ is an unambiguously defined ring homomorphism that maps K injectively onto K'. If we view K and L as D-algebras via inclusion, and we see that the map ρ is in fact a D-algebra homomorphism. Thus, K and K' are isomorphic as D-algebras. It is in this sense that the field of fractions K is the smallest field that contains D as a subring.

From now on, we shall simply write an element $[a, b]$ of K as the fraction a/b. In this notation, the above rules for addition, multiplication, and testing equality in K now look quite familiar:

$$\frac{a_1}{b_1} + \frac{a_2}{b_2} = \frac{a_1b_2 + a_2b_1}{b_1b_2}, \frac{a_1}{b_1} . \frac{a_2}{b_2} = \frac{a_1a_2}{b_1b_2}, \frac{a_1}{b_1} = \frac{a_2}{b_2} \Leftrightarrow a_2b_1.$$

FUNCTION FIELDS

An important special case of the above construction for the field of fractions of D is when $D = F[X]$, where F is a field. In this case, the field of fractions is denoted $F(X)$, and is called the field of rational functions (over F). This terminology is a bit unfortunate, since just as with polynomials, although the elements of $F(X)$ define functions, they are not (in general) in one-to-one correspondence with these functions.

Since $F[X]$ is a subring of $F(X)$, and since F is a subring of $F[X]$, we see that F is a subfield of $F(X)$.

More generally, we may apply the above construction to $D = F[X_1,..., X_n]$, the ring of multi-variate polynomials over the field F, in which case the field of fractions is denoted $F(X_1,..., X_n)$, and is also called the field of rational functions (over F, in the variables $X_1,..., X_n$).

EXERCISE

1. Show that the field of fractions of $\mathbb{Z}[i]$ within $\mathbb{C}$ is $\mathbb{Q}[i]$.
2. Let F be a field of characteristic zero. Show that F contains an isomorphic copy of $\mathbb{Q}$.

ALGEBRAS

Throughout this section, R denotes a ring (i.e., a commutative ring with unity).

Sometimes, a ring may also be naturally viewed as an R-module, in which case, both the theory of rings and the theory of modules may be brought to bear to study its properties.

Definition : An R-algebra is a set E, together with addition and multiplication operations on E, and a function $\mu : R \times E \rightarrow E$, such that

(i) With respect to addition and multiplication, E forms a ring;

(ii) With respect to addition and the scalar multiplication map m, E forms an R-module;

(iii) For all $c \in R$, and a, b $\in E$, we have

$$\mu(c, \alpha)\beta = \mu(c, \alpha\beta) = \alpha\mu(c, \beta).$$

An R-algebra E may also be called an algebra over R. As we usually do for R-modules, we shall write $c\alpha$ (or $c \cdot \alpha$) instead of $\mu(c, \alpha)$. When we do this, part (iii) of the definition states that

$$(c\alpha)\beta = c(\alpha\beta) = \alpha(c\beta)$$

for all $c \in R$ and $\alpha, \beta \in E$. In particular, we may write $c\alpha\beta$ without any ambiguity. Note that there are two multiplication operations at play here: scalar multiplication (such as $c\alpha$), and ring multiplication (such as $\alpha\beta$).

Also note that since we are assuming E is commutative, the second equality in part (iii) is already implied by the first. A simple consequence of the definition is that for all $c, d \in R$ and $\alpha, \beta \in E$, we have $(c\alpha)(d\beta) = (cd)(\alpha\beta)$. From this, it follows that for all $c \in R$, $\alpha \in E$, and $k \geq 0$, we have $(c\alpha)^k = c^k\alpha^k$.

Example.

1. Suppose E is a ring and $\tau : R \rightarrow E$ is a ring homomorphism. With scalar multiplication defined by $c := \tau(c)$ for $c \in R$ and $\alpha \in E$, one may easily check that E is indeed an R-algebra. In this case, we say that E is an R-algebra via the map τ.
2. If R is a subring of E, then with $\tau : R \rightarrow E$ being the inclusion map, we can view E as an R-algebra as in the previous example. In this case, we say that E is an R-algebra via inclusion.
3. If $\tau : R \rightarrow E$ is a natural embedding of rings, then by a slight abuse of terminology, just as we sometimes say that R is a subring of E, we shall also say that E is an R-algebra via inclusion.

Theorem: If R is a field, then an R-algebra is either the trivial ring or contains an isomorphic copy of R as a subring.

Theorem: If E is an R-algebra, then the map

$$\tau : R \rightarrow E$$
$$c \mapsto c.1_E,$$

is a ring homomorphism, and $c\alpha = \tau(c)$ for all $c \in R$ and $\alpha \in E$.

Proof: Exercise.

In the special situation where R is a field, we can say even more. In this situation, and with τ as in the above theorem, then either E is trivial or τ is injective. In the latter case, E contains an isomorphic copy of R as a subring.

Important Constructions of *R*-algebras

Example.

1. Let E be an R-algebra and let I be an ideal of E. Then it is easily verified that I is also a submodule of E. This means that the quotient ring E/I may also be viewed as an R-module, and indeed, it is an R-algebra, called the quotient algebra (over R) of E modulo I. For $\alpha \in E$ and $c \in R$, addition, multiplication, and scalar multiplication in E are defined as follows:

$$[\alpha]_I + [\beta]_I := [\alpha + \beta]_I, \ [\alpha]_I . [\beta]_I := [\alpha.\beta]_I, \ c . [\alpha]_I := [c . \alpha]_I.$$

2. If $E_1, \ldots, E_k$ are R-algebras, then their direct product E_1 X ... X E_k is an R-algebra as well, where addition, multiplication, and scalar multiplication are defined component-wise. As usual, if $E = E_1 = \ldots = E_k$, we write this as E^{Xk}.
3. If I is an arbitrary set, and E is an R-algebra, then Map(I, E), which is the set of all functions $f : I \rightarrow E$, may be naturally viewed as an R-algebra, with addition, multiplication, and scalar multiplication defined point-wise.

POLYNOMIAL EVALUATION

Let E be an R-algebra. Consider the ring of polynomials $R[X]$ (which is an R-algebra via inclusion). Any polynomial $g \in R[X]$ naturally defines a function on E: if $g = \sum_i a_i X^i$, with each $a_i \in R$, and $\alpha \in E$, then

$$g(\alpha) := \sum_i a_i \alpha^i .$$

Just as for rings, we say that is a root of g if $g(\alpha) = 0_E$.

For fixed $\alpha \in E$, the polynomial evaluation map

$$\rho : R[X] \to E$$
$$g \mapsto g(\alpha)$$

is easily seen to be an R-algebra homomorphism. The image of ρ is denoted $R[\alpha]$, and is a subalgebra of E. Indeed, $R[\alpha]$ is the smallest subalgebra of E containing α, and is called the subalgebra (over R) generated by α.

Theorem: Let us consider $\rho : E \to E'$ be an R-algebra homomorphism. Then for all $g \in R[X]$ and $\alpha \in E$, we have

$$\rho(g(\alpha)) = g(\rho(\alpha)).$$

Proof: Let $g = \sum_i a_i X^i \in R[X]$. Then we have

$$\rho(g(\alpha)) = \rho\left(\sum_i a_i \alpha^i\right) = \sum_i \rho(a_i \alpha^i) = \sum_i a_i \rho(\alpha^i) = \sum_i a_i \rho(\alpha)^i$$

If $E = R[\alpha]$ for some $\alpha \in E$, then every element of E can be expressed as $g(\alpha)$ for some $g \in R[X]$, and $\rho(g(\alpha)) = g(\rho(\alpha))$; hence, the action of ρ is completely determined by its action on α.

Example.

1. Let us consider $f \in R[X]$ be a non-zero polynomial with $\mathrm{lc}(f) \in R^*$.

Let $\xi := [X]_f \in E$. Then $E = R[\xi]$, and moreover, every element of E can be expressed uniquely as $g(\xi)$, where $g \in R[X]$ and $\deg(g) < \deg(f)$. In addition, ξ is a root of f. If $\deg(f) > 0$, and otherwise, they are trivial.

Now let E' be any R-algebra, and suppose that ρ: $E \to E'$ is an R-algebra homomorphism, and let $\xi' := \rho(\xi)$. By the previous theorem, ρ sends $g(\xi)$ to $g(\xi')$, for each $g \in R[X]$. Thus, the image of ρ is $R[\xi']$. Also, we have

$f(\xi') = f(\rho(\xi)) = \rho(f(\xi)) = \rho(0_E) = 0_E$. Therefore, ξ' must be a root of f.

Conversely, suppose that $\xi' \in E'$ is a root of f. Then the polynomial evaluation map from $R[X]$ to $E0$ that sends $g \in R[X]$ to $g(\xi') \in E'$ is an R-algebra homomorphism whose kernel contains f. Using the generalized versions of the first isomorphism theorems for rings and R-modules, we obtain the R-algebra homomorphism

$$\rho : E \to E'$$
$$g(\xi) \mapsto g(\xi').$$

EXERCISE

1. Show that the only $\mathbb{R}$-algebra homomorphisms from $\mathbb{C}$ into itself are the identity map and the complex conjugation map.
2. Let E be an R-algebra. For $\alpha \in E$, consider the α-multiplication map on E, which sends $\beta \in E$ to $\alpha\beta \in E$. Show that this map is an R-linear map.
3. Show that every ring may be viewed in a unique way as a $\mathbb{Z}$-algebra, and that subrings are subalgebras, and ring homomorphisms are $\mathbb{Z}$-algebra homomorphisms.

SUBALGEBRAS

Let E be an R-algebra. A subset S of E is called a subalgebra (over R) of E if it is both a subring of E and a submodule of E. This means that S contains 1_E, and is closed under addition, multiplication, and scalar multiplication; restricting these operations to S, we may view S as an R-algebra in its own right.

Theorem: If E is an R-algebra via inclusion, and S is a subring of E, then S is a subalgebra if and only if S contains R. More generally, if E is an arbitrary R-algebra, and S is a subring of E, then S is a subalgebra of E if and only if S contains $c \,.\, 1_E$ for all $c \in R$.

R-ALGEBRA HOMOMORPHISMS

Let us suppose E and E' be R-algebras. A function $\rho : E \to E'$ is called an R-algebra homomorphism if ρ is both a ring homomorphism and an R-linear map. This means that $\rho(1_E) = 1_{E'}$, and

$$\rho(\alpha + \beta) = \rho(\alpha) + \rho(\beta),\ \rho(\alpha\beta) = \rho(\alpha)\tau(\beta),\ \text{and}\ \rho(c\alpha) = c\rho(\alpha)$$

for all $\alpha, \beta \in E$ and all $c \in R$. As usual, if ρ is bijective, then it is called an R-algebra isomorphism, and if, in addition, $E = E'$, it is called an R-algebra automorphism.

Theorem: Let E be an R-algebra, and let $\rho : E \to E$ be an R-algebra homomorphism. Then the set $S := \{\alpha \in E : \rho(\alpha) = \alpha\}$ is a subalgebra of E, called the subalgebra of E fixed by ρ. Moreover, if E is a field, then so is S.

Proof: Let us verify that S is closed under addition. If $\alpha, \beta \in S$, then we have

$$\rho(\alpha + \beta) = \rho(\alpha) + \rho(\beta) \text{ (since } \rho \text{ is a group homomorphism)}$$
$$= \alpha + \beta \text{ (since } \alpha, \beta \in S).$$

Using the fact that ρ is a ring homomorphism, one can similarly show that *S* is closed under multiplication, and that $1_E \in S$. Likewise, using the fact that ρ is an *R*-linear map, one can also show that *S* is closed under scalar multiplication.

This shows that *S* is a subalgebra, proving the first statement. For the second statement, suppose that *E* is a field. Let be a non-zero element of *S*, and suppose $\beta \in E$ is its multiplicative inverse, so that $\alpha\beta = 1_E$. We want to show that β lies in *S*. Again, using the fact that ρ is a ring homomorphism, we have

$$\alpha\beta = 1_E = \rho(1_E) = \rho(\alpha\beta) = \rho(\alpha)\rho(\beta) = \alpha\rho(\beta),$$

and hence $\alpha\beta = \alpha\rho(\beta)$; canceling ρ, we obtain $\beta = \rho(\beta)$, and so $\beta \in S$.

Theorem: If *E* and *E*′ are *R*-algebras via inclusion, and $\rho : E \to E'$ is a ring homomorphism, then ρ is an *R*-algebra homomorphism if and only if the restriction of to *R* is the identity map. More generally, if *E* and *E*′ are arbitrary *R*-algebras and $\rho : E \to E'$ is a ring homomorphism, then is an *R*-algebra homomorphism if and only if $\rho(c \,.\, 1_E) = c \,.\, 1_{E'}$, for all $c \in R$.

Example.

1. Since $\mathbb{C}$ contains $\mathbb{R}$ as a subring, we may naturally view $\mathbb{C}$ as an $\mathbb{R}$-algebra via inclusion. The complex conjugation map on $\mathbb{C}$ that sends $a + bi$ to $a - bi$, for $a, b \in$ R, is an $\mathbb{R}$-algebra automorphism on $\mathbb{C}$.

Many simple facts about *R*-algebra homomorphisms can be obtained by combining corresponding facts for ring and *R*-module homomorphisms. For example, the composition of two *R*-algebra homomorphisms is again an *R*-algebra homomorphism, since the composition is both a ring homomorphism and an *R*-linear map. As another example, if $\rho : E \to E'$ is an *R*-algebra homomorphism, then its image *S*′ is both a subring and a submodule of *E*′, and hence, *S*′ is a subalgebra of *E*′. The kernel *K* of ρ is an ideal of *E*, and we may form the quotient algebra *E/K*. The first isomorphism theorems for rings and modules tell us that *E/K* and *S*′ are isomorphic both as rings and as *R*-modules, and hence, they are isomorphic as *R*-algebras. Specifically, the map

$$\bar{\rho} : E/K \to E'$$

$$[\alpha]_K \mapsto \rho(\alpha)$$

is an injective *R*-algebra homomorphism whose image is *S*′.

2. If *E* is an *R*-algebra and` *I* is an ideal of *E*, then as observed in Example 16.6, *I* is also a submodule of *E*, and we may form the quotient algebra *E/I*. The natural map

$$\rho : E \to E/I$$

$$\alpha \mapsto [\alpha]_I$$

is both a ring homomorphism and an *R*-linear map, and hence is an *R*-algebra homomorphism.

The following theorem isolates an important subalgebra associated with any R-algebra homomorphism $\rho : E \to E$.

UNIQUE FACTORIZATION

Let us consider D denotes an integral domain. We call $a, b \in D$ associate if $a = ub$ for some $u \in D^*$. Equivalently, a and b are associate if and only if $a \mid b$ and $b \mid a$. A non-zero element $p \in D$ is called irreducible if it is not a unit, and all divisors of p are associate to 1 or p. Equivalently, a non-zero, non-unit $p \in D$ is irreducible if and only if it cannot be expressed as $p = ab$ where neither a nor b are units.

Definition: We call D a unique factorization domain (UFD) if

- Every non-zero element of D that is not a unit can be written as a product of irreducibles in D, and
- Such a factorization into irreducibles is unique up to associates and the order in which the factors appear.

Another way to state part (ii) of the above definition is that if $p_1 \dots p_r$ and $p_1' \dots p_s'$ are two factorizations of some element as a product of irreducibles, then $r = s$, and there exists a permutation π on the indices $\{1,\dots, r\}$ such that p_i and $p'_{\pi(i)}$ are associate. We have seen, both $\mathbb{Z}$ and F[X] are UFDs. In both of those cases, we chose to single out a distinguished irreducible element among all those associate to any given irreducible: for $\mathbb{Z}$, we always chose positive primes, and for F[X], we chose monic irreducible polynomials.

Example.

1. Having already seen two examples of UFDs, it is perhaps a good idea to 1ook at an example of an integral domain that is not a UFD. Consider the subring $\mathbb{Z}$[p–3] of the complex numbers, which consists of all complex numbers of the form $a + b\sqrt{-3}$, where $a, b \in \mathbb{Z}$. As this is a subring of the field $\mathbb{C}$, it is an integral domain (one may also view $\mathbb{Z}[\sqrt{-3}]$ as the quotient ring $\mathbb{Z}[X]/(X^2 + 3)$).

Let us first determine the units in $\mathbb{Z}[\sqrt{-3}]$. For $a, b \in \mathbb{Z}$, we have $N(a + b\sqrt{-3}) = a^2 + 3b^2$, where N is the usual norm map on C. If $\alpha \in \mathbb{Z}[\sqrt{-3}]$ is a unit, then there exists $\alpha' \in \mathbb{Z}[\sqrt{-3}]$ such that 0 = 1. Taking norms, we obtain

$$1 = N(1) = N(\alpha\alpha') = N(\alpha)N(\alpha').$$

Since the norm of an element of $\mathbb{Z}[\sqrt{-3}]$ is a non-negative integer, this implies that $N(\alpha) = 1$. If $\alpha = a + b\sqrt{-3}$, with $a, b \in \mathbb{Z}$, then $N(\alpha) = a^2 + 3b^2$, and it is clear that $N(\alpha) = 1$ if and only if $\alpha = \pm 1$. We conclude that the only units in $\mathbb{Z}[\sqrt{-3}]$ are ± 1.

Now consider the following two factorizations of 4 in $\mathbb{Z}[\sqrt{-3}]$:

$$4 = 2 \,.\, 2 = (1 + \sqrt{-3})(1 - \sqrt{-3}).$$

We claim that 2 is irreducible. For suppose, say, that $2 = \alpha\alpha'$, for $\alpha, \alpha' \in \mathbb{Z}[\sqrt{-3}]$, with neither a unit. Taking norms, we have $4 = N(2) = N(\alpha)N(\alpha')$, and therefore, $N(\alpha) = N(\alpha') = 2$ – but this is impossible, since there are no integers a and b such that $a^2 + 3b^2 = 2$. By the same reasoning, since $N(1 + \sqrt{-3}) = N(1 - \sqrt{-3}) = 4$, we see that $1 + \sqrt{-3}$ and $1 - \sqrt{-3}$ are both irreducible. Further, it is clear that 2 is not associate to either $1 + \sqrt{-3}$ or $1 - \sqrt{-3}$, and so the two factorizations of 4 in are fundamentally different. For $a, b \in D$, we call $d \in D$ a common divisor of a and b if $d \mid a$ and $d \mid b$; moreover, we call such a d a greatest common divisor of a and b if all other common divisors of a and b divide d. We say that a and b are relatively prime if the only common divisors of a and b are units.

It is immediate from the definition of a greatest common divisor that it is unique, up to multiplication by units, if it exists at all. Unlike in the case of $\mathbb{Z}$ and $F[X]$, in the general setting, greatest common divisors need not exist; moreover, even when they do, we shall not attempt to "normalize" greatest common divisors, and we shall speak only of "a" greatest common divisor, rather than "the" greatest common divisor. Just as for integers and polynomials, we can generalize the notion of a greatest common divisor in an arbitrary integral domain D from two to any number of elements of D, and we can also define a least common multiple of any number of elements as well. Although these greatest common divisors and least common multiples need not exist in an arbitrary integral domain D, if D is a UFD, they will always exist.

The existence question easily reduces to the question of the existence of a greatest common divisor and least common multiple of a and b, where a and b are non-zero elements of D. So assuming that D is a UFD, we may write

$$a = u\prod_{i=1}^{r} p_i^{e_i} \text{ and } b = v\prod_{i=1}^{r} p_i^{f_i},$$

where u and v are units, $p_1, \ldots, p_r$ are non-associate irreducibles, and $e_1, \ldots, e_r$ and $f_1, \ldots, f_r$ are non-negative integers, and it is easily seen that

$$\prod_{i=1}^{r} p_i^{\min(e_i, f_i)}$$

is a greatest common divisor of a and b, while

$$\prod_{i=1}^{r} p_i^{\max(e_i, f_i)}$$

is a least common multiple of a and b.

It is also evident that in a UFD D, if $c \mid ab$ and c and a are relatively prime, then $c \mid b$. In particular, if p is irreducible and $p \mid ab$, then $p \mid a$ or $p \mid b$. This is equivalent to saying that if p is irreducible, then the quotient ring D/pD is an integral domain .

EXERCISE

1. Let D be a UFD and F its field of fractions. Show that
   (*a*) If $x = a/b$ for $a, b \in D$ relatively prime, then for any other $a', b' \in D$ with $x = a'/b'$, we have $a' = ca$ and $b' = cb$ for some $c \in D$.
   (*b*) Every element $x \in F$ can be expressed as $x = a/b$, where $a, b \in D$ are relatively prime.
2. (*a*) Consider an equivalence class C induced by the "is associate to" relation. Show that if C contains an irreducible element, then all elements of C are irreducible.
   (*a*) Show that the "is associate to" relation is an equivalence relation.
3. Show that the ring $\mathbb{Z}[\sqrt{-5}]$ is not a UFD.
4. Let D be a UFD and let $p \in D$ be irreducible. Show that there is no prime ideal Q of D with $\{0_D\} \subseteq Q \subseteq pD$.

UNIQUE FACTORIZATION IN *D*[X]

Theorem: Let D be a UFD and F its field of fractions. If $f \in D[X]$ with $\deg(f) > 0$ is irreducible, then f is also irreducible in $F[X]$.

Proof: Suppose that f is not irreducible in $F[X]$, so that $f = gh$ for non-constant polynomials $g, h \in F[X]$, both of degree strictly less than that of f. We may write $g = (c/d)g'$, where $c, d \in D$ and $g' \in D[X]$ is primitive. Set $h' := (c/d)h$, so that $f = gh = g'h'$. By Theorem, we have $h' \in D[\mathbf{X}]$, and this shows that f is not irreducible in $D[X]$.

Theorem: If D is a UFD, then so is D[X].

This theorem implies, for example, that $\mathbb{Z}[X]$ is a UFD. Applying the theorem inductively, one also sees that $\mathbb{Z}[X_1,..., X_n]$ is a UFD, as is $F[X_1,..., X_n]$ for every field F.

We begin with some simple observations. First, recall that for an integral domain D, $D[X]$ is an integral domain, and the units in $D[X]$ are precisely the units in D. Second, it is easy to see that an element of D is irreducible in D if and only if it is irreducible in $D[X]$. Third, for $c \in D$ and $f = \sum_i c_i X^i \in D[X]$, we have c j f if and only if c j ci for all i.

We call a non-zero polynomial $f \in D[X]$ primitive if the only elements of D that divide f are units. If D is a UFD, then given any non-zero polynomial $f \in D[X]$, we can write it as $f = cf'$, where $c \in D$ and $f' \in D[X]$ is a primitive polynomial: just take c to be a greatest common divisor of all the coefficients of f.

Example. In $\mathbb{Z}[X]$, the polynomial $f = 4X^2 + 6X + 20$ is not primitive, but we can write $f = 2f$, where $f = 2X^2 + 3X + 10$ is primitive.

Theorem: Let D be a UFD. The product of two primitive polynomials in $D[X]$ is also primitive.

Proof: Let $g, h \in D[X]$ be primitive polynomials, and let $f := gh$. If f is not primitive, then $c \mid f$ for some non-zero, non-unit $c \in D$, and as D is a UFD, there is some irreducible element $p \in D$ that divides c, and therefore, divides f as well. By Theorem, it follows that $p \mid g$ or $p \mid h$, which implies that either g is not primitive or h is not primitive.

Suppose that D is a UFD and that F is its field of fractions. Any non-zero polynomial $f \in F[X]$ can always be written as $f = (c/d)f'$, where $c, d \in D$, with $d \neq 0$, and $f' \in D[X]$ is primitive. To see this, clear the denominators of the coefficients of f, writing $df = f''$, where $0 \neq d \in D$ and $f'' \in D[X]$. Then take c to be a greatest common divisor of the coefficients of f'', so that $f'' = cf'$, where $f' \in D[X]$ is primitive. Then we have $f = (c/d)f$, as required. Of course, we may assume that c and d are relatively prime – if not, we may divide c and d by a greatest common divisor.

Example. Let $f = (3/5)X^2 + 9X + 3/2 \in \mathbb{Q}[X]$. Then we can write $f = (3/10)f'$, where $f' = 2X^2 + 30X + 5 \in \mathbb{Z}[X]$ is primitive.

Eisenstein's Criterion

Theorem: Let D be a UFD and F its field of fractions. Let $f = c_nX^n + c_{n-1}X^{n-1} + ... + c_0 \in D[X]$. If there exists an irreducible $p \in D$ such that

$$p \mid c_n, p \mid c_{n-1}, ..., p \mid c_0, p^2 \mid c_0,$$

then f is irreducible over F.

Proof: Let f be as above, and suppose it were not irreducible in $F[X]$. Then by Theorem, we could write $f = gh$, where $g, h \in D[X]$, both of degree strictly less than that of f. Let us write

$$g = a_kX^k + ... + a_0 \text{ and } h = b_\ell X^\ell + ... + b_0,$$

where $a_k \neq 0$ and $b_\ell \neq 0$, so that $0 < k < n$ and $0 < \ell < n$. Now, since $c_n = a_kb_\ell$, and $p \mid c_n$, it follows that $p \mid a_k$ and $p \mid b_\ell$. Further, since $c_0 = a_0b_0$, and $p \mid c_0$ but $p_2 \mid c_0$, it follows that p divides one of a_0 or b_0, but not both – for concreteness, let us assume that $p \mid a_0$ but $p \mid b_0$. Also, let m be the smallest positive integer such that $p \mid am$ – note that $0 < m \leq k < n$.

Now let us consider the natural map that sends $a \in D$ to $\bar{a} := [a]p \in D/pD$, which we can extend coefficient-wise to a ring homomorphism from $D[X]$ to $(D/pD)[X]$. Because D is a UFD and p is irreducible, D/pD is an integral domain. Since $f = gh$, we have

$$\bar{c}_nX^n = \bar{f} = \bar{g}\bar{h} = (\bar{a}_kX^k + ... + \bar{a}_mX^m)\bar{b}_\ell X^\ell(+ ... + \bar{b}_0).$$

Theorem: Let D be a UFD and let F be its field of fractions. Suppose that $f, g \in D[X]$ and $h \in F[X]$ are non-zero polynomials such that $f = gh$ and g is primitive. Then $h \in D[X]$.

Proof: Write $h = (c/d)h'$, where $c, d \in D$ and $h' \in D[X]$ is primitive. Let us assume that c and d are relatively prime. Then we have

$$d \,.\, f = c \,.\, gh'.$$

We claim that $d \in D$. To see this, note that implies that $d \mid (c \,.\, gh')$, and the assumption that c and d are relatively prime implies that $d \mid gh'$. But by Theorem, gh' is primitive, from which it follows that d is a unit. That proves the claim.

It follows that $c/d \in D$, and hence $h = (c/d)h' \in D[X]$.

Theorem: Let D be a UFD. Every non-zero, non-unit element of $D[X]$ can be expressed as a product of irreducibles in $D[X]$.

Proof: Let f be a non-zero, non-unit polynomial in $D[X]$. If f is a constant, then because D is a UFD, f factors into irreducibles in D. So assume f is not constant. If f is not primitive, we can write $f = cf'$, where c is a non-zero, non-unit in D, and f' is a primitive, non-constant polynomial in $D[X]$. Again, as D is a UFD, c factors into irreducibles in D.

From the above discussion, it suffices to prove the theorem for non-constant, primitive polynomials $f \in D[X]$. If f is itself irreducible, we are done. Otherwise, we can write $f = gh$, where $g, h \in D[X]$ and neither g nor h are units. Further, by the assumption that f is a primitive, non-constant polynomial, both g and h must also be primitive, non-constant polynomials; in particular, both g and h have degree strictly less than deg(f), and the theorem follows by induction on degree.

Theorem: Let D be a UFD. Let $f \in D[X]$ with $\deg(f) > 0$ be irreducible, and let $g, h \in D[X]$. If f divides gh in $D[X]$, then f divides either g or h in $D[X]$.

Proof: Suppose that $f \in D[X]$ with $\deg(f) > 0$ is irreducible. This implies that f is a primitive polynomial. By Theorem 16.48, f is irreducible in $F[X]$, where F is the field of fractions of D. Suppose f divides gh in $D[X]$. Then because $F[X]$ is a UFD, f divides either g or h in $F[X]$.

In the proof of Theorem 16.43, there is a clear connection between factorization in $D[X]$ and $F[X]$, where F is the field of fractions of D. We should perhaps make this connection more explicit. Let $f \in D[X]$ be a non-zero polynomial. We may write f as

$$f = up_1^{a_1} \ldots p_r^{a_r} f_1^{b_1} \ldots f_s^{b_s}.$$

where $u \in D$, the pi's are non-associate, irreducible elements of D, and the f_j's are non-associate, irreducible, non-constant polynomials over D (and in particular, primitive). For $j = 1, \ldots, s$, let $g_j := \mathrm{lc}(f_j)^{-1} f_j$ be the monic associate of f_j in $F[X]$. Then in $F[X]$, f factors as

$$f = cg_1^{b_1} \ldots g_s^{b_s},$$

where

$$c := u.\prod_i p_i^{a_i}.\prod_i lc(f_j)^{b_j} \in F,$$

and the g_j's are distinct, irreducible, monic polynomials over F.

Example. Consider the polynomial $f = 4X^2 + 2X - 2 \in \mathbb{Z}[X]$. Over $\mathbb{Z}[X]$, f factors as $2(2X-1)(X+1)$, where each of these three factors is irreducible in $\mathbb{Z}[X]$. However, over $\mathbb{Q}[X]$, f factors as $4(X - 1/2)(X + 1)$, where 4 is a unit, and the other two factors are irreducible.

The following theorem provides a useful criterion for establishing that a polynomial is irreducible.

Theorem: Let D be a UFD, let p be an irreducible in D, and let $g, h \in D[X]$. Then $p \mid gh$ implies $p \mid g$ or $p \mid h$.

Proof: Consider the quotient ring D/pD, which is an integral domain (because D is a UFD), and the corresponding ring of polynomials $(D/pD)[X]$, which is also an integral domain. Also consider the natural map that sends $a \in D$ to $\bar{a} := [a]_p \in D/pD$, which we can extend coefficient-wise to a ring homomorphism from $D[X]$ to $(D/pD)[X]$. If $p \mid gh$, then we have

$$0 = \overline{gh} = \bar{g}\bar{h},$$

and since $(D/pD)[X]$ is an integral domain, it follows that $\bar{g}` = 0$ or $\bar{g} = 0$, which means that $p \mid g$ or $p \mid h$.

Theorem: For every prime number q, the qth cyclotomic polynomial

$$\Phi_q := \frac{X^q - 1}{X - 1} = X^{q-1} + X^{q-2} + \ldots + 1$$

is irreducible over $\mathbb{Q}$.

Proof: Let

$$f := \Phi_q(X+1) = \frac{(X+1)^q - 1}{(X+1) - 1}.$$

It is easy to see that

$$f := \sum_{i=0}^{q-1} c_i X^i \text{, where } c_i = \binom{q}{i+1} \ (i = 0, \ldots, q-1).$$

Thus, $c_{q-1} = 1$, $c_0 = q$, and for $0 < i < q - 1$, we have $q \mid c_i$. Theorem therefore applies, and we conclude that f is irreducible over $\mathbb{Q}$. It follows that Πq is irreducible over $\mathbb{Q}$, since if $\Phi_q = gh$ were a non-trivial factorization of Φ_q, then $f = \Phi_q(X+1) = g(X+1).\ h(X+1)$ would be a non-trivial factorization of f.

EXERCISE

1. Let a be a non-zero, square-free integer, with $a \notin \{\pm 1\}$, and let n be a positive integer. Show that the polynomial $X^n - a$ is irreducible in $\mathbb{Q}[X]$.
2. Let D be a UFD, let p be an irreducible element of D, and consider the natural map that sends $a \in D$ to $\bar{a} := [a]_p \in D/pD$, which we extend coefficient-wise to a ring homomorphism from $D[X]$ to $(D/pD)[X]$. Show that if $f \in D[X]$ is a primitive polynomial such that $p \mid \mathrm{lc}(f)\ \bar{f}$ and $f \in (D/pD)[X]$ is irreducible, then f is irreducible.
3. Let $g, h \in \mathbb{Z}[X]$ be non-zero polynomials with $d := \gcd(g, h) \in \mathbb{Z}[X]$. Show that for every prime p not dividing $\mathrm{lc}(g)\ \mathrm{lc}(h)$, we have $\bar{d} \mid \gcd(\bar{g}, \bar{h})$, and except for finitely many primes p, we have $\bar{d} = \gcd(\bar{g}, \bar{h})$. Here, $\bar{d}, \bar{g}$ and $\bar{h}$ denote the images of d, g, and h in $\mathbb{Z}_p[X]$ under the coefficient-wise extension of the natural map from $\mathbb{Z}$ to $\mathbb{Z}_p$.
4. Show that the polynomial $X^4 + 1$ is irreducible in $\mathbb{Q}[X]$.
5. Design and analyze an efficient algorithm for the following problem. The input is a pair of polynomials $g, h \in \mathbb{Z}[X]$, along with their greatest common divisor d in the ring $\mathbb{Q}[X]$. The output is the greatest common divisor of g and h in the ring $\mathbb{Z}[X]$.
6. Show that neither $\mathbb{Z}[X]$ nor $F[X, Y]$ (where F is a field) are PIDs (even though they are UFDs).
7. Let $f \in \mathbb{Z}[X]$ be a monic polynomial. Show that if f has a root $x \in \mathbb{Q}$, then $x \in \mathbb{Z}$, and x divides the constant term of f.
8. Let F be a field, and let $g, h \in F[X, Y]$. Define $V(g, h) := \{(x, y) \in F X F : g(x, y) = h(x, y) = 0\}$. Show that if g and h are relatively prime, then $V(g, h)$ is a finite set.
9. Let F be a field, and consider the ring of bivariate polynomials $F[X, Y]$. Show that in this ring, the polynomial $X^2 + Y^2 - 1$ is irreducible, provided F does not have characteristic 2. What happens if F has characteristic 2?

UNIQUE FACTORIZATION IN EUCLIDEAN AND PRINCIPAL IDEAL DOMAINS

Definition: D is said to be a Euclidean domain if there is a "size function" S mapping the non-zero elements of D to the set of non-negative integers; such that for all $a, b \in D$ with $b \neq 0$, there exist $q, r \in D$, with the property that $a = bq + r$ and either $r = 0$ or $S(r) < S(b)$.

Example.

1. Recall again the ring

$$\mathbb{Z}[i] = \{a + bi : a, b \in \mathbb{Z}\}$$

of Gaussian integers. Let us show that this is a Euclidean domain, using the usual norm map N on complex numbers for the size function. Let $\alpha, \alpha \in \mathbb{Z}[i]$, with $\beta \neq 0$. We want to show the existence of $\kappa, \rho, \in \mathbb{Z}[i]$ such that $\alpha = \beta\kappa + \rho$, where $N(\rho) < N(\beta)$. Suppose that in the field C, we compute $\alpha\beta^{-1} = r + si$, where $r, s \in \mathbb{Q}$. Let m, n be integers such that $|m - r| \leq 1/2$ and $|n - s| \leq 1/2$ – such integers m and n always exist, but may not be uniquely determined. Set $\kappa := m + ni \in \mathbb{Z}[i]$ and $\rho := \alpha - \beta\kappa$. Then we have

$$\alpha\beta^{-1} = \kappa + \delta,$$

where $\delta \in C$ with $N(\delta) \leq 1/4 + 1/4 = 1/2$, and

$$\rho = \alpha - \beta\kappa = \alpha - \beta(\alpha\beta)^{-1} - \delta) = \delta\beta,$$

and hence

$$N(\rho) = N(\delta\beta) = N(\delta)N(\beta) \leq \frac{1}{2}N(\beta).$$

2. Both $\mathbb{Z}$ and $F[X]$ are Euclidean domains. In $\mathbb{Z}$, we can take the ordinary absolute value function jj as a size function, and for $F[X]$.

Theorem: Let D be a PID. Let $p \in D$ be irreducible, and let $a, b \in D$. Then $p \mid ab$ implies that $p \mid a$ or $p \mid b$.

Theorem: If D is a Euclidean domain and I is an ideal of D, then there exists $d \in D$ such that $I = dD$.

Proof: If $I = \{0\}$, then $d = 0$ does the job, so let us assume that $I \neq \{0\}$. Let d be any non-zero element of I such that $S(d)$ is minimal, where S is a size function that makes D into a Euclidean domain. We claim that $I = dD$.

It will suffice to show that for all $c \in I$, we have $d \mid c$. Now, we know that there exists q, $r \in D$ such that $c = dq + r$, where either $r = 0$ or $S(r) < S(d)$. If $r = 0$, we are done; otherwise, r is a non-zero element of I with $S(r) < S(d)$, contradicting the minimality of $S(d)$.

Recall that an ideal of the form $I = dD$ is called a principal ideal. If all ideals of D are principal, then D is called a principal ideal domain (PID).

Theorem: If D is a PID, then D is a UFD.

Theorem: Let D be a PID. For all $a, b, c \in D$ such that $c \mid ab$ and a and c are relatively prime, we have $c \mid b$.

Theorem: If D is a PID, and $I_1 \subseteq I_2 \subseteq \ldots$ are ideals of D, then there exists an integer k such that $I_k = I_{k+1} = \ldots..$

Proof: Let $I := \bigcup_{i=1}^{\infty} I_i$, which is an ideal of D. Thus, $I = dD$ for some $d \in D$. But $d \in \bigcup_{i=1}^{\infty} I_i$ implies that $d \in I_k$ for some k, which shows that $I = dD \subseteq I_k$. It follows that $I = I_k = I_{k+1} = \ldots.$

Theorem: Let D be a PID. For all $a, b \in D$, there exists a greatest common divisor d of a and b, and moreover, $aD + bD = dD$.

Theorem: If D is a PID, then every non-zero, non-unit element of D can be expressed as a product of irreducibles in D.

Proof: Let $c \in D$, $c \neq 0$, and c not a unit. If c is irreducible, we are done. Otherwise, we can write $c = ab$, where neither a nor b are units. As ideals, we have $cD \subseteq aD$ and $cD \subseteq bD$. If we continue this process recursively, building up a "factorization tree" where c is at the root, a and b are the children of c, and so on, then the recursion must stop, since any infinite path in the tree would give rise to ideals

$$cD = I_1 \subseteq I_2 \subseteq \ldots,$$

contradicting previous Theorem.

The proof of the uniqueness part of Theorem is essentially the same as for proofs we gave for $\mathbb{Z}$ and $F[X]$.

EXERCISE

1. Show that in a PID, all non-zero prime ideals are maximal.
2. Consider the polynomial

$$X^3 - 1 = (X - 1)(X^2 + X + 1).$$

Over $\mathbb{C}$, the roots of $X^3 - 1$ are $1, (-1 \pm \sqrt{-3})/2$. Let $\omega := (-1 + \sqrt{-3})/2$, and note that $\omega^2 = -1 - \omega = (-1 - \sqrt{-3})/2$, and $\omega^3 = 1$.

(a) Show that the ring $\mathbb{Z}[\omega]$ consists of all elements of the form $a + b\omega$, where $a, b \in \mathbb{Z}$, and is an integral domain. This ring is called the ring of Eisenstein integers.

(*b*) Show that the only units in $\mathbb{Z}[\omega]$ are ± 1, $\pm\omega$, and $\pm\omega^2$.

(*c*) Show that $\mathbb{Z}[\omega]$ is a Euclidean domain.

3. Show that $\mathbb{Z}[\sqrt{-2}]$ is a Euclidean domain.
4. Let $\alpha, \beta \in \mathbb{C}$. Prove the following statements.
   (*a*) $N(\alpha + \beta) \leq 2(N(\alpha) + N(\beta))$.
   (*b*) $M(\alpha) \leq |\alpha| \leq \sqrt{2}\, M(\alpha)$.
   (*c*) $|\alpha\beta| = |\alpha||\beta|$.
   (*d*) $|\alpha + \beta| \leq |\alpha| + |\beta|$.

   The following exercises develop algorithms for computing with Gaussian integers. For computational purposes, we assume that a Gaussian integer $\alpha = a + bi$, with a, $b \in \mathbb{Z}$, is represented as the pair of integers (a, b).
5. Let $\alpha, \beta \in \mathbb{Z}[i]$.
   (*a*) Show how to compute $\alpha \,.\, \beta$ in time $O(\text{len}(M(\alpha)).\ \text{len}(M(\beta)))$.
   (*b*) Show how to compute $M(\alpha)$ in time $O(\text{len}(M(\alpha)))$ and $N(\alpha)$ in time $O(\text{len}(M(\alpha))^2)$.
   (*c*) Assuming $\beta \neq 0$, show how to compute $\kappa, \rho \in \mathbb{Z}[i]$ such that $\alpha = \beta\kappa + \rho$, $N(\rho) \leq \frac{1}{2} N(\beta)$, and $N(\kappa) \leq 4N(\alpha)/N(\beta)$. Your algorithm should run in time $O(\text{len}(M(\alpha)).\ \text{len}(M(\beta)))$.
   (*d*) Show how to compute $\alpha + \beta$ in time $O(\text{len}(M(\alpha)) + \text{len}(M(\beta)))$.
6. Using the division with remainder algorithm from part (d) of the previous exercise, adapt the Euclidean algorithm for (ordinary) integers to work with Gaussian integers. On inputs $\alpha, \beta \in \mathbb{Z}[i]$, your algorithm should compute a greatest common divisor $\delta \in \mathbb{Z}[i]$ of α and β in time $O(\ell^3)$, where $\ell := \max\{\text{len}(M(\alpha)), \text{len}(M(\beta))\}$.
7. Extend the algorithm of the previous exercise, so that it computes $\sigma, \tau \in \mathbb{Z}[i]$ such that $\alpha\sigma + \beta\tau = \delta$. Your algorithm should run in time $O(\ell^2)$, and it should also be the case that $\text{len}(M(\sigma))$ and $\text{len}(M(\tau))$ are $O(\ell)$.
8. In Exercise 16.41, we saw that 2 factors as $-i(1 + i)^2$ in $\mathbb{Z}[i]$, where $1 + i$ is irreducible. This exercise examines the factorization in $\mathbb{Z}[i]$ of prime numbers $p > 2$. Show that:
   (*a*) All prime numbers $p \equiv 3 \pmod 4$ are irreducible in $\mathbb{Z}[i]$.
   (*b*) For every irreducible $\pi \in \mathbb{Z}[i]$, there exists a unique prime number p such that 'divides p;
   (*c*) For all prime numbers $p \equiv 1 \pmod 4$, we have $p = \pi\bar{\pi}$, where $\pi \in \mathbb{Z}[i]$ is irreducible, and the complex conjugate $\bar{\pi}$ of π is also irreducible and not associate to π;

9. Let $\pi := 1 + i \in \mathbb{Z}[i]$.

   (a) Show that for all $\alpha \in \mathbb{Z}[i]$ with $\alpha \equiv 1 \pmod{\pi}$, there exists a unique $\varepsilon \in \{\pm 1, \pm i\}$ such that $\alpha \equiv \varepsilon \pmod{2\pi}$.

   (b) Show that $2 = \pi\overline{\pi} = -i\pi^2$, that $N(\pi) = 2$, and that π is irreducible in $\mathbb{Z}[i]$.

   (c) Let $\alpha \in \mathbb{Z}[i]$, with $\alpha = a + bi$ for $a, b \in \mathbb{Z}$. Show that $\pi \mid \alpha$ if and only if $a - b$ is even, in which case

   $$\frac{\alpha}{\pi} = \frac{a+b}{2} + \frac{b-a}{2}i.$$

   (*d*) Show that for all $\alpha, \beta \in \mathbb{Z}[i]$ with $\alpha \equiv \beta \equiv 1 \pmod{\pi}$, there exists a unique $\varepsilon \in \{\pm 1, \pm i\}$ such that $\alpha \equiv \varepsilon\beta \pmod{2\pi}$.

   (*e*) Show that for all $\alpha \in \mathbb{Z}[i]$, we have $\alpha \equiv 0 \pmod{\pi}$ or $\alpha \equiv 1 \pmod{\pi}$.

   (*f*) Show that the quotient ring $\mathbb{Z}[i]/\pi\mathbb{Z}[i]$ is isomorphic to the ring $\mathbb{Z}_2$.

FORMAL DERIVATIVES

Let us suppose R denotes a ring. Consider a polynomial $g \in R[X]$. If Y is another indeterminate, we may evaluate g at $X + Y$, and collecting monomials of like degree in Y, we may write

$$g(X + Y) = g_0 + g_1Y + g_2Y^2 + ...$$

where $g_i \in R[X]$ for $i = 0, 1, 2,....$ Evidently, $g_0 = g$ (just substitute 0 for Y in , and we may write

$$g(X + Y \equiv g + g_1Y \pmod{Y^2}).$$

We define the formal derivative of g, denoted $D(g)$, to be the unique polynomial $g_1 \in R[X]$ satisfying. We stress that unlike the "analytical" notion of derivative from calculus, which is defined in terms of limits, this definition is purely "symbolic." Nevertheless, some of the usual rules for derivatives still hold:

Theorem: We have:

(*i*) $D(c) = 0$ for all $c \in R$;

(*ii*) $D(X) = 1$;

(*iii*) $D(g + h) = D(g) + D(h)$ for all $g, h \in R[X]$;

(*iv*) $D(gh) = D(g)h + gD(h)$ for all $g, h \in R[X]$.

Proof: Parts (i) and (ii) are immediate from the definition. Parts (iii) and (iv) follow from the definition by a simple calculation. Suppose

$$g(X + Y \equiv g + g_1Y \pmod{Y^2}) \text{ and } h(X + Y \equiv h + h_1Y \pmod{Y^2})$$

where $g_1 = D(g)$ and $h_1 = D(h)$. Then

$$(g + h)(X + Y \equiv g(X + Y + h(X + Y \equiv (g + h) + (g_1 + h_1)Y \pmod{Y^2}),$$

and

$$(gh)(X + Y \equiv g(X + Y)\, h(X + Y) \equiv gh + (g_1 h + g h_1)Y \pmod{Y^2}).$$

Combining parts (i) and (iv) of this theorem, we see that $D(cg) = cD(g)$ for all $c \in R$ and $g \in R[X]$. This fact can also be easily derived directly from the definition of the derivative.

Combining parts (ii) and (iv) of this theorem, together with a simple induction argument, we see that $D(X^n) = nX^{n-1}$ for all positive integers n. This fact can also be easily derived directly from the definition of the derivative by considering the binomial expansion of $(X + Y)^n$.

Combining part (iii) of this theorem and the observations in the previous two paragraphs, we see that for any polynomial $g = \sum_{i=0}^{k} a_i X^i \in R[X]$, we have

$$D(g) = \sum_{i=1}^{k} i a_i X^{i-1},$$

which agrees with the usual formula for the derivative of a polynomial.

The notion of a formal derivative can be generalized to multi-variate polynomials. Let $g \in R[X_1,\ldots, X_n]$. For any $i = 1,\ldots, n$, we can view g as a polynomial in the variable X_i, whose coefficients are elements of $R[X_1,\ldots, X_{i-1}, X_{i+1},\ldots, X_n]$.

Then if we formally di–erentiate with respect to the variable X_i, we obtain the formal "partial" derivative $DX_i\,(g)$.

EXERCISE

1. Suppose p is a prime, $g \in \mathbb{Z}[X]$, and $x \in \mathbb{Z}$, such that $g(x) \equiv 0 \pmod{p}$ and $D(g)(x) \not\equiv 0 \pmod{p}$. Show that for every positive integer e, there exists an integer $\hat{x}$ such that $g(\hat{x}) \equiv 0 \pmod{p^e}$, and give an efficient procedure to compute such an $\hat{x}$, given p, g, x, and e.

2. Show that for $g_1,\ldots, g_n \in R[X]$, we have

$$D\left(\prod_i g_i\right) = \sum_i D(g_i) \prod_{j \neq i} g_j$$

and that for $g \in R[X]$, and $n \geq 1$, we have

$$D(g^n) = n g^{n-1} D(g).$$

3. Let $g \in R[X]$ with $\deg(g) = k \geq 0$, and let $x \in R$. Show that if we evaluate g at $X + x$, writing

$$g(X+x) = \sum_{i=0}^{k} b_i X^i ,$$

with $b_0, \ldots, b_k \in R$, then we have

$$i!.\ b_i = (D^i(g))(x) \text{ for } i = 0, \ldots, k.$$

4. Prove the "chain rule" for formal derivatives: if $g, h \in R[X]$ and $f = g(h) \in R[X]$, then $D(f) = D(g)(h).\ D(h)$; more generally, if $g \in R[X_1, \ldots, X_n]$, and $h_1, \ldots, h_n \in R[X]$, and $f = g(h_1, \ldots, h_n) \in R[X]$, then

$$D_X(f) = \sum_{i=1}^{n} D_{X_i}(g)(h_1, \ldots, h_n) D_X(h_i).$$

5. Let $g \in R[X]$, and let $x \in R$ be a root of g. Show that x is a multiple root of g if and only if x is also a root of $D(g)$.

POLYNOMIAL CONGRUENCES

Let us consider F denotes a field. We state these results here – the proofs of these results are essentially the same as in the integer case, and as such, are omitted for the most part.

Theorem: Let $\{f_i\}_{i=1}^k$ be a pairwise relatively prime family of non-zero polynomials in $F[X]$, and let $g_1, \ldots, g_k$ be arbitrary polynomials in $F[X]$. Then there exists a solution $g \in F[X]$ to the system of congruences

$$g \equiv g_i \ (\text{mod } f_i) \ (i = 1, \ldots, k).$$

Moreover, any $g' \in F[X]$ is a solution to this system of congruences if and only if $g \equiv g' \ (\text{mod } f)$, where $f := \prod_{i=1}^k f_i$.

Let us recall the formula for the solution g. We have

$$g := \sum_{i=1}^{k} g_i e_i,$$

where

$$e_i := f_i^* t_i,\ f_i^* := f / f_i,\ t_i := (f_i^*)^{-1} \text{ mod } fi \ (i = 1, \ldots, k).$$

Now, let us consider the special case of the Chinese remainder theorem where $f_i = X - x_i$ with $x_i \in F$, and $g_i = y_i \in F$, for $i = 1, \ldots, k$. The condition that $\{f_i\}_{i=1}^k$ is pairwise relatively prime is equivalent to the condition that the x_i's are distinct. Observe that a polynomial $g \in F[X]$ satisfies the system of congruences

$$g \equiv g_i \ (\text{mod } f_i) \ (i = 1, \ldots, k)$$

if and only if

$$g(x_i) = y_i \; (i = 1,..., k).$$

Moreover, we have $f_i^* = \prod_{j \neq i}(X - x_j)$ and $t_i = 1/\prod_{j \neq i}(x_i - x_j) \in F$. So we get

$$g = \sum_{i=1}^{k} y_i \frac{\prod_{j \neq i}(X - x_j)}{\prod_{j \neq i}(x_i - x_j)}.$$

The reader will recognize this as the usual Lagrange interpolation formula. Thus, the Chinese remainder theorem for polynomials includes Lagrange interpolation as a special case.

Theorem: Let $g, f \in F[X]$ with $f \neq 0$, and let $d := \gcd(g, f)$.

(*i*) For every $h \in F[X]$, the congruence $gz \equiv h \pmod f$ has a solution $z \in F[X]$ if and only if $d \mid h$.

(*ii*) For every $z \in F[X]$, we have $gz \equiv 0 \pmod f$ if and only if $z \equiv 0 \pmod{f/d}$.

(*iii*) For all $z, z' \in F[X]$, we have $gz \equiv gz' \pmod f$ if and only if $z \equiv z' \pmod{f/d}$.

Let $g, f \in F[X]$ with $f \neq 0$. Part (iii) of Theorem gives us a cancellation law for polynomial congruences:

if $\gcd(g, f) = 1$ and $gz \equiv gz' \pmod f$, then $z \equiv z' \pmod f$.

We say that $z \in F[X]$ is a multiplicative inverse of g modulo f if $gz \equiv 1 \pmod f$. Part (i) of Theorem says that g has a multiplicative inverse modulo f if and only if $\gcd(g, f) = 1$. Moreover, part (iii) of Theorem says that the multiplicative inverse of g, if it exists, is uniquely determined modulo f.

As for integers, we may generalize the "mod" operation as follows. Suppose $g, h, f \in F[X]$, with $f \neq 0$, $g \neq 0$, and $\gcd(g, f) = 1$. If s is the rational function $h/g \in F(X)$, then we define $s \bmod f$ to be the unique polynomial $z \in F[X]$ satisfying

$$gz \equiv h \pmod f \text{ and } \deg(z) < \deg(f).$$

With this notation, we can simply write $g^{-1} \bmod f$ to denote the unique multiplicative inverse of g modulo f of degree less than $\deg(f)$.

Theorem: Let $g, f \in F[X]$, where $f \neq 0$. Then there exists a unique $z \in F[X]$ such that $z \equiv g \pmod f$ and $\deg(z) < \deg(f)$, namely, $z := g \bmod f$.

POLYNOMIAL QUOTIENT ALGEBRAS

Letus suppose $f \in F[X]$ be a polynomial of degree $\ell \geq 0$, and consider the quotient ring $E := F[X]/(f)$. As discussed in Example 16.7, we may naturally view E as an F-algebra. Moreover, if we set $\xi := [X]_f \in E$, then $E = F[\xi]$, and viewing E as a vector space over F, we see that $\{\xi^{i-1}\}_{i=1}^{\ell}$ is a basis for E.

Now suppose $\alpha \in E$. We have $\alpha = [g]_f = g(\xi)$ for some $g \in F[X]$, and from the above discussion about polynomial congruences, we see that is a unit if and only if $\gcd(g, f) = 1$.

If $\ell = 0$, then E is trivial. If f is irreducible, then E is a field, since $g \not\equiv 0 \pmod f$ implies $\gcd(g, f) = 1$. If f is reducible, then E is not a field, and indeed, not even an integral domain: for any non-trivial factor $g \in F[X]$ of f, $[g]f \in E$ is a zero divisor.

The Chinese remainder theorem for polynomials also has a more algebraic interpretation. Namely, if $\{f_i\}_{i=1}^k$ is a pairwise relatively prime family of non-zero polynomials in $F[X]$, and $f := \prod_{i=1}^k f_i$, then the map

$$\theta : F[X]/(f) \to F[X]/(f_1) \times \ldots \times F[X]/(fk)$$

$$[g]_f \mapsto ([g]_{f_1}, \ldots, [g]_{f_k})$$

is unambiguously defined, and is in fact an F-algebra isomorphism.

Example

1. The polynomial $X^2 + 1$ is irreducible over $\mathbb{R}$, since if it were not, it would have a root in $\mathbb{R}$, which is clearly impossible, since -1 is not the square of any real number. It follows immediately that $\mathbb{C} = \mathbb{R}[X]/(X^2 + 1)$ is a field, without having to explicitly calculate a formula for the inverse of a nonzero complex number.

2. Consider the polynomial $f := X^4 + X^3 + 1$ over $\mathbb{Z}_2$. We claim that f is irreducible. It suffices to show that f has no irreducible factors of degree 1 or 2.

If f had a factor of degree 1, then it would have a root; however, $f(0) = 0 + 0 + 1 = 1$ and $f(1) = 1 + 1 + 1 = 1$. So f has no factors of degree 1.

Does f have a factor of degree 2? The polynomials of degree 2 are X_2, $X^2 + X$, $X^2 + 1$, and $X^2 + X + 1$. The first and second of these polynomials are divisible by X, and hence not irreducible, while the third has a 1 as a root, and hence is also not irreducible. The last polynomial, $X^2 + X + 1$, has no roots, and hence is the only irreducible polynomial of degree 2 over $\mathbb{Z}_2$. So now we may conclude that if f were not irreducible, it would have to be equal to

$$(X^2 + X + 1)^2 = X^4 + 2X^3 + 3X^2 + 2X + 1 = X^4 + X^2 + 1,$$

which it is not.

Thus, $E := \mathbb{Z}_2[X]/(f)$ is a field with $2^4 = 16$ elements. We may think of elements E as bit strings of length 4, where the rule for addition is bit-wise "exclusiveor." The rule for multiplication is more complicated: to multiply two given bit strings, we interpret the bits as coefficients of polynomials (with the left-most bit the coefficient of X^3), multiply the polynomials, reduce the product modulo f, and write down the bit string corresponding to the reduced product polynomial. For example, to multiply 1001 and 0011, we compute

$$(X^3 + 1)(X + 1) = X^4 + X^3 + X + 1,$$

and

$$(X^4 + X^3 + X + 1) \bmod (X^4 + X^3 + 1) = X.$$

Hence, the product of 1001 and 0011 is 0010.

Theorem says that E^* is a cyclic group. Indeed, the element $\xi := 0010$ (i.e., $\xi = [X]_f$) is a generator for E^*, as the following table of powers shows:

i	ξ^i	i	ξ^i
1	0010	8	1110
2	0100	9	0101
3	1000	10	1010
4	1001	11	1101
5	1011	12	0011
6	1111	13	0110
7	0111	14	1100
		15	0001

Such a table of powers is sometimes useful for computations in small finite fields such as this one. Given $\alpha \in E^*$, we can compute $\alpha\beta$ by obtaining (by table lookup) i, j such that $\alpha = \xi_i$ and $= \alpha_j$, computing $k := (i + j) \bmod 15$, and then obtaining $\alpha\beta = \xi^k$ (again by table lookup).

MINIMAL POLYNOMIALS

Let us consider F denotes a field. Suppose that E is an arbitrary F-algebra, and let a be an element of E. Consider the polynomial evaluation map

$$\rho : F[X] \to E$$
$$g \mapsto g(\alpha),$$

which is an F-algebra homomorphism. By definition, the image of ρ is $F[\alpha]$. The kernel of ρ is an ideal of $F[X]$, and since every ideal of $F[X]$ is principal, it follows that Ker $\rho = \phi F[X]$ for some polynomial $\phi \in F[X]$; moreover, we can make the choice of ϕ unique by insisting that it is monic or zero. The polynomial ϕ is called the minimal polynomial of α (over F).

On the one hand, suppose $\phi \neq 0$. Since any polynomial that is zero at α is a polynomial multiple of ϕ, we see that ϕ is the unique monic polynomial of smallest degree that vanishes at α. Moreover, the first isomorphism theorems for rings and modules tell us that $F[\alpha]$ is isomorphic (as an F-algebra) to $F[X]/(\phi)$, via the isomorphism

$$\bar{\rho}: F[X]/(\phi) \to F[\alpha]$$

$$[g]_\phi \text{ 7! } g(\alpha).$$

Under this isomorphism, $[X]_\phi \in F[X]/(\phi)$ corresponds to $\alpha \in F[\alpha]$, and we see that $\{\alpha^{i-1}\}_{i=1}^{m}$ is a basis for $F[\alpha]$ over F, where $m = \deg(\phi)$. In particular, every element of $F[\alpha]$ can be written uniquely as $\sum_{i=1}^{m} c_i\alpha^{i-1}$, where $c_1,..., c_m \in F$.

On the other hand, suppose $\phi = 0$. This means that no non-zero polynomial vanishes at α. Also, it means that the map ρ is injective, and hence $F[\alpha]$ is isomorphic (as an F-algebra) to $F[X]$; in particular, $F[\alpha]$ is not finitely generated as a vector space over F.

Note that if $\alpha \in E$ has a minimal polynomial $\phi \neq 0$, then $\deg(\phi) > 0$, unless E is trivial (i.e., $1_E = 0_E$), in which case $\phi = 1$.

Example.

1. Consider the real numbers $\sqrt{2}$ and $\sqrt[3]{2}$.

We claim that $X^2 - 2$ is the minimal polynomial of $\sqrt{2}$ over $\mathbb{Q}$. To see this, first observe that $\sqrt{2}$ is a root of $X^2 - 2$. Thus, the minimal polynomial of $\sqrt{2}$ divides $X^2 - 2$. However, as we saw in Example, the polynomial $X^2 - 2$ is irreducible over $\mathbb{Q}$, and hence must be equal to the minimal polynomial of $\sqrt{2}$ over $\mathbb{Q}$.

A similar argument shows that X3 $-\in$is the minimal polynomial of $\sqrt[3]{2}$ over $\mathbb{Q}$.

We also see that $\mathbb{Q}[\sqrt{2}]$ is isomorphic (as a $\mathbb{Q}$-algebra) to $\mathbb{Q}[X]/(X^2 - 2)$, and since $X^2 - 2$ is irreducible, it follows that the ring $\mathbb{Q}[\sqrt{2}]$ is actually a field. As a vector space over $\mathbb{Q}$, $\mathbb{Q}[\sqrt{2}]$ has dimension 2, and every element of $\mathbb{Q}[\sqrt{2}]$ may be written uniquely as $a + b\sqrt{2}$ for $a, b \in \mathbb{Q}$. Indeed, for all $a, b \in \mathbb{Q}$, not both zero, the multiplicative inverse of $a + b\sqrt{2}$ is $(a/c) + (b/c)\sqrt{2}$, where $c := a^2 - 2b^2$.

Similarly, $\mathbb{Q}[\sqrt[3]{2}]$ is a field and has dimension 3 as a vector space over $\mathbb{Q}$, and every element of $\mathbb{Q}[\sqrt[3]{2}]$ may be written uniquely as $a + b\sqrt[3]{2} + c\sqrt[3]{4}$ for $a, b, c \in \mathbb{Q}$.

A simple but important fact is the following:

Theorem: Suppose E is an F-algebra, and that as an F-vector space, E has finite dimension n. Then every $\alpha \in E$ has a non-zero minimal polynomial of degree at most n.

Proof: Indeed, the family of elements

$$1_E, \alpha, ..., \alpha^n$$

must be linearly dependent (as must any family of $n + 1$ elements of a vector space of dimension n), and hence there exist $c_0, ..., cn \in F$, not all zero, such that

$$c_0 1_E + c_1\alpha + ... + c_n\alpha_n = 0_E,$$

and therefore, the non-zero polynomial $f := \sum_i c_i X^i$ vanishes at α.

Example. Let $f \in F[X]$ be a monic polynomial of degree ℓ, and consider the F-algebra $E := F[X]/(f) = F[\xi]$, where $\xi := [X]_f \in E$. Clearly, the minimal polynomial of ξ over F is f. Moreover, as a vector space over F, E has dimension ℓ, with $\{\xi^{i-1}\}_{i=1}^{\ell}$ being a basis. Therefore, every $\alpha \in E$ has a non-zero minimal polynomial of degree at most ℓ.

EXERCISE

1. Show that if the factorization of f over $F[X]$ into monic irreducibles is $f = f_1^{e_1} ... f_r^{e_r}$, and if a = $[h]_f$ Ξ $F[X]/(f)$, then the minimal polynomial f of a over F is lcm(f_1,..., f_r), where each f_i is the minimal polynomial of $[h]_{f_i^{e_i}}$ Ξ $F[X]/(f_i^{e_i})$ over F.
2. Let r : E ® $EÄ$ be an F-algebra homomorphism, let a Ξ E, let f be the minimal polynomial of over F, and let fÄ be the minimal polynomial of r(a) over F. Show that fÄ | fÄ , and that fÄ = f if is injective.

FACTORIZATION OF POLYNOMIALS

Let us suppose F denotes a field. Like the ring $\mathbb{Z}$, the ring $F[X]$ of polynomials is an integral domain, and because of the division with remainder property for polynomials, $F[X]$ has many other properties in common with $\mathbb{Z}$. Recall that the units of $F[X]$ are precisely the units F^* of F, that is, the nonzero constants. We call two polynomials $g, h \in F[X]$ associate if $g = ch$ for some $c \in F^*$. It is easy to see that g and h are associate if and only if $g \mid h$ and $h \mid g$ – indeed, this follows as a special case of part (i) of Theorem.

Clearly, any nonzero polynomial g is associate to a unique monic polynomial (i.e., a polynomial with leading coefficient 1), called the monic associate of g; indeed, the monic associate of g is $\text{lc}(g)^{-1}.g$ (where, as usual, lc(g) denotes the leading coefficient of g). We called polynomial $f \in F[X]$ irreducible if it is non-constant and all divisors of f are associate to 1 or f. Conversely, we call f reducible if it is non-constant and is not irreducible.

Equivalently, a non-constant polynomial f is reducible if and only if there exist polynomials $g, h \in F[X]$ of degree strictly less than that of f such that $f = gh$. Clearly, if g and h are associate polynomials, then g is irreducible if and only if h is irreducible. The irreducible polynomials play a role similar to that of the prime numbers. Just as it is convenient to work with only positive prime numbers, it is also convenient to restrict attention to monic irreducible polynomials.

Theorem: Let us suppose $p \in F[X]$ be irreducible, and let $g, h \in F[X]$. Then $p \mid gh$ implies that $p \mid g$ or $p \mid h$.

Proof: Assume that $p \mid gh$. The only divisors of p are associate to 1 or p. Thus, gcd(p, g) is either 1 or the monic associate of p. If $p \mid g$, we are done; otherwise, if $p \mid g$, we must have gcd(p, g) = 1, and by the previous theorem, we conclude that $p \mid h$.

Now to prove the uniqueness part of Theorem. Suppose we have

$$p_1 \cdots p_r = q_1 \cdots q_s,$$

where $p_1,..., p_r$ and $q_1,..., q_s$ are monic irreducible polynomials (with duplicates allowed among the p_i's and among the q_j's). If $r = 0$, we must have $s = 0$ and we are done. Otherwise, as p_1 divides the right-hand side, by inductively applying Theorem, one sees that p_1 is equal to q_j for some j. We can cancel these terms and proceed inductively (on r).

Theorem: Every non-zero polynomial $f \in F[X]$ can be expressed as

$$f = c.p_1^{e_1} \ldots p_r^{e_r},$$

where $c \in F^*$, $p_1,..., p_r$ are distinct monic irreducible polynomials, and $e_1,..., e_r$ are positive integers. Moreover, this expression is unique, up to a reordering of the irreducible polynomials.

To prove this theorem, we may assume that f is monic, since the non-monic case trivially reduces to the monic case. If f is 1 or a monic irreducible, we are done. Otherwise, there exist $g, h \in F[X]$ of degree strictly less than that of f such that $f = gh$, and again, we may assume that g and h are monic. By induction on degree, both g and h can be expressed as a product of monic irreducible polynomials, and hence, so can f.

Theorem: For all $g, h \in F[X]$, there exists a unique greatest common divisor d of g and h, and moreover, $gF[X] + hF[X] = dF[X]$.

Proof: We apply the previous theorem to the ideal $I := gF[X] + hF[X]$. Let $d \in F[X]$ with $I = dF[X]$, as in that theorem. Note that $g, h, d \in I$ and d is monic or zero. It is clear that d is a common divisor of g and h. Moreover, there exist $s, t \in F[X]$ such that $gs + ht = d$. If $d' \mid g$ and $d' \mid h$, then clearly $d' \mid (gs + ht)$, and hence $d' \mid d$.

Finally, for uniqueness, if e is a greatest common divisor of g and h, then $d \mid e$ and $e \mid d$, and hence e is associate to d, and the requirement that e is monic or zero implies that $e = d$. For $g, h \in F[X]$, we denote by gcd(g, h) the greatest common divisor of g and h. Note that as we have defined it, lc(g) gcd(g, 0) = g. Also note that when at least one of g or h are non-zero, gcd(g, h) is the unique monic polynomial of maximal degree that divides both g and h. An immediate consequence of Theorem is that for all $g, h \in F[X]$, there exist $s, t \in F[X]$ such that $gs + ht$ = gcd(g, h), and that when at least one of g or h are non-zero, gcd(g, h) is the unique monic polynomial of minimal degree that can be expressed as $gs +$

ht for some $s, t \in F[X]$.

We say that $g, h \in F[X]$ are relatively prime if $\gcd(g, h) = 1$, which is the same as saying that the only common divisors of g and h are units. It is immediate from Theorem that g and h are relatively prime if and only if $gF[X] + hF[X] = F[X]$, which holds if and only if there exist $s, t \in F[X]$ such that $gs + ht = 1$.

Theorem: Let us suppose that I be an ideal of $F[X]$. Then there exists a unique polynomial $d \in F[X]$ such that $I = dF[X]$ and d is either zero or monic.

Proof: We first prove the existence part of the theorem. If $I = \{0\}$, then $d = 0$ does the job, so let us assume that $I \neq \{0\}$. Since I contains non-zero polynomials, it must contain monic polynomials, since if g is a non-zero polynomial in I, then its monic associate $\mathrm{lc}(g)^{-1} g$ is also in I. Let d be a monic polynomial of minimal degree in I. We want to show that $I = dF[X]$. We first show that $I \subseteq dF[X]$. To this end, let g be any element in I.

It suffices to show that $d \mid g$. Using Theorem, we may write $g = dq + r$, where $\deg(r) < \deg(d)$. Then by the closure properties of ideals, one sees that $r = g - dq$ is also an element of I, and by the minimality of the degree of d, we must have $r = 0$. Thus, $d \mid g$. We next show that $dF[X] \subseteq I$. This follows immediately from the fact that $d \in I$ and the closure properties of ideals. That proves the existence part of the theorem. As for uniqueness, note that if $dF[X] = eF[X]$, we have $d \mid e$ and $e \mid d$, from which it follows that d and e are associate, and so if d and e are both either monic or zero, they must be equal.

For $g, h \in F[X]$, we call $d \in F[X]$ a common divisor of g and h if $d \mid g$ and $d \mid h$; moreover, we call such a d a greatest common divisor of g and h if d is monic or zero, and all other common divisors of g and h divide d.

Theorem: For $f, g, h \in F[X]$ such that f j gh and $\gcd(f, g) = 1$, we have $f \mid h$.

Proof: Suppose that $f \mid gh$ and $\gcd(f, g) = 1$. Then since $\gcd(f, g) = 1$, by Theorem we have $fs + gt = 1$ for some $s, t \in F[X]$. Multiplying this equation by h, we obtain $fhs + ght = h$. Since $f \mid f$ by definition, and $f \mid gh$ by hypothesis, it follows that $f \mid h$.

Theorem: There are infinitely many monic irreducible polynomials in $F[X]$.

If F is infinite, then this theorem is true simply because there are infinitely many monic, linear polynomials; in any case, one can easily prove this theorem by mimicking the proof of Theorem.

For a monic irreducible polynomial p, we may define the function v_p, mapping non-zero polynomials to non-negative integers, as follows: for every polynomial $f \neq 0$, if $f = p^e g$, where $p \mid g$, then $v_p(f) := e$. We may then write the factorization of f into irreducibles as

$$f = c \prod_p p^{v_p(f)},$$

where the product is over all monic irreducible polynomials p, with all but finitely many of the terms in the product equal to 1.

Just as for integers, we may extend the domain of definition of v_p to include 0, defining $v_p(0) := \infty$. For all polynomials g, h, we have

$$v_p(g \,.\, h) = v_p(g) + v_p(h) \text{ for all } p.$$

From this, it follows that for all polynomials g, h, we have

$$h \mid g \Leftrightarrow v_p(h) \leq v_p(g) \text{ for all } p,$$

and

$$v_p(\gcd(g, h)) = \min(v_p(g), v_p(h)) \text{ for all } p.$$

For $g, h \in F[X]$, a common multiple of g and h is a polynomial m such that $g \mid m$ and $h \mid m$; moreover, such an m is the least common multiple of g and h if m is monic or zero, and m divides all common multiples of g and h. In light of Theorem, it is clear that the least common multiple exists and is unique, and we denote the least common multiple of g and h by lcm(a, b).

Note that as we have defined it, lcm(g, 0) = 0, and that when both g and h are non-zero, lcm(g, h) is the unique monic polynomial of minimal degree that is divisible by both g and h. Also, for all $g, h \in F[X]$, we have

$$v_p(\text{lcm}(g, h)) = \max(v_p(g), v_p(h)) \text{ for all } p.$$

The notions of greatest common divisor and least common multiple generalize naturally from two to any number of polynomials. We also say that a family of polynomials $\{g_i\}_{i=1}^{k}$ is pairwise relatively prime if $\gcd(g_i, g_j) = 1$ for all indices i, j with $i \neq j$.

Any rational function $g/h \in F(X)$ can be expressed as a fraction g_0/h_0 in lowest terms, that is, $g/h = g_0/h_0$ and $\gcd(g_0, h_0) = 1$, and this representation is unique up to multiplication by units.

GENERAL PROPERTIES OF EXTENSION FIELDS

We now discuss a few general notions related to extension fields. These are all quite simple applications of the theory developed so far. Recall that if F and E are fields, with F being a subring of E, then F is called a subfield of E, and E is called an extension field of F. As usual, we shall blur the distinction between a subring and a natural embedding; that is, if $\tau : F \to E$ is a natural embedding, we shall simply identify elements of F with their images in E under τ, and in so doing, we may view E as an extension field of F.

Usually, the map τ will be clear from context; for example, if $E = F[X]/(f)$ for some irreducible polynomial $f \in F[X]$, then we shall simply say that E is an extension field of F, although strictly speaking, F is embedded in E via the map that sends $c \in F$ to $[c]_f \in E$. We start with some definitions. Let E be an extension field of a field F.

Then E is an F-algebra via inclusion, and in particular, an F-vector space. If E is a finite dimensional F-vector space, then we say that E is a finite extension of F, and $\dim_F(E)$ is called the degree (over F) of the extension, and is denoted $(E : F)$; otherwise, we say that E is an infinite extension of F. An element $\alpha \in E$ is called algebraic over F if there exists a non-zero polynomial $g \in F[X]$ such that $g(\alpha) = 0$, and in this case, we define the degree of α (over F) to be the degree of its minimal polynomial over F; otherwise, α is called transcendental over F.

If all elements of E are algebraic over F, then we call E an algebraic extension of F. Suppose E is an extension field of a field F. For $\alpha \in E$, we define

$$F(\alpha) := \{g(\alpha)/h(\alpha) : g, h \in F[X], h(\alpha) \neq 0\}.$$

It is easy to see that $F(\alpha)$ is a subfield of E, and indeed, it is the smallest subfield of E containing F and α. Clearly, the ring $F[\alpha] = \{g(\alpha): g \in F[X]\}$, which is the smallest subring of E containing F and α, is a subring of $F(\alpha)$. We derive some basic properties of $F(\alpha)$ and $F[\alpha]$. The analysis naturally breaks down into two cases, depending on whether α is algebraic or transcendental over F.

On the one hand, suppose is algebraic over F. Let ϕ be the minimal polynomial of over F, so that $\deg(\phi) > 0$, and the quotient ring $F[X]/(\phi)$ is isomorphic (as an F-algebra) to the ring $F[\phi]$. Since $F[\phi]$ is a subring of a field, it must be an integral domain, which implies that $F[X]/(\phi)$ is an integral domain, and so is irreducible.

This in turn implies that $F[X]/(\phi)$ is a field, and so $F[\alpha]$ is not just a subring of E, it is a subfield of E. Since $F[\alpha]$ is itself already a subfield of E containing F and α, it follows that $F(\alpha) = F[\alpha]$. Moreover, $F[\alpha]$ is a finite extension of F; indeed $(F[\alpha]: F) = \deg(\phi) =$ the degree of α over F, and the elements $1, \alpha, \ldots, \alpha^{m-1}$, where $m := \deg(\phi)$, form a basis for $F[\alpha]$ over F.

On the other hand, suppose that α is transcendental over F. In this case, the minimal polynomial of α over F is the zero polynomial, and the ring $F[\alpha]$ is isomorphic (as an F-algebra) to the ring $F[X]$, which is definitely not a field. But consider the "rational function evaluation map" that sends $g/h \in F(X)$ to $g(\alpha)/h(\alpha) \in F(\alpha)$.

Since no non-zero polynomial over F vanishes at α, it is easy to see that this map is well defined, and is in fact an F-algebra isomorphism. Thus, we see that $F(\alpha)$ is isomorphic (as an F-algebra) to $F(X)$. It is also clear that $F(\alpha)$ is an infinite extension of F.

Theorem: If E is a finite extension of a field F, and K is a subfield of E containing F, then E is a finite extension of K, K is a finite extension of F, and $(E : F) = (E : K)(K : F)$.

Again, suppose that E is a finite extension of a field F. Theorem implies that E is algebraic over F, and indeed, that each element of E has degree over F bounded by $(E : F)$. However, we can say a bit more about these degrees. Suppose $\alpha \in E$. Then the degree of α over F is equal to $(F[\alpha]: F)$, and by the previous theorem, applied to $K := F[\alpha]$, we have $(E : F) = (E : F[\alpha])(F[\alpha]: F)$. In particular, the degree of α over F divides $(E : F)$.

Theorem: Suppose E is a finite extension of a field K, with a basis $\{\beta_j\}_{j=1}^m$ over K, and K is a finite extension of F, with a basis $\{\alpha_i\}_{i=1}^n$ over F. Then the elements

$$\alpha_i \beta_j \ (i = 1,..., n; j = 1,..., m)$$

form a basis for E over F. In particular, E is a finite extension of F and

$$(E : F) = (E : K)(K : F).$$

Now suppose that E is a finite extension of a field F. Let K be an intermediate field, that is, a subfield of E containing F. Then evidently, E is a finite extension of K (since any basis for E over F also spans E over K), and K is a finite extension of F (since as F-vector spaces, K is a subspace of E). The previous theorem then implies that $(E : F) = (E : K)(K : F)$.

Theorem: Let F be a field, and $f \in F[X]$ a non-zero polynomial of degree n. Then there exists a finite extension E of F over which f factors as

$$f = c(X - \alpha_1)(X - \alpha_2)... (X - \alpha_n),$$

where $c \in F$ and $\alpha_1,..., \alpha_n \in E$.

Proof: We may assume that f is monic. We prove the existence of E by induction on the degree n of f. If $n = 0$, then the theorem is trivially true. Otherwise, let h be an irreducible factor of f, and set $K := F[X]/(h)$, so that $\xi := [X]h \in K$ is a root of h, and hence of f. So over K, which is a finite extension of F, the polynomial f factors as

$$f = (X - \xi)g,$$

where $g \in K[X]$ is a monic polynomial of degree $n - 1$. Applying the induction hypothesis, there exists a finite extension E of K over which g splits into linear factors. Thus, over E, f splits into linear factors, and by Theorem, E is a finite extension of F.

Theorem: If E is a finite extension of a field F, then it is an algebraic extension, and for each $\alpha \in E$, the degree of α over F divides $(E : F)$.

Theorem: Let E be an extension field of a field F.

(*i*) If $\alpha \in E$ is transcendental over F, then $F(\alpha)$ is isomorphic (as an F-algebra) to the rational function field $F(X)$, while the subring $F[\alpha]$ is isomorphic (as an F-algebra) to the ring of polynomials $F[X]$; moreover, $F(\alpha)$ is an infinite extension of F.

(*ii*) If $\alpha \in E$ is algebraic over F, then $F(\alpha) = F[\alpha]$, and $F[\alpha]$ is isomorphic (as an F-algebra) to $F[X]/(\phi)$, where ϕ is the minimal polynomial of α over F, which is irreducible; moreover, $F[\phi]$ is a finite extension of F, and $(F[\alpha]: F) = \deg(\phi)$ = the degree of α over F, and the elements $1, \alpha,..., \alpha^{m-1}$, where $m := \deg(\phi)$, form a basis for $F[\alpha]$ over F.

Suppose E is an extension field of a field K, which itself is an extension of a field F. Then E is also an extension field of F. The following theorem examines the relation between the degrees of these extensions, in the case where E is a finite extension of K, and K is a finite extension of F. The proof is a simple calculation, which we leave to the reader to verify.

Example.

1. Suppose $f \in F[X]$ is a monic irreducible polynomial of degree ℓ, so that $E := F[X]/(f) = F[\xi]$, where $\xi := [X]_f \in E$, is an extension field of F. The element ξ is algebraic of degree ℓ over F. Moreover, E is a finite extension of F, with $(E : F) = \ell$; in particular, E is an algebraic extension of F, and for each $\alpha \in E$, the degree of over F divides ℓ.

2. We see that the real numbers $\sqrt{2}$ and $\sqrt[3]{2}$ are algebraic over $\mathbb{Q}$. The fields $\mathbb{Q}[\sqrt{2}]$ and $\mathbb{Q}[\sqrt[3]{2}]$ are extension fields of $\mathbb{Q}$, where $(\mathbb{Q}[\sqrt{2}]:\mathbb{Q}) = 2$ = the degree of $\sqrt{2}$ over $\mathbb{Q}$, and $(\mathbb{Q}[\sqrt[3]{2}]:\mathbb{Q}) = 3$ = the degree of $\sqrt[3]{2}$ over $\mathbb{Q}$. As both of these fields are finite extensions of $\mathbb{Q}$, they are algebraic extensions as well. Since their degrees over $\mathbb{Q}$ are prime numbers, it follows that they have no subfields other than themselves and $\mathbb{Q}$. In particular, if $\alpha \in \mathbb{Q}[\sqrt{2}] \setminus \mathbb{Q}$, then $\mathbb{Q}[\alpha] = \mathbb{Q}[\sqrt{2}]$. Similarly, if $\in \mathbb{Q}[\sqrt[3]{2}] \setminus \mathbb{Q}$, then $\mathbb{Q}[\alpha] = \mathbb{Q}[\sqrt[3]{2}]$.

EXERCISE

1. Consider a field F and its field of rational functions $F(X)$. Let $\alpha \in F(X) \setminus F$. Show that X is algebraic over $F(\beta)$, and that is transcendental over F.

2. Consider the real numbers p2 and p3. Show that

$$(\mathbb{Q}[\sqrt{2},\sqrt{3}]:\mathbb{Q}) = (\mathbb{Q}\ \sqrt{2}\,[+\sqrt{3}]:\mathbb{Q}) = 4.$$

3. Let E be an extension field of a field F. Suppose $\alpha \in E$ is transcendental over F, and that E is algebraic over $F(\alpha)$. Show that for every $\beta \in E$, β is transcendental over F if and only if E is algebraic over $F(\beta)$.

4. Let E be an extension field of a field F, and let $\alpha_1, \ldots, n \in E$ be algebraic over F. Show that the ring $F[\alpha_1, \ldots, \alpha_n]$ is in fact a field, and that $F[\alpha_1, \ldots, \alpha_n]$ is a finite (and hence algebraic) extension of F.

5. Consider the real numbers $\sqrt{2}$ and $\sqrt[3]{2}$. Show that

$$(\mathbb{Q}[\sqrt{2},\sqrt[3]{2}]:\mathbb{Q}) = (\mathbb{Q}\ \sqrt{2}\,[+\sqrt[3]{2}]:\mathbb{Q}) = 6.$$

6. Let E be an extension of F. Show that the set of all elements of E that are algebraic over F is a subfield of E containing F.

7. Show that if E is an algebraic extension of K, and K is an algebraic extension of F, then E is an algebraic extension of F.

FORMAL POWER SERIES AND LAURENT SERIES

Earlier we have discussed generalizations of polynomials that allow an infinite number of nonzero coefficients. Although we are mainly interested in the case where the coefficients come from a field F, we develop the basic theory for general rings R.

REVERSED LAURENT SERIES

While formal Laurent series are useful in some situations, in many others, it is more useful and natural to consider reversed Laurent series over R. These are formal expressions of the form

$$g = \sum_{i=-\infty}^{m} a_i X^i ,$$

where $a_m, a_{m-1}, \ldots \in R$. Thus, in a reversed Laurent series, we allow an infinite number of terms involving negative powers of X, but only a finite number of terms involving positive powers of X. Formally, such a reversed Laurent series is a doubly infinite sequence $\{a_i\}_{i=-\infty}^{\infty}$, with the restriction that for some integer m, we have $a_i = 0$ for all $i > m$. We may again use the usual formulas to define addition and multiplication – and again, only finitely many non-zero terms.

The ring of all reversed Laurent series is denoted $R((X^{-1}))$, and as the notation suggests, the map that sends X to X^{-1} (and acts as the identity on R) is an R-algebra isomorphism of $R((X))$ with $R((X^{-1}))$. Also, one may naturally view R[X] as a subring of $R((X^{-1}))$.

For $g = \sum_{i=-\infty}^{m} a_i X^i \in R((X^{-1}))$ with $a_m \neq 0$, let us define the degree of g, denoted deg(g), to be the value m, and the leading coefficient of g, denoted lc(g), to be the value a_m. As for ordinary polynomials, we define the degree of 0 to be $-\infty$, and the leading coefficient of 0 to be 0. Note that if g happens to be a polynomial, then these definitions of degree and leading coefficient agree with that for ordinary polynomials.

Theorem: Let $g, h \in R[X]$ with $h \neq 0$ and lc(h) $\in R^*$, and using the usual division with remainder property for polynomials, write $g = hq + r$, where $q, r \in R[X]$ with deg(r) < deg(h). Let h^{-1} denote the multiplicative inverse of h in $R((X^{-1}))$. Then $q = \lfloor gh^{-1} \rfloor$.

Proof: Multiplying the equation $g = hq + r$ by h^{-1}, we obtain $gh^{-1} = q + rh^{-1}$, and deg (rh^{-1}) < 0, from which it follows that $\lfloor gh^{-1} \rfloor = q$.

Let F be a field, so that $F((X^{-1}))$ is also field. Now, $F((X^{-1}))$ contains $F[X]$ as a subring, and hence contains (an isomorphic copy of) the rational function field $F(X)$. Just as $F(X)$ corresponds to the field of rational numbers, $F((X^{-1}))$ corresponds to the field real numbers.

Theorem: For $g, h \in R((X^{-1}))$, we have $\deg(gh) \leq \deg(g) + \deg(h)$, where equality holds unless both $\mathrm{lc}(g)$ and $\mathrm{lc}(h)$ are zero divisors. Furthermore, if $h \neq 0$ and $\mathrm{lc}(h)$ is a unit, then h is a unit, and we have $\deg(gh^{-1}) = \deg(g) - \deg(h)$.

It is also natural to define a floor function for reversed Laurent series: for $g \in R((X^{-1}))$ with $g = \sum_{i=-\infty}^{m} a_i X^i$, we define

$$\lfloor g \rfloor := \sum_{i=0}^{m} a_i X^i \in R[X]$$

that is, we compute the floor function by simply throwing away all terms involving negative powers of X.

EXERCISE

1. Let R be a ring, and let $a \Xi R$. Show that the multiplicative inverse of $X - a$ in $R((X^{-1}))$ is $\sum_{j=1}^{\infty} a^{j-1} X^{-j}$.
2. Let R be an arbitrary ring, let $a_1, \ldots, a_\ell \in R$, and let

   $$f := (X - a_1)(X - a_2) \ldots (X - a_\ell) \in R[X].$$

   For $j \geq 0$, define the "power sum"

   $$s_j := \sum_{i=1}^{\ell} a_i^j.$$

   Show that in the ring $R((X^{-1}))$, we have

   $$\frac{D(f)}{f} = \sum_{i=1}^{\ell} \frac{1}{(X - a_i)} = \sum_{j=1}^{\infty} s_{j-1} X^{-j},$$

   where $D(f)$ is the formal derivative of f.
3. Continuing with the previous exercise, derive Newton's identities, which state that if $f = X^\ell + c_1 X^{\ell-1} + \ldots + c_\ell$, with $c_1, \ldots, c_\ell \in R$, then

   $$s_1 + c_1 = 0$$
   $$s_2 + c_1 s_1 + 2c_2 = 0$$
   $$s_3 + c_1 s_2 + c_2 s_1 + 3c_3 = 0$$
   $$\vdots$$
   $$s_\ell + c_1 s_{\ell-1} + \ldots + c_{\ell-1} s_1 + \ell c_\ell = 0$$
   $$s_{j+\ell} + c_1 s_{j+\ell-1} + \ldots + c_{\ell-1} s_{j+1} + c_\ell s_j = 0 \ (j \geq 1).$$

4. Write down the rule for determining the multiplicative inverse of an element of $R((X^{-1}))$ whose leading coefficient is a unit in R.
5. Let F be a field of characteristic other than 2. Show that a non-zero $g \in F((X^{-1}))$ has a square-root in $F((X^{-1}))$ if and only if deg(g) is even and lc(g) has a square-root in F.

FORMAL POWER SERIES

The ring $R[[X]]$ of formal power series over R consists of all formal expressions of the form

$$g = a_0 + a_1X + a_2X^2 + ...,$$

where $a_0, a_1, a_2,... \in R$. Unlike ordinary polynomials, we allow an infinite number of non-zero coefficients. We may write such a formal power series as

$$g = \sum_{i=0}^{\infty} a_i X^i.$$

Formally, such a formal power series is an infinite sequence $\{a_i\}_{i=0}^{\infty}$, and the rules for addition and multiplication are exactly the same as for polynomials. Indeed, the formulas for addition and multiplication may be applied directly – all of the relevant sums are finite, and so everything is well defined. We leave it to the reader to verify that with addition and multiplication so defined, $R[[X]]$ indeed forms a ring. We shall not attempt to interpret a formal power series as a function, and therefore, "convergence" issues shall simply not arise.

Clearly, $R[[X]]$ contains $R[X]$ as a subring. Let us consider the group of units of $R[[X]]$.

Theorem: Let $g = \Sigma_{i=0}^{\infty} a_i X^i \in R[[X]]$. Then $g \in (R[[X]])^*$ if and only if $a_0 \in R^*$.

Proof: If a_0 is not a unit, then it is clear that g is not a unit, since the constant term of a product of formal power series is equal to the product of the constant terms.

Conversely, if a_0 is a unit, we show how to define the coefficients of the inverse $h = \Sigma_{i=0}^{\infty} b_i X^i$ of g. Let $f = gh = \Sigma_{i=0}^{\infty} c_i X^i$. We want $f = 1$, which means that $c_0 = 1$ and $c_i = 0$ for all $i > 0$. Now, $c_0 = a_0 b_0$, so we set $b_0 := a_0^{-1}$. Next, we have $c_1 = a_0 b_1 + a_1 b_0$, so we set $b_1 := -a_1 b_0 . a_0^{-1}$. Next, we have $c_2 = a_0 b_2 + a_1 b_1 + a_2 b_0$, so we set $b_2 := -(a_1 b_1 + a_2 b_0). a_0^{-1}$. Continuing in this way, we see that if we define $b_i := -(a_1 b_{i-1} + ... + a_i b_0). a_0^{-1}$ for $i \geq 1$, then $gh = 1$.

EXERCISE

1. Let F be a field. Show that every non-zero ideal of $F[[X]]$ is of the form (X^m) for some uniquely determined integer $m \geq 0$.
2. In the ring $R[[X]]$, the multiplicative inverse of $1 - X$ is $\sum_{i=0}^{\infty} X^i$.

FORMAL LAURENT SERIES

One may generalize formal power series to allow a finite number of negative powers of X. The ring $R((X))$ of formal Laurent series over R consists of all formal expressions of the form

$$g = a_m X^m + a_{m+1} X^{m+1} + \dots,$$

where m is allowed to be any integer (possibly negative), and $a_m, a_{m+1}, \dots \in R$.

Thus, elements of $R((X))$ may have an infinite number of terms involving positive powers of X, but only a finite number of terms involving negative powers of X. We may write such a formal Laurent series as

$$g = \sum_{i=m}^{\infty} a_i X^i$$

Formally, such a formal Laurent series is a doubly infinite sequence $\{a_i\}_{i=-\infty}^{\infty}$, with the restriction that for some integer m, we have $a_i = 0$ for all $i < m$. We may again use the usual formulas to define addition and multiplication (where the indices i, j, and k now range over all integers, not just the non-negative integers). Note that while the sum in has an infinite number of terms, only finitely many of them are non-zero.

One may naturally view $R[[X]]$ as a subring of $R((X))$, and of course, $R[X]$ is a subring of $R[[X]]$ and so also a subring of $R((X))$.

Theorem: If F is a field, then $F((X))$ is a field.

Theorem: If D is an integral domain, then $D((X))$ is an integral domain.

Proof: Let $g = \sum_{i=m+n}^{\infty} c_i X^i$ and $h = \sum_{i=n}^{\infty} b_i X^i$, where $a_m \neq 0$ and $b_n \neq 0$. Then $gh = \sum_{i=m+n}^{\infty} c_i X^i$, where $c_{m+n} = a_m b_n \neq 0$.

Theorem: Let $g \in R((X))$, and suppose that $g \neq 0$ and $g = \sum_{i=m}^{\infty} a_i X^i$ with $a_m \in R^*$. Then g has a multiplicative inverse in $R((X))$.

Proof: We can write $g = X^m g'$, where g' is a formal power series whose constant term is a unit, and hence there is a formal power series h such that $g'h = 1$. Thus, $X^{-m} h$ is the multiplicative inverse of g in $R((X))$.

EXERCISE

Let F be a field. Show that $F((X))$ is the field of fractions of $F[[X]]$; that is, there is no proper subfield $E \subseteq F((X))$ that contains $F[[X]]$.

Chapter 5

Finite Fields Arithmetic

Finite fields are the general starting point for the constructions of many combinatorial structures. It will be important to know the fundamentals concerning these fields in order to investigate combinatorial structures and related areas of combinatorial interest. Unfortunately, the area of field theory is rather large and it would be impossible for us to cover it in detail and still have time to work with the results.

In the interest of conserving time, we will present the elements of general field theory without proofs and only prove statements when we turn our attention specifically to finite fields. The material that we will gloss over is usually covered in a graduate (or good undergraduate) course in abstract algebra.

It will be assumed that you are familiar with the definition of a field, the definition and basic properties of vector spaces and the fact that the integers modulo a prime p form a field (a finite one), which we will denote by GF(p) (standing for the Galois Field of order p).

GENERAL FIELD THEORY

Let L be a field containing a subset K, which is itself a field under the operations inherited from L. Then L is called an *extension* of K, and K a *subfield* of L. Every field has a smallest subfield, called the *prime subfield*, which is isomorphic to either the field of rationals Q, in which case we say that it has *characteristic zero*, or to a GF(p) for some prime p, in which case we say that it has *characteristic p*. We shall denote the characteristic of an arbitrary field K by char K.

If L is an extension of K (denoted L > K), and þ is an element of L but not an element of K, then the smallest field containing both K and þ will be denoted by K(þ) and is a subfield of L. Similarly, the smallest field containing K and the elements $þ_1,þ_2,\dots,þ_n$ in L > K will be written as $K(þ_1,þ_2,\dots,þ_n)$.

Any extension field L of K can be viewed as a vector space over K and the dimension of this vector space is called the *degree* of L over K, and is denoted by [L:K]. If the vector space is finite dimensional we say that L is a *finite extension* of K. If L is a finite extension of K and M is a finite extension of L, then [M:K] = [M:L][L:K].

Denote by K[x] the ring of polynomials over K in the variable x. K[x] is a principal ideal domain (an ideal is a subring that is closed under multiplication, a principal ideal is an ideal generated by a single element, and a principal ideal domain is a commutative ring with unity all of whose ideals are principal). A polynomial in K[x] is said to be *monic* if the coefficient of the highest power in x is unity. It is *irreducible* if it is not the product of two nonscalar polynomials in K[x].

If f(x) is an irreducible polynomial in K[x], then any zero (i.e., root) of f(x) is not in K and so there is a smallest extension field L of K that contains it. Furthermore, L is isomorphic to the quotient field K[x]/<f(x)>, where <f(x)> denotes the principal ideal of K[x] generated by f(x). If the irreducible polynomial f(x) is of degree n, then [L:K] = n. Furthermore, if þ is the zero of f(x) in question, then L = K(þ) and the elements $1, þ, þ^2, \ldots, þ^{n-1}$ form a basis of L over K.

Before we continue, let us illustrate these ideas with a well-known example. The field C of complex numbers is usually described in one of two ways, either as the set of numbers {a + bi} where a and b are real numbers and i = sqrt(-1), or if you prefer not to mention the imaginary number i, C can be described as the set of all pairs (a,b) of real numbers where addition of two pairs is the usual componentwise addition and multiplication of two pairs is defined by (a,b)(c,d) = (ac - bd, ad + bc).

The second version is of course viewing C as a vector space of dimension 2 over the real numbers R. Let us see how we would construct C starting with the subfield R. Now K = R and K[x] is the ring of all polynomials with real coefficients. The polynomial $x^2 + 1$ is monic since the coefficient of x^2 is 1 and it is irreducible over R since it cannot be factored into polynomials with only real coefficients.

The principal ideal <x^2+1> consists of all polynomials that have x^2+1 as a factor. The quotient field structure, R[x]/<x^2+1> is obtained by taking each polynomial in R[x] and dividing it by x^2+1. The polynomials that have the same remainder after division form equivalence classes, which are the elements of the quotient field. The different possible remainders are the polynomials a + bx, where a,b in R, and we identify the equivalence classes with these remainders.

The association a + bx iff a + bi clearly shows that the quotient field and C are isomorphic. Now, let þ be a zero of $x^2 + 1$, i.e. $þ^2 + 1 = 0$, so þ = sqrt(-1) = i which is not an element of R. {1,i} forms a basis for C over R since every element of C can be written as a(1) + b(i) with a,b in R and 1 and i are easily seen to be linearly independent. Using this basis, we can identify the elements of C with their coefficient vectors, i.e. a + bi iff (a,b) to get the second representation as a vector space of dimension 2 (notice the highest power of x in the irreducible polynomial).

Let f(x) be in K[x], then the smallest field containing K and all the zeros of f(x) is called the *splitting field* of f(x) over K. We may call it "the" splitting field due to the following theorem:

Theorem: Let *f*(*x*) be an irreducible polynomial in K[x], then a splitting field for f(x) over K exists and any two such splitting fields are isomorphic.

If f(x) is a polynomial in K[x] of degree n, then its splitting field over K is at most of degree n! and this bound may or may not be obtained, depending on the polynomial and the field. We now consider some elementary properties of polynomials over a field. If f(x) in K[x] is given by

$$f(x) = \beta_n x^n + \beta_{n-1} x^{n-1} + ... + \beta_1 x + \beta_0, \text{ where } \beta_i \text{ in K}$$

then the formal derivative of f(x), denoted by f'(x) is the polynomial

$$f'(x) = n\beta_n x^{n-1} + (n-1)\beta_{n-1} x^{n-2} + ... + \beta_1$$

which is a polynomial of degree at most n – 1. Note that the formal derivative of a polynomial may be zero even though the polynomial is not a constant, for example if K = GF(3), where 3 = 0, the polynomial $f(x) = x^3 + 2$ has a zero formal derivative. A polynomial f(x) in K[x] is said to have a zero þ of *multiplicity* m in some extension field L of K if m is the largest positive integer for which

$$(x-þ)^m \mid f(x)$$

in L[x], where the vertical bar indicates division with no remainder. The zeros of an irreducible polynomial f(x) in K[x] in the splitting field for f(x) over K are called *conjugates*. Let f(x), g(x) in K[x] be two polynomials such that deg f > deg g. By the Euclidean algorithm there exist two polynomials q(x) and r(x) such that

$$f(x) = q(x)g(x) + r(x), \text{ where deg r} < \text{deg g}.$$

By repeated application of the algorithm, the greatest common divisor (or gcd) d(x) of f(x) and g(x) (denoted by (f(x), g(x))) can be expressed as

$$d(x) = (fs(x),g(x)) = a(x)f(x) + b(x)g(x), \text{ for some } a(x),b(x) \text{ in K[x]}.$$

For further reference we collect some elementary properties of polynomials in the following theorem.

Theorem: *Let f(x), g(x) in K[x] and let L be any extension of K. Then:*

- If (f(x), g(x)) = d(x) in K[x] then (f(x), g(x)) = d(x) in L[x].
- If(x) | g(x) in K[x] iff f(x) | g(x) in L[x].
- *f(x) has a multiple zero iff (f(x), f'(x)) is not 1.*

It is significant that two polynomials have a common root in some extension field if they have a common divisor over the original field. The question of multiplicities of roots of irreducible polynomials can be settled precisely. If f(x) in K[x], char K = 0 and f(x) is irreducible, then f(x) cannot have multiple zeros.

If char K = p, then an irreducible polynomial f(x) having multiple zeros must be of the form f(x) = g(x^p) for some polynomial g(x) in K[x]. In this case each zero of f(x) has the same multiplicity. It can be shown that if f(x) is an irreducible polynomial over a finite field, then it has only simple (multiplicity one) zeros.

Theorem: A polynomial *f(x)* in *K[x]* of degree n has at most n zeros in any extension of K.

We consider now the concept of field isomorphism, which will be useful in the investigation of finite fields. An *isomorphism* of the field K_1 onto the field K_2 is a one-to-one onto map that preserves both field operations, i.e.,

$$\mu(þ + \beta) = \mu(þ) + \mu(\beta),\ \mu(þ\beta) = \mu(þ)\mu(\beta) \text{ for all } þ, \beta \text{ in K1}.$$

An *automorphism* of K is an isomorphism of K onto itself. The set of all automorphisms of a field forms a group under composition. If $\mu_1,\mu_2,\ldots,\mu_n$ are distinct isomorphisms of K_1onto K_2, then these isomorphisms are linearly independent over K_2 in the sense that if

Sum $a_i\ \mu_i$ (b) = 0 with a_i in K_2

for all b in K_1, then all the a_i must = 0. In particular, distinct automorphisms of a field are linearly independent.

A field automorphism leaves the elements of the prime subfield fixed. The automorphism may leave a larger subfield fixed. More generally, if

$$K' = \{ a \text{ in } K \mid \mu_i (a) = a, i = 1,\ldots,n \}$$

then K′ is a subfield of K, called the *fixed field* of K with respect to the automorphisms μ_i. It can be shown that [K:K'] > = n. We will be interested in subgroups of the group of all automorphisms that fix certain subfields. We denote the group of all automorphisms of a field L by G(L) and the subgroup of G(L) that fixes all elements of the subfield K of L by G(L/K). It is important to note that the fixed field of G(L/K) may properly contain K. It is easily shown that G(L/K) is a subgroup of G(L), conversely, if H is any subgroup of G(L), then the set of elements of L fixed by H is a subfield of L.

FUNDAMENTAL PROPERTIES OF FINITE FIELDS

We now restrict our attention to finite fields, i.e., fields with a finite number of elements. A finite field, since it cannot contain Q, must have a prime subfield of the form GF(p) for some prime p, also:

Theorem: ***Any finite field with characteristic p has p^n elements for some positive integer n.***

Proof: Let L be the finite field and K the prime subfield of L. The vector space of L over K is of some finite dimension, say n, and there exists a basis $þ_1,þ_2,\ldots,þ_n$ of L over K. Since every element of L can be expressed uniquely as a linear combination of the $þ_i$ over K, i.e., every a in L can be written as, a = Sum $\beta_i\ þ_i$, with β_i in K, and since K has p elements, L must have p^n elements.

This theorem, while it does restrict the size of a finite field, does not say that one will exist for a particular power of a prime, nor does it specify how many finite fields can exist of a particular order. The answers to these questions can be deduced from the following theorem.

Theorem: *Let L be a field with characteristic p and prime subfield K. Then L is the splitting field for $f(x) = xp^n - x$ iff L has p^n elements.*

Proof: Suppose that L is the splitting field for $f(x) = xp^n - x$ over *K*. Since $(f(x), f'(x)) = 1$, the roots of $f(x)$ are distinct and so L has at least p^n elements. Consider the subset

$$E = \{ þ \text{ in } L \mid þ^{pn} = þ \}$$

of L. Clearly E contains p^n elements since it consists of the roots of f(x). Suppose that þ,í in E; then $(þ\beta)^{pn} = (þ)^{pn} (\beta)^{pn} = þ\beta$ and hence, $þ\beta$ in *E*. Also,

$$(Þ + \beta)^{p^n} = \sum Þ^i \beta^{p^n - 1} = Þ^{p^n} + \beta^{p^n} = Þ + \beta$$

since $p \mid C(p^n, i)$ for $0 < i < p^n$, and hence, (þ + í) in E. The existence of additive and multiplicative inverses is easy to show, so E is a subfield of L and also a splitting field for f(x). Thus by Thm II.1.1 E = L and L contains p^n elements.

Suppose now that L contains p^n elements. The multiplicative group of L, which we will denote by L*, forms a group of order p^n - 1 and hence the order of any element of L* divides p^n - 1. Thus $þ^{pn} = þ$ for all þ in L* and the relation is trivially true for þ = 0. Thus f(x) splits in L.

The following important theorem is useful in establishing the subfield structure of the Galois Fields among other things.

Theorem: $GF(p^n)^*$ *is cyclic.*

Proof: The multiplicative group $GF(p^n)^*$ is, by definition, abelian and of order $p^n - 1$. If $p^n - 1 = p_1^{n_1} \ldots p_k^{e_k}$, then, factoring $GF(p^n)^*$ into a direct product of its Sylow subgroups, we have

$$GF(p^n)^* = S(p_1) \text{ X} \ldots \text{ X } S(p_k)$$

where $S(p_i)$ is the Sylow subgroup of order $(p_i)^e{}_i$. The order of every element in $S(p_i)$ is a power of p_i and let a_i in $S(p_i)$ have the maximal order, say $(p_i)^{e'}{}_i$, $e'_i <= e_i$, for i = 1,...,k. S i n c e $(p_i, p_j) = 1$, i not equal j, the element $a = a_1 a_2 \ldots a_k$ has maximal order $m = (p_1)^{e'-1} \ldots (p_k)^{e'}{}_k$ in $GF(p^n)^*$. Furthermore every element of $GF(p^n)^*$ satisfies the polynomial $x^m - 1$, implying that $m >= p^n - 1$. Since a in $GF(p^n)^*$ has order m, m divides $p^n - 1$ and so, $m = p^n - 1$. Thus the element a is a generator and $GF(p^n)^*$ is cyclic.

A generator of $GF(p^n)^*$ is called a *primitive element* of $GF(p^n)$.

The following theorem has some useful consequences.

Theorem: *Over any field K, $(x^m - 1) \mid (x^n - 1)$ iff $m \mid n$.*

Proof: If $n = qm + r$, with $r < m$, then by direct computation

$$x^n + 1 = x^r \left(\sum_{i=0}^{q-1} x^{im} \right) (x^m - 1) + (x^r - 1)$$

It follows that $(x^m - 1) \mid (x^n - 1)$ iff $x^r - 1 = 0$, i.e., $r = 0$.

Corollary: *For any prime integer p, $(p^m - 1) \mid (p^n - 1)$ iff $m \mid n$.*

Proof: Basically the same as that of the theorem, do for homework.

Theorem: $GF(p^m)$ is a subfield of $GF(p^n)$ *iff $m \mid n$.*

Proof: Suppose $GF(p^m)$ is a subfield of $GF(p^n)$; then $GF(p^n)$ may be interpreted as a vector space over $GF(p^m)$ with dimension, say, k. Hence, $p^n = p^{km}$ and $m \mid n$.

Now suppose $m \mid n$, which from the previous theorem and its corollary implies that $(x^{pm-1} - 1) \mid (x^{pn-1} - 1)$. Thus every zero of $x^{pm} - x$ that is in $GF(p^m)$ is also a zero of $x^{pn} - x$ and hence in $GF(p^n)$. It follows that $GF(p^m)$ is contained in $GF(p^n)$. Notice that there is precisely one subfield of $GF(p^n)$ of order p^m, otherwise $x^{pm} - x$ would have more than p^m roots.

Although we will not prove it, the automorphism group of a finite field is cyclic. The standard generator of this group is the so-called *Frobenius automorphism* defined for a finite field of characteristic p as the map $x \to x^p$ for all x in $GF(p^n)$.

PRELIMINARIES

We begin by stating some simple but useful divisibility criteria for polynomials over an arbitrary field. These will play a crucial role in the development of the theory.

Let F be a field. A polynomial $f \in F[X]$ is called square-free if it is not divisible by the square of any polynomial of degree greater than zero. Using formal derivatives, we obtain the following useful criterion for establishing that a polynomial is square-free:

Theorem: Let $a \geq 2$ be an integer and let k, ℓ be positive integers. Then $a^k - 1$ divides $a^\ell - 1$ if and only if k divides ℓ.

Theorem: If F is a field, and $f \in F[X]$ with $\gcd(f, D(f)) = 1$, then f is square-free.

Proof: Suppose f is not square-free, and write $f = g^2h$, for $g, h \in F[X]$ with $\deg(g) > 0$. Taking formal derivatives, we have

$$D(f) = 2gD(g)h + g^2D(h),$$

and so clearly, g is a common divisor of f and $D(f)$.

Theorem: Let F be a field, and let k, ℓ be positive integers. Then $X^k - 1$ divides $X^\ell - 1$ in $F[X]$ if and only if k divides ℓ.

Proof: Let $\ell = kq + r$, with $0 \leq r < k$. We have

$$X^\ell \equiv X^{kq} X^r \equiv X^r \pmod{X^k - 1},$$

and $X^r \equiv 1 \pmod{X^k - 1}$ if and only if $r = 0$.

Theorem: Let $a \geq 2$ be an integer, k, ℓ be positive integers, and F a field. Then $X^{a^k} - X$ divides $X^{a^\ell} - X$ in $F[X]$ if and only if k divides ℓ.

Proof: Now, $X^{a^k} - X$ divides $X^{a^\ell} - X$ if and only if $X^{a^k - 1} - 1$ divides $X^{a^\ell - 1} - 1$. By Theorem, this happens if and only if $a^k - 1$ divides $a^\ell - 1$. By Theorem, this happens if and only if k divides ℓ.

Let us suppose F is a field, and E is an extension field of F; that is, F is a subfield of E (or, more generally, F is embedded in E via some canonical embedding, and we identify elements of F with their images in E under this embedding). We may view E as an F-algebra via inclusion, and in particular, as an F-vector space. If E' is also an extension field of F, and $\rho: E \to E'$ is a ring homomorphism, then is an F-algebra homomorphism if and only if $\rho(a) = a$ for all $a \in F$.

Let us further assume that as an F-vector space, E has finite dimension ℓ. This dimension ℓ is called the degree of E over F, and is denoted $(E: F)$, and E is called a finite extension of F. Now consider an element $\alpha \in E$. Then α is algebraic over F, which means that there exists a non-zero polynomial $g \in F[X]$ such that $g(\alpha) = 0$.

The monic polynomial $\phi \in F[X]$ of least degree such that $\phi(\alpha) = 0$ is called the minimal polynomial of α over F. The polynomial ϕ is irreducible over F, and its degree $m := \deg(\phi)$ is called the degree of α over F. The ring $F[\alpha] = \{g(\alpha): g \in F[X]\}$, which is the smallest subring of E containing F and α, is actually a field, and is isomorphic, as an F-algebra, to $F[X] / (\phi)$, via the map that sends $g(\alpha) \in F[\alpha]$ to $[g]_\phi \in F[X] / (\phi)$. In particular, $(F[\alpha]: F) = m$, and the elements $1, \alpha, \ldots, \alpha^{m-1}$ form a basis for $F[\alpha]$ over F. Moreover, m divides ℓ.

CONJUGATES, NORMS AND TRACES

Throughout this section, F denotes a finite field of cardinality q, E denotes an extension of degree ℓ over F, and σ denotes the Frobenius map on E over F.

Let us suppose an element $\alpha \in E$. We say that $\beta \in E$ is conjugate to α (over F) if $\beta = \sigma^i(\alpha)$ for some $i \in \mathbb{Z}$. The reader may verify that the "conjugate to" relation is an equivalence relation. We call the equivalence classes of this relation conjugacy classes, and we call the elements of the conjugacy class containing α the conjugates of α.

Starting with α, we can start listing conjugates:

$\alpha, \sigma(\alpha), \sigma^2(\alpha), \ldots.$

As σ^{ℓ} is the identity map, this list will eventually start repeating. Let k be the smallest positive integer such that $\sigma^k(\alpha) = \sigma^i(\alpha)$ for some $i = 0, \ldots, k-1$. It must be the case that $i = 0$ – otherwise, applying σ^{-1} to the equation $\sigma^k(\alpha) = \sigma^i(\alpha)$ would yield $\sigma^{k-1}(\alpha) = \sigma^{i-1}(\alpha)$, and since $0 \leq i-1 < k-1$, this would contradict the minimality of k.

Thus, $\alpha, \sigma(\alpha), \ldots, \sigma^{k-1}(\alpha)$ are all distinct, and $\sigma^k(\alpha) = \sigma$. Moreover, for every integer i, we have $\sigma^i(\alpha) = \sigma^j(\sigma)$, where $j = i \bmod k$. Therefore, the k distinct elements $\alpha, \sigma(\alpha), \ldots, \sigma^{k-1}(\alpha)$ are all the conjugates of α. Also, $\sigma^i(\alpha) = \alpha$ if and only if k divides i, and since $\sigma^{\ell}(\alpha) = \alpha$, it must be the case that k divides ℓ. In addition, the conjugates of α are powers of α, and in particular, they all belong to $F[\alpha]$.

With α and k as above, consider the polynomial

$$\phi := \prod_{i=0}^{k=1} (X - \sigma^i(\alpha)).$$

The coefficients of ϕ obviously lie in E, but we claim that in fact, they lie in F. This is easily seen as follows. Extend the domain of definition of σ from E to $E[X]$ by applying σ coefficient-wise to polynomials; this yields a ring homomorphism from $E[X]$ into $E[X]$, which we also denote by σ. Applying σ to ϕ, we obtain

$$\sigma(\phi) = \prod_{i=0}^{k-1} \sigma(X - \sigma^i(\alpha)) = \prod_{i=0}^{k-1} (X - \sigma^{i+1}(\alpha)) = \prod_{i=0}^{k-1} (X - \sigma^i(\alpha)),$$

since $\sigma^k(\alpha) = \alpha$. Thus we see that $\sigma(\phi) = \phi$. Writing $\phi = \sum_i c_i X^i$,, it follows that $\sigma(c_i) = c_i$ for all i, and hence by Theorem, $c_i \in F$ for all i. Hence $\phi \in F[X]$. We further claim that ϕ is the minimal polynomial of α. To see this, let $f \in F[X]$ be any polynomial over F for which α is a root. Then for every integer i, by Theorem, we have

$$0 = \alpha^i(0) = \alpha^i(f(\alpha)) = f(\sigma^i(\alpha)).$$

Thus, all the conjugates of α are also roots of f, and so ϕ divides f. That proves that ϕ is the minimal polynomial of α. Since ϕ is the minimal polynomial of α and $\deg(\phi) = k$, it follows that the number k is none other than the degree of α over F.

Theorem: The function $N_{E/F}$, restricted to E^*, is a group homomorphism from E^* onto F^*.

Proof: We have

$$N_{E/F}(\alpha) = \prod_{i=0}^{\ell-1} \alpha^{q^i} = \alpha^{\sum_{i=0}^{\ell-1} q^i} = \alpha^{(q^{\ell}-1)/(q-1)}$$

Since E^* is a cyclic group of order $q^{\ell} - 1$, the image of the $(q^{\ell} - 1)/(q - 1)$-power map on E^* is the unique subgroup of E^* of order $q - 1$. Since F^* is a subgroup of E^* of order $q - 1$, it follows that the image of this power map is F^*.

Theorem: Let $\alpha \in E$ be of degree k over F, and let ϕ be the minimal polynomial of α over F. Then k is the smallest positive integer such that $\sigma^k(\alpha) = \alpha$, the distinct conjugates of are α, $\sigma(\alpha),..., \alpha^{k-1}(\alpha)$, and ϕ factors over E (in fact, over $F[\alpha]$) as

$$\phi = \prod_{i=0}^{k-1}(X - \sigma^i(\alpha)).$$

Another useful way of reasoning about conjugates is as follows. First, if $\alpha = 0$, then the degree of α over F is 1, and there is nothing more to say, so let us assume that $\alpha \in E$. If r is the multiplicative order of α, then note that every conjugate $\sigma^i(\alpha)$ also has multiplicative order r – this follows from the fact that for every positive integer s, $\alpha^s = 1$ if and only if $(\sigma^i(\alpha))^s = 1$. Also, note that we must have $r \,||\, E^* \,| = q^{\ell} - 1$, or equivalently, $q^{\ell} \equiv 1 \pmod r$. Focusing now on the fact that σ is the q-power map, we see that the degree k of α is the smallest positive integer such that $\alpha^{q^k} = \alpha$, which holds if and only if $\alpha^{q^k - 1} = 1$, which holds if and only if $q^k \equiv 1 \pmod r$. Thus, the degree of α over F is simply the multiplicative order of q modulo r.

Theorem: The function $Tr_{E/F}$ is an F-linear map from E onto F.

Proof. The fact that $Tr_{E/F}$ is an F-linear map is a simple consequence of the fact that σ is an F-linear map. $Tr_{E/F}$ maps into F. Since the image of $Tr_{E/F}$ is a subspace of F, the image is either $\{0\}$ or F, and so it suffices to show that $Tr_{E/F}$ does not map all of E to zero. But an element $\alpha \in E$ is in the kernel of $Tr_{E/F}$ if and only if is a root of the polynomial

$$X + X^q + ... + X^{q^{\ell-1}},$$

which has degree $q^{\ell-1}$. Since E contains q^{ℓ} elements, not all elements of E can lie in the kernel of $Tr_{E/F}$.

Example. As an application of some of the above theory, let us investigate the factorization of the polynomial $X^r - 1$ over F, a finite field of cardinality q. Let us assume that $r > 0$ and is relatively prime to q. Let E be a splitting field of $X^r - 1$, so that E is a finite extension of F in which $X^r - 1$ splits into linear factors:

$$X^r - 1 = \prod_{i=0}^{r}(X - \alpha_i).$$

We claim that the roots α_i of $X^r - 1$ are distinct – this follows from the Theorem and the fact that $\gcd(X^r - 1, {}_rX^{r-1}) = 1$.

Next, observe that the r roots of $X^r - 1$ in E actually form a subgroup of E^*, and since E^*

is cyclic, this subgroup must be cyclic as well. So the roots of $X^r - 1$ form a cyclic subgroup of E^* of order r. Let ζ be a generator for this group. Then all the roots of $X^r - 1$ are contained in $F[\zeta]$, and so we may as well assume that $E = F[\zeta]$.

Let us compute the degree of ζ over F. By Theorem, the degree ℓ of ζ over F is the multiplicative order of q modulo r. Moreover, the $\varphi(r)$ roots of $X^r - 1$ of multiplicative order r are partitioned into $\varphi(r)/\ell$ conjugacy classes, each of size ℓ (here, φ is Euler's phi function); indeed, as the reader is urged to verify, these conjugacy classes are in one-to-one correspondence with the cosets of the subgroup of $\mathbb{Z}_r^*$ generated by $[q]_r$, where each such coset $C \subseteq \mathbb{Z}_r^*$ corresponds to the conjugacy class $\{\zeta^a: [a]_r \in C\}$.

More generally, for every $s \mid r$, every root of $X^r - 1$ whose multiplicative order is s has degree k over F, where k is the multiplicative order of q modulo s. As above, the $\varphi(s)$ roots of multiplicative order s are partitioned into $\varphi(s)/k$ conjugacy classes, which are in one-to-one correspondence with the cosets of the subgroup of $\mathbb{Z}_s^*$ generated by $[q]_s$.

This tells us exactly how $X^r - 1$ splits into irreducible factors over F. Things are a bit simpler when r is prime, in which case, from the above discussion, we see that

$$X^r - 1 = (X - 1) \prod_{i=1}^{(r-1)/\ell} f_i,$$

where the f_i's are distinct monic irreducible polynomials, each of degree ℓ, and ℓ is the multiplicative order of q modulo r.

Theorem: If $\alpha \in E$ has multiplicative order r, then the degree of α over F is equal to the multiplicative order of q modulo r.

For $\alpha \in E$, define the polynomial

$$\chi := \prod_{i=0}^{\ell-1} (X - \sigma^i(\alpha)).$$

It is easy to see, using the same type of argument as was used to prove Theorem, that $\chi \in F[X]$, and indeed, that

$$\chi = \phi^{\ell/k},$$

where k is the degree of α over F. The polynomial χ is called the characteristic polynomial of α (from E to F).

Two functions that are often useful are the "norm" and "trace." The norm of (from E to F) is defined as

$$N_{E/F}(\alpha) := \prod_{i=0}^{\ell-1} \sigma^i(\alpha),$$

while the trace of α (from E to F) is defined as

$$Tr_{E/F}(\alpha) := \prod_{i=0}^{\ell-1} \sigma^i(\alpha).$$

It is easy to see that both the norm and trace of α are elements of F, as they are fixed by σ; alternatively, one can see this by observing that they appear, possibly with a minus sign, as coefficients of the characteristic polynomial χ– indeed, the constant term of χ is equal to $(-1)^{\ell}$ $N_{E/F}(\alpha)$, and the coefficient of $X^{\ell-1}$ in χ is $-Tr_{E/F}(\alpha)$.

The following two theorems summarize the most important facts about the norm and trace functions.

EXERCISE

1. Let E be an extension of degree ℓ over a finite field F. Let be the Frobenius map on E over F.
   (*a*) Show that the minimal polynomial of s over F is $X^{\ell} - 1$.
   (*b*) Show that there exists $\beta \in E$ such that the minimal polynomial of β under σ is $X^{\ell} - 1$.
   (*c*) Conclude that $\beta, \sigma(\beta),..., \sigma^{\ell-1}(\beta)$ form a basis for E over F. This type of basis is called a normal basis.
2. Let F be a finite field, and let $f \in F[X]$ be a monic irreducible polynomial of degree ℓ. Let $E = F[X]/(f) = F[\xi]$, where $\xi := [X]_f$
   (*a*) Show that
   $$\frac{D(f)}{f} = \sum_{j=1}^{\infty} Tr_{E/F}(\xi^{j-1})X^{-j}.$$
   (*b*) From part (a), deduce that the sequence of elements
   $Tr_{E/F}(\xi^{j-1})$ $(j = 1, 2,...)$
   is linearly generated over F with minimal polynomial f.
   (*c*) Show that one can always choose a polynomial f so that sequence in part (b) is purely periodic with period $q^{\ell} - 1$.
3. Show that for all primes p, the polynomial $x^4 + 1$ is reducible in $\mathbb{Z}_p[X]$.
4. Let E be a finite extension of a finite field F of characteristic p. Show that if $\alpha \in E$ and $0 \neq a \in F$, and if α and $\alpha + a$ are conjugate over F, then p divides the degree of α over F.

5. Let E be a finite extension of a finite field F. Show that every F-algebra automorphism on E must be a power of the Frobenius map on E over F.

6. Let E be an extension of degree ℓ over a finite field F. Show that for $a \in F$, we have $N_{E/F}(a) = a^{\ell}$ and $Tr_{E/F}(a) = \ell a$.

7. Let F be a finite field of characteristic p. For $a \in F$, consider the polynomial $f := X^p - X - a \in F[X]$.

   (*a*) Show that if $F = \mathbb{Z}_p$ and $a \neq 0$, then f is irreducible.

   (*b*) More generally, show that if $Tr_{F/\mathbb{Z}_p}(a) \neq 0$, then f is irreducible, and otherwise, f splits into distinct monic linear factors over F.

8. Let E be a finite extension of a finite field F. Let K be an intermediate field, $F \subseteq K \subseteq E$. Show that for all $\alpha \in E$

   (*a*) $N_{E/F}(\alpha) = N_{K/F}(N_{E/K}(\alpha))$, and

   (*b*) $Tr_{E/F}(\alpha) = Tr_{K/F}(Tr_{E/K}(\alpha))$.

9. Let F be a finite field, and $f \in F[X]$ a monic irreducible polynomial of degree k over F. Let E be an extension of degree ℓ over F. Show that over E, f factors as the product of d distinct monic irreducible polynomials, each of degree k/d, where $d := \gcd(k, \ell)$.

THE EXISTENCE OF FINITE FIELDS

Let us consider F be a finite field. F must have cardinality p^w, where p is prime and w is a positive integer, and p is the characteristic of F. However, we can say a bit more than this. The field $\mathbb{Z}_p$ is embedded in F, and so we may simply view $\mathbb{Z}_p$ as a subfield of F. Moreover, it must be the case that w is equal to $(F: \mathbb{Z}_p)$.

We want to show that there exist finite fields of every prime-power cardinality. Actually, we shall prove a more general result:

If F is a finite field, then for every integer $\ell \geq 1$, there exists an extension field E of degree ℓ over F.

For the remainder of this section, F denotes a finite field of cardinality $q = p^w$, where p is prime and $w \geq 1$.

Suppose for the moment that E is an extension of degree ℓ over F. Let us derive some basic facts about E. First, observe that E has cardinality q^{ℓ}. By Theorem, E^* is cyclic, and the order of E^* is $q^{\ell} - 1$. If $\gamma \in E^*$ is a generator for E^*, then every non-zero element of E can be expressed as a power of γ; in particular, every element of E can be expressed as a polynomial in with coefficients in F; that is, $E = F[\gamma]$. Let $\phi \in F[X]$ be the minimal polynomial of γ over F, which is an irreducible polynomial of degree ℓ. It follows that E is isomorphic (as an F-algebra) to $F[X]/(\phi)$.

So we have shown that every extension of degree ℓ over F must be isomorphic, as an F-algebra, to $F[X]/(f)$ for some irreducible polynomial $f \in F[X]$ of degree ℓ. Conversely, given any irreducible polynomial f over F of degree ℓ, we can construct the finite field $F[X]/(f)$, which has degree ℓ over F. Thus, the question of the existence of a finite field of degree ℓ over F reduces to the question of the existence of an irreducible polynomial over F of degree ℓ.

Theorem: Let E be a extension of degree ℓ over F, and let σ be the Frobenius map on E over F. Then for all integers i and j, we have $\sigma^i = \sigma^j$ if and only if $i \equiv j \pmod{\ell}$.

Proof: We may assume $i \geq j$. We have

$$\sigma^i = \sigma^j \Leftrightarrow \sigma^{i-j} = \sigma^0 \Leftrightarrow \alpha^{q^{i-j}} - \alpha = 0 \text{ for all } \alpha \in E$$

$$\Leftrightarrow \left(\prod_{\alpha \in E}(X-\alpha)\right) \mid X^{q^{i-j}} - \alpha = 0 \text{ (by Theorem 7.13)}$$

$$\Leftrightarrow (X^{q^{\ell}} - X) \mid (X^{q^{i-j}} - X) \text{ (by Theorem, applied to } E)$$

$$\Leftrightarrow \ell \mid (i-j) \text{ (by Theorem)}$$

$$\Leftrightarrow i \equiv j \pmod{\ell}$$

From the above theorem, it follows that every power of the Frobenius map σ can be written uniquely as σ^i for some $i = 0, \ldots, \ell - 1$.

Theorem: For every $a \in F$, we have $a^q = a$.

Proof. The multiplicative group of units F^* of F has order $q - 1$, and hence, every $a \in F^*$ satisfies the equation $a^{q-1} = 1$. Multiplying this equation by a yields $a^q = a$ for all $a \in F^*$, and this latter equation obviously holds for $a = 0$ as well.

Theorem: Let E be an F-algebra. Then the map $\sigma: E \to E$ that sends $\alpha \in E$ to α^q is an F-algebra homomorphism.

Proof. By Theorem, either E is trivial or contains an isomorphic copy of F as a subring. In the former case, there is nothing to prove. So assume that E contains an isomorphic copy of F as a subring. It follows that E must have characteristic p.

Since $q = p^w$, we see that $\sigma = \tau^w$, where $\tau(\alpha) := \alpha^p$. The map τ is a ring homomorphism, and hence so is σ. Moreover, by Theorem, we have

$$\sigma(c1_E) = (c1_E)^q = c^q 1_E^q = c1_E$$

for all $c \in F$. Thus, σ is an F-algebra homomorphism.

Theorem: We have

$$X^q - X = \prod_{a \in F}(X - a).$$

Proof: Since each $a \in F$ is a root of $X^q - X$, by Theorem, the polynomial $\prod_{a \in F}(X - a)$ divides the polynomial $X^q - X$. Since the degrees and leading coeffocients of these two polynomials are the same, the two polynomials must be equal.

Theorem: For all $\ell \geq 1$, we have

$$q^\ell = \sum_{k|\ell} k\Pi_F(k).$$

Theorem: Let E be a finite extension of F, and let σ be the Frobenius map on E over F. Then σ is an F-algebra automorphism on E. Moreover, for all $\alpha \in E$, we have $\sigma(\alpha) = \alpha$ if and only if $\alpha \in F$.

Proof: The fact that σ is an F-algebra homomorphism follows from the previous theorem. Any ring homomorphism from a field into a field is injective. Surjectivity follows from injectivity and finiteness.

For the second statement, observe that $\sigma(\alpha) = \alpha$ if and only if α is a root of the polynomial $X^q - X$, and since all q elements of F are already roots, there can be no other roots.

As the Frobenius map on finite fields plays a fundamental role in the study of finite fields, let us develop a few simple properties right away.

Suppose E is a finite extension of F, and let be the Frobenius map on E over F. Since the composition of two F-algebra automorphisms is also an F-algebra automorphism, for every $i \geq 0$, the i-fold composition σ^i, which sends $\alpha \in E$ to $\alpha^{q^i} \in E$, is also an F-algebra automorphism. Since σ is an F-algebra automorphism, the inverse function σ^{-1} is also an F-algebra automorphism.

Hence, i is an F-algebra automorphism for all $i \in \mathbb{Z}$. If E has degree ℓ over F, then applying Theorem to the field E, we see that σ^ℓ is the identity map.

Theorem: For $k \geq 1$, let P_k denote the product of all the monic irreducible polynomials in $F[X]$ of degree k. For all positive integers ℓ, we have

$$X^{q^\ell} - X = \prod_{k|\ell} P_k,$$

where the product is over all positive divisors k of ℓ.

Proof: First, we claim that the polynomial $X^{q^\ell} - X$ is square-free. This follows immediately from Theorem, since $DX^{q^\ell}(-X) = q^\ell X^{q^\ell - 1} - 1 = -1$.

Thus, we have reduced the proof to showing that if f is a monic irreducible polynomial of degree k, then f divides $X^{q^\ell} - X$ if and only if k divides ℓ.

So let f be a monic irreducible polynomial of degree k. Let $E := F[X]/(f) = F[\xi]$, where $\xi := [X]_f \in E$. Observe that E is an extension field of degree k over F. Let σ be the Frobenius map on E over F.

First, we claim that f divides $X^{q^\ell} - X$ if and only if $\sigma^\ell(\xi) = \xi$. Indeed, f is the minimal polynomial of ξ over F, and so f divides $X^{q^\ell} - X$ if and only if ξ is a root of $X^{q^\ell} - X$, which is the same as saying $\xi^{q^\ell} = \xi$, or equivalently, $\sigma^\ell(\xi) = \xi$.

Second, we claim that $\sigma^\ell(\xi) = \xi$ if and only if $\sigma^\ell(\alpha) = \alpha$ for all $\alpha \in E$. To see this, first suppose that $\sigma^\ell(\alpha) = \alpha$ for all $\alpha \in E$. Then in particular, this holds for $\alpha = \xi$. Conversely, suppose that $\sigma^\ell(\xi) = \xi$. Every $\alpha \in E$ can be written as $\alpha = g(\xi)$ for some $g \in F[X]$, and since α^ℓ is an F-algebra homomorphism, by

Theorem we have

$$\sigma^\ell(\alpha) = \sigma^\ell(g(\xi)) = g(\sigma^\ell(\xi)) = g(\xi) = \alpha.$$

Finally, we see that $\sigma(\alpha) = \alpha$ for all $\alpha \in E$ if and only if $\sigma' = \sigma^0$, which by Theorem holds if and only if $k \mid \ell$.

For $\ell \geq 1$, let $\prod_F \ell$ denote the number of monic irreducible polynomials of degree ℓ in $F[X]$.

Theorem: For all $\ell \geq 1$, we have

$$\frac{q^\ell}{2\ell} \leq \Pi_F(\ell) \leq \frac{q^\ell}{\ell},$$

and

$$\Pi_F(\ell) = \frac{q^\ell}{\ell} + O\left(\frac{q^{\ell/2}}{\ell}\right).$$

EXERCISE

1. How many irreducible polynomials of degree 30 over $\mathbb{Z}_2$ are there?
2. Show that

$$\Pi_F(\ell) = \ell^{-1} \sum_{k \mid \ell} \mu(k) q^{\ell/k},$$

where μ is the Mφbius function.

UNIQUENESS OF FINITE FIELDS

Let E be an extension of degree ℓ over a field F. If K is an intermediate field, that is, a subfield of E containing F, then Theorem says that $(E: F) = (E: K)(K: F)$, and so in particular, the degree of K over F divides ℓ.

In the case where F is a finite field, we can say much more about such intermediate fields. Recall that if $\rho: E \rightarrow E$ be an F-algebra homomorphism, then the subalgebra of E fixed by ρ is defined as $K := \{\alpha \in E: \rho(\alpha) = \alpha\}$. Not only is K a subalgebra of E, but it is also a field, and so K is itself an intermediate field.

Theorem: Let E and E' be finite extensions of the same degree over a finite field F. Then E and E' are isomorphic as F-algebras.

Proof: Let q be the cardinality of F, and let ℓ be the degree of the extensions. As we have argued before, we have $E' = F[\alpha']$ for some $\alpha' \in E'$, and so E' is isomorphic as an F-algebra to $F[X]/(\phi)$, where ϕ is the minimal polynomial of α' over F. As ϕ is an irreducible polynomial of degree ℓ, by Theorem, ϕ divides $X^{q^\ell} - X$, and by Theorem (applied to E), $X^{q^\ell} - X = \prod_{\alpha \in E}(X - \alpha)$, from which it follows that ϕ has a root $\alpha \in E$. Since ϕ is irreducible, ϕ is the minimal polynomial of α over F, and hence $F[\alpha]$ is isomorphic as an F-algebra to $F[X]/(\phi)$. Since α has degree ℓ over F, we must have $E = F[\alpha]$. Thus, $E = F[\alpha] \cong F[X]/(\phi) \cong F[\alpha'] = E'$.

Theorem:. Let E be an extension of degree ℓ over a finite field F. Let σ be the Frobenius map on E over F. Then the intermediate fields K, with $F \subseteq K \subseteq E$, are in one-to-one correspondence with the divisors k of ℓ, where the divisor k corresponds to the subalgebra of E fixed by k, which has degree k over F.

Proof: Let q be the cardinality of F.

Suppose k is a divisor of ℓ. By Theorem (applied to E), the polynomial $X^{q^\ell} - X$ splits into distinct monic linear factors over E. By Theorem, the polynomial $X^{q^k} - X$ divides $X^{q^\ell} - X$. Hence, $X^{q^k} - X$ also splits into distinct monic linear factors over E. This says that the subalgebra of E fixed by k, which consists of the roots of $X^{q^k} - X$, has precisely q^k elements, and hence is an extension of degree k over F.

Now let K be an arbitrary intermediate field, and let k be the degree of K over F. As already mentioned, we must have $k \mid \ell$. Also, by Theorem (applied with K in place of F), K is the subalgebra of E fixed by σ^k.

EXERCISE

1. Let E be a finite extension of a finite field F, and suppose $\alpha, \beta \in E$, where has degree a over F, β has degree b over F, and $\gcd(a, b) = 1$. Show that β has degree b over $F[\alpha]$, that α has degree a over $F[\beta]$, and that $\alpha + \beta$ has degree ab over F.
2. This exercise develops an alternative proof for the existence of finite fields – however, it does not yield a density result for irreducible polynomials. Let F be a finite field of cardinality q, and let $\ell \geq 1$ be an integer. Let E be a splitting field for the polynomial X^{q^ℓ} $- X \in F[X]$, and let be the Frobenius map on E over F. Let K be the subalgebra of E fixed by σ^ℓ. Show that K is an extension of F of degree ℓ.
3. Let E be an extension of degree ℓ over a finite field F of cardinality q. Show that at least half the elements of E have degree ℓ over F, and that the total number of elements of degree ℓ over F is $q^\ell + O(q^{\ell/2})$.

ALGORITHMS FOR FINITE FIELDS

Factoring Polynomials: Square-free Decomposition

In the remaining sections of this chapter, we develop efficient algorithms for factoring polynomials over the finite field F. We begin in this section with a simple and efficient preprocessing step. Recall that a polynomial is called square-free if it is not divisible by the square of any polynomial of degree greater than zero. This preprocessing algorithm takes the polynomial to be factored, and partially factors it into a product of square-free polynomials. Given this algorithm, we can focus our attention on the problem of factoring square-free polynomials.

Let $f \in F[X]$ be a monic polynomial of degree $\ell > 0$. Suppose that f is not square-free. According to Theorem, $d := \gcd(f, D(f)) \neq 1$, where $D(f)$ is the formal derivative of f; thus, we might hope to get a non-trivial factorization of f by computing d. However, we have to consider the possibility that $d = f$. Can this happen? The answer is "yes," but if it does happen that $d = f$.

Theorem: Algorithm SFD correctly computes a square-free decomposition of f using $O(\ell^2 + \ell(w-1)\,\mathrm{len}(p)/p)$ operations in F.

Proof: Let $f = \prod_i f_i^{e_i}$ be the factorization of the input f into irreducibles. Let S be the set of indices i such that $e_i \not\equiv 0 \pmod p$, and let $S0$ be the set of indices i such that $e_i \equiv 0 \pmod p$. Also, for $j \geq 1$, let $S_{\geq j} := \{i \in S : e_i \geq j\}$ and $S_{=j} := \{i \in S : e_i = j\}$.

Consider the first iteration of the main loop. By Theorem the value first assigned to g is $\prod_{i \in S} f_i$. It is straightforward to prove by induction on j that at the beginning of the jth iteration of the inner while loop, the value assigned to g is $\prod_{i \in S_{\geq j}} f_i$, and the value assigned to f is $\prod_{i \in S_{\geq j}} f_i^{e_i - j+1} \cdot \prod_{i \in S'} f_i^{e_i}$. Moreover, in the jth loop iteration, the value assigned to m is $\prod_{i \in S_{\geq j}} f_i$. It follows that when the while loop terminates, the value assigned to f is $\prod_{i \in S_{=j}} f_i$, and the value assigned to L is a square-free decomposition of $\prod_{i \in S} f_i^{e_i}$; if f does not equal 1 at this point, then subsequent iterations of the main loop will append to L a square-free decomposition of $\prod_{i \in S'} f_i^{e_i}$.

That proves the correctness of the algorithm. Now consider its running time. Again, consider just the first iteration of the main loop. The cost of computing $f/\gcd(f, D(f))$ is at most $C_1 \ell^2$ operations in F, for some constant C_1. Now consider the cost of the inner while loop. It is not hard to see that the cost of the jth iteration of the inner while loop is at most

$$D_2 \ell = \sum_{i \in S_{\geq j}} \deg(f_i)$$

operations in F, for some constant C_2. This follows from the observation in the previous paragraph that the value assigned to g is $\prod_{i \in S_{\geq j}} f_i$, along with our usual cost estimates for division and Euclid's algorithm. Therefore, the total cost of all iterations of the inner while loop is at most

$$C_2 \ell \sum_{j \geq 1} \sum_{i \in S_{\geq j}} \deg(f_i)$$

operations in F. In this double summation, for each $i \in S$, the term $\deg(f_i)$ is counted exactly e_i times, and so we can write this cost estimate as

$$C_2 \ell \sum_{i \in S} e_i \deg(f_i) \leq C_2 \ell^2.$$

Finally, it is easy to see that in the if-then statement at the end of the main loop body, if the algorithm does in fact compute a pth root, then this takes at most

$$C_3 \ell (w - 1) \operatorname{len}(p)/p$$

operations in F, for some constant C_3. Thus, we have shown that the total cost of the first iteration of the main loop is at most

$$(C_1 + C_2) \ell^2 + C_3 \ell (w - 1) \operatorname{len}(p) / p$$

operations in F. If the main loop is executed a second time, the degree of f at the start of the second iteration is at most ℓ / p, and hence the cost of the second loop iteration is at most

$$(C_1 + C_2)(\ell / p)^2 + C_3(\ell / p)(w - 1) \operatorname{len}(p)/p$$

operations in F. More generally, for $t = 1, 2, \ldots$, the cost of loop iteration t is at most

$$(C_1 + C_2)(\ell/p^{t-1})^2 + C_3(\ell/p^{t-1})(w-1)\ \mathrm{len}(p)/p,$$

operations in F, and summing over all $t \geq 1$ yields the stated bound.

Theorem: Suppose that $f \in F[X]$ is a monic polynomial of degree $\ell > 0$, and that $\gcd(f, D(f)) = f$. Then $f = g(X^p)$ for some $g \in F[X]$. Moreover, if $g = \sum_i a_i X^i$, then $f = h^p$, where

$$h = \sum_i a_i^{p^{(w-1)}} X^i.$$

Proof. Since $\deg(D(f)) < \deg(f)$ and $\gcd(f, D(f)) = f$, we must have $D(f) = 0$. If $f = \sum_i c_i X^i$, then $D(f) = \sum_i i c_i X^{i-1}$. Since this derivative must be zero, it follows that all the coefficients c_i with $i \not\equiv 0 \pmod p$ must be zero to begin with. That proves that $f = g(X^p)$ for some $g \in F[X]$. Furthermore, if h is defined as above, then

$$h^p = \left(\sum_i a_i^{p^{(w-1)}} X^i\right)^p = \sum_i a_i^{p^w} X^{ip} \sum_i a_i (X^p)^i = g(X^p) = f.$$

Our goal now is to design an efficient algorithm that takes as input a monic polynomial $f \in F[X]$ of degree $\ell > 0$, and outputs a list of pairs $((g_1, s_1), \ldots, (g_t, s_t))$, where

- Each $g_i \in F[X]$ is monic, non-constant, and square-free,
- Each s_i is a positive integer,
- The family of polynomials $\{g_i\}_{i=1}^t$ is pairwise relatively prime, and
- $f = \prod_{i=1}^t g_i^{s_i}$.

We call such a list a square-free decomposition of f. There are a number of ways to do this.

Theorem: Let $f \in F[X]$ be a monic polynomial of degree $\ell > 0$. Suppose that the factorization of f into irreducibles is $f = f_1^{e_1} \ldots f_r^{e_r}$. Then

$$\frac{f}{\gcd(f, D(f))} = \prod_{\substack{1 \leq i \leq r \\ e_i \not\equiv 0 \pmod p}} f_i.$$

Proof: The theorem can be restated in terms of the following claim: for each $i = 1, \ldots, r$, we have

- $f_i^{e_i} \mid D(f)$ if $e_i \equiv 0 \pmod p$, and
- $f_i^{e_i - 1} \mid D(f)$ but $f_i^{e_i} \mid D(f)$ if $e_i \not\equiv 0 \pmod p$.

To prove the claim, we take formal derivatives using the usual rule for products, obtaining

$$D(f) = \sum_j e_j f_j^{e_j - 1} D(f_i) \prod_{k \neq j} f_k^{e_k}.$$

Consider a fixed index i. With the possible exception of the term with $j = i$. In the case where $e_i \equiv 0 \pmod p$, the term with $j = i$ vanishes, and that proves the claim in this case. So assume that $e_i \not\equiv 0 \pmod p$.

By the previous theorem, and the fact that f_i is irreducible, and in particular, not the pth power of any polynomial, we see that $D(f_i)$ is non-zero, and (of course) has degree strictly less than that of f_i. From this, and (again) the fact that f_i is irreducible, it follows that the term with $j = i$ is divisible by $f_i^{e_i - 1}$, but not by $f_i^{e_i}$, from which the claim follows.

This theorem provides the justification for the following square-free decomposition algorithm.

ALGORITHM SFD

On input f, where $f \in F[X]$ is a monic polynomial of degree $\ell > 0$, compute a square-free decomposition of f as follows:

Initialize an empty list L

$s \leftarrow 1$

repeat

$j \leftarrow 1, g \leftarrow f / \gcd(f, D(f))$

while $g \neq 1$ do

$f \leftarrow f/g,\ h \leftarrow \gcd(f, g),\ m \leftarrow g/h$

if $m \neq 1$ then append (m, js) to L

$g \leftarrow h, j \leftarrow j + 1$

if $f \neq 1$ then // f is a pth power

// compute a pth root as in

$f \leftarrow f^{1/p}, s \leftarrow ps$

until $f = 1$

output L

BERLEKAMP'S ALGORITHM

We now develop an alternative algorithm, due to Berlekamp, for factoring a polynomial over the finite field F into irreducibles. We shall assume that the input polynomial is square-free, using Algorithm SFD as a preprocessing step, if necessary.

Let us now assume we have a monic square-free polynomial f of degree $\ell > 0$ that we want to factor into irreducibles. We first present the mathematical ideas underpinning the algorithm.

Let E be the F-algebra $F[X]/(f)$. Let σ be the Frobenius map on E over F, which maps $\alpha \in E$ to $\alpha^q \in E$. We know that σ is an F-algebra homomorphism. Consider the subalgebra B of E fixed by σ. Thus,

$$B = \{\alpha \in E: \alpha^q = \alpha\}.$$

The subalgebra B is called the Berlekamp subalgebra of E. Let us take a closer look at it. Suppose that f factors into irreducibles as

$$f = f_1 \cdots f_r,$$

and let

$$\theta: E \to E_1 \times \cdots \times E_r$$

$$[g]_f \mapsto ([g]_{f_1}, \ldots, [g]_{f_r})$$

be the F-algebra isomorphism from the Chinese remainder theorem, where $E_i := F[X] / (f_i)$ is an extension field of F of finite degree for $i = 1, \ldots, r$. Now, for $\alpha \in E$, if $\theta(\alpha) = (\alpha_1, \ldots, \alpha_r)$, then we have $q =$ if and only if $\alpha_i^q = \alpha_i$ for $i = 1, \ldots, r$; moreover, by Theorem, we know that for all $\alpha_i \in E_i$, we have $\alpha_i^q = \alpha_i$ if and only if $\alpha_i \in F$. Thus, we may characterize B as follows:

$$B = \{\theta^{-1}(c_1, \ldots, c_r): c_1, \ldots, c_r \in F\}.$$

Since B is a subalgebra of E, then as F-vector spaces, B is a subspace of E. Of course, E has dimension ℓ over F, with the natural basis $\{\xi^{i-1}\}_{i=1}^{\ell}$, where $\xi := [X]_f$. As for the Berlekamp subalgebra, from the above characterization of B, it is evident that the elements

$$\theta^{-1}(1, 0, \ldots, 0), \theta^{-1}(0, 1, 0, \ldots, 0), \ldots, \theta^{-1}(0, \ldots, 0, 1)$$

form a basis for B over F, and hence, B has dimension r over F.

Now we come to the actual factoring algorithm.

CONSTRUCT A BASIS FOR B

The first stage of Berlekamp's factoring algorithm constructs a basis for B over F. We can easily do this using Gaussian elimination, as follows. Let $\rho: E \to E$ be the map that sends $\alpha \in E$ to $\sigma(\alpha) - \alpha = \alpha^q - \alpha$. Since σ is an F-linear map, the map ρ is also F-linear. Moreover, the kernel of is none other than the Berlekamp subalgebra B. So to find a basis for B, we simply need to find a basis for the kernel of ρ using Gaussian elimination over F.

To perform the Gaussian elimination, we need to choose a basis S for E over F, and construct the matrix $Q := \mathrm{Mat}_{S,S}(\rho) \in F^{\ell \times \ell}$, that is, the matrix of ρ with respect to this basis, so that evaluation of ρ corresponds to multiplying a row vector on the right by Q. We are free to choose a basis in any convenient way, and the most convenient basis, of course, is $S := \{\xi^{i-1}\}_{i=1}^{\ell}$, since for computational purposes, we already represent an element $\alpha \in E$ by its coordinate vector $\mathrm{Vec}_S(\alpha)$. The matrix Q, then, is the $\ell \times \ell$ matrix whose ith row, for $i = 1, \ldots, \ell$, is $\mathrm{Vec}_S(\rho(\xi^{i-1}))$.

Note that if $\alpha = \xi^q$, then $\rho(\xi^{i-1}) = (\xi^{i-1})^q - \xi^{i-1} = (\xi^q)^{i-1} - \xi^{i-1} = \alpha^{i-1} - \xi^{i-1}$. This observation allows us to construct the rows of Q by first computing ξ^q via repeated squaring, and then just computing successive powers of ξ^q.

After we construct the matrix Q, we apply Gaussian elimination to get row vectors $v_1, \ldots, v_r$ that form a basis for the row null space of Q. It is at this point that our algorithm actually discovers the number r of irreducible factors of f. Our basis for B is $\{\beta_i\}_{i=1}^{r}$, where $\mathrm{Vec}_S(\beta_i) = v_i$ for $i = 1, \ldots, r$.

Putting this altogether, we have the following algorithm to compute a basis for the Berlekamp subalgebra.

Algorithm B1

On input f, where $f \in F[X]$ is a monic square-free polynomial of degree $\ell > 0$, do the following, where $E := F[X]/(f)$, $\xi := [X]_f \in E$, and $S := \{\xi^{i-1}\}_{i=1}^{\ell}$:

let Q be an $\ell \times \ell$ matrix over F (initially with undefined entries)

compute $\alpha \leftarrow \xi^q$ using repeated squaring

$\beta \leftarrow 1_E$

for $i \leftarrow 1$ to ℓ do // invariant: $\beta = \alpha^{i-1} = (\xi^{i-1})^q$

$\mathrm{Row}_i(Q) \leftarrow \mathrm{Vec}_S(\beta)$, $Q(i, i) \leftarrow Q(i, i) - 1$, $\beta \leftarrow \beta\alpha$

compute a basis $\{v_i\}_{i=1}^{r}$ of the row null space of Q using

Gaussian elimination

for $i = 1, \ldots, r$ do $\beta_i \leftarrow Vec_S^{-1}(v_i)$

(*vi*)output $\{\beta_i\}_{i=1}^{r}$

The correctness of Algorithm B1 is clear from the above discussion. As for the running time:

Theorem: Algorithm B1 uses $O(\ell^3 + \ell^2 \operatorname{len}(q))$ operations in F.

Proof: This is just a matter of counting. The computation of α takes $O(\operatorname{len}(q))$ operations in E using repeated squaring, and hence $O(\ell^2 \operatorname{len}(q))$ operations in F. To build the matrix Q, we have to perform an additional $O(\ell)$ operations in E to compute the successive powers of α, which translates into $O(\ell^3)$ operations in F. Finally, the cost of Gaussian elimination is an additional $O(\ell^3)$ operations in F.

SPLITTING WITH A BASIS FOR B

The second stage of Berlekamp's factoring algorithm is a probabilistic procedure that factors f using a basis $\{\beta_i\}_{i=1}^r$ for B. As we did with Algorithm EDF, we begin by discussing how to efficiently split f into two non-trivial factors, and then we present a somewhat more elaborate algorithm that completely factors f.

Let $M_1 \in F[X]$ be the polynomial that is,

$$M_1 := \begin{cases} \sum_{j=0}^{w-1} X^{2^j} & \text{if } p = 2, \\ X^{(q-1)/2} - 1 & \text{if } p > 2 \end{cases}$$

Using our basis for B, we can easily generate a random element β of B by simply choosing $c_1,\ldots, c_r$ at random, and computing $\beta := \sum_i c_i\beta_i$. If $\theta(\beta) = (b_1,\ldots, b_r)$, then the family of random variables $\{b_i\}_{i=1}^r$ is mutually independent, with each b_i uniformly distributed over F. Just as in Algorithm EDF, $\gcd(\operatorname{rep}(M_1(\beta)), f)$ will be a non-trivial factor of f with probability at least 1/2, if $p = 2$, and probability at least 4/9, if $p > 2$.

That is the basic splitting strategy. We turn this into an algorithm to completely factor f using the same technique of iterative refinement that was used in Algorithm EDF. That is, at any stage of the algorithm, we have a partial factorization $f = \prod_{h \in H} h$, which we try to refine by attempting to split each $h \in H$ using the strategy outlined above. One technical difficulty is that to split such a polynomial h, we need to efficiently generate a random element of the Berlekamp subalgebra of $F[X]/(h)$.

A particularly efficient way to do this is to use our basis for the Berlekamp subalgebra of $F[X]/(f)$ to generate a random element of the Berlekamp subalgebra of $F[X]/(h)$ for all $h \in H$ simultaneously. Let $g_i := \operatorname{rep}(i)$ for $i = 1,\ldots, r$. If we choose $c_1,\ldots, c_r \in F$ at random, and set $g := c_1g_1 + \ldots + c_rg_r$, then $[g]_f$ is a random element of the Berlekamp subalgebra of $F[X]/(f)$, and by the Chinese remainder theorem, it follows that the family of random variables $\{[g]_h\}_{h\in H}$ is mutually independent, with each $[g]_h$ uniformly distributed over the Berlekamp subalgebra of $F[X]/(h)$.

Here is the algorithm for completely factoring a polynomial, given a basis for the corresponding Berlekamp subalgebra.

Algorithm B2

On input $f, \{\beta_i\}_{i=1}^r$, where $f \in F[X]$ is a monic square-free polynomial of degree $\ell > 0$, and $\{\beta_i\}_{i=1}^r$ is a basis for the Berlekamp subalgebra of $F[X]/(f)$, do the following, where $g_i := \text{rep}(\beta_i)$ for $i = 1, \ldots, r$:

$H \leftarrow \{f\}$

while $|H| < r$ do

choose $c_1, \ldots, c_r \in F$ at random

$g \leftarrow c_1 g_1 + \ldots + c_r g_r \in F[X]$

$H' \leftarrow \phi$;

for each $h \in H$ do

$\beta \leftarrow [g]_h \in F[X]/(h)$

$d \leftarrow \gcd(\text{rep}(M_1(\beta)), h)$

if $d = 1$ or $d = h$

then $H' \leftarrow H' \cup \{h\}$

else $H' \leftarrow H' \cup \{d, h/d\}$

$H \leftarrow H'$

output H

The correctness of the algorithm is clear. As for its expected running time, we can get a quick-and-dirty upper bound as follows:

- The cost of generating g in each loop iteration is $O(r\ell)$ operations in F. For a given h, the cost of computing $\beta := [g]_h \in F[X]/(h)$ is $O\ell(\deg(h))$ operations in F, and the cost of computing $M_1(\beta)$ is $O(\deg(h)^2 \text{ len}(q))$ operations in F. Therefore, the number of operations in F performed in each iteration of the main loop is at most a constant times

$$r\ell + \ell \sum_{h \in H} \deg(h) + len(q) \sum_{h \in H} \deg(h)^2$$

$$\leq 2\ell^2 + len(q)\left(\sum_{h \in H} \deg(h)\right)^2 = O(\ell^2 len(q)).$$

- The respected number of iterations of the main loop until we get some nontrivial split is $O(1)$.
- The algorithm finishes after getting $r - 1$ non-trivial splits.
- Therefore, the total expected cost is $O(r \ell^2 \operatorname{len}(q))$ operations in F.

A more careful analysis reveals:

Theorem: Algorithm B2 uses an expected number of

$$O(\operatorname{len}(r) \ell^2 \operatorname{len}(q))$$

operations in F.

Proof: The proof follows the same line of reasoning as the analysis of Algorithm EDF. Indeed, using the same argument as was used there, the expected number of iterations of the main loop is $O(\operatorname{len}(r))$. As discussed in the paragraph above this theorem, the cost per loop iteration is $O(\ell^2 \operatorname{len}(q))$ operations in F. The theorem follows.

The bound in the above theorem is tight: unlike Algorithm EDF, we cannot make the multiplicative factor of len(r) go away.

Theorem: Berlekamp's factoring algorithm uses an expected number of $O(\ell^3 + \ell^2 \operatorname{len}(\ell) \operatorname{len}(q))$ operations in F.

We have assumed the input to Berlekamp's algorithm is a square-free polynomial. However, we may use Algorithm SFD as a preprocessing step to ensure that this is the case. Even if we include the cost of this preprocessing step, the running time estimate in Theorem remains valid.

So we see that Berlekamp's algorithm is faster than the Cantor–Zassenhaus algorithm, whose expected operation count is $O(\ell^3 \operatorname{len}(q))$. The speed advantage of Berlekamp's algorithm grows as q gets large. The one disadvantage of Berlekamp's algorithm is space: it requires space for $\Theta(\ell^2)$ elements of F, while the Cantor–Zassenhaus algorithm requires space for only $O(\ell)$ elements of F. One can in fact implement the Cantor–Zassenhaus algorithm so that it uses $O(\ell^3 + \ell^2 \operatorname{len}(q))$ operations in F, while using space for only $O(\ell^{1.5})$ elements of F.

EXERCISE

1. This exercise develops a variant of the Cantor–Zassenhaus algorithm that uses $O(\ell^3 + \ell^2 \operatorname{len}(q))$ operations in F, while using space for only $O(\ell^{1.5})$ elements of F. Our problem is reduced to that of implementing Algorithm DDF within the stated time and space bounds, assuming that the input polynomial is square-free.

   (*a*) Let $f \in F[X]$ be a monic polynomial of degree $\ell > 0$, and let $m = O(\ell^{1/2})$. Let $\xi := [X]_f \in E$, where $E := F[X] / (f)$. Show how to compute

   $$\xi^q, \xi^{q^2}, \ldots, \xi^{q^{m-1}} \in E \text{ and } \xi^{q^m}, \xi^{q^{2m}}, \ldots, \xi^{q^{(m-1)m}} \in E$$

   using $O(\ell^3 + \ell^2 \operatorname{len}(q))$ operations in F, and space for $O(\ell^{1.5})$ elements of F.

(*b*) Show that for all non-negative integers i, j, with $i \neq j$, the irreducible polynomials in $F[X]$ that divide $X^{q^i} - X^{q^j}$ are precisely those whose degree divides $i - j$.

2. Give an example of a family of input polynomials that cause Algorithm B2 to use an expected number of at least $\Omega(\ell^2 \text{len}(\ell) \text{len}(q))$ operations in F. Assume that computing $M_1(\beta)$ for $\beta \in F[X]/(h)$ takes $\Omega(\deg(h)^2 \text{len}(q))$ operations in F.
3. Using the ideas behind Berlekamp's factoring algorithm, devise a deterministic irreducibility test that, given a monic polynomial of degree ℓ over F, uses $O(\ell^3 + \ell^2 \text{len}(q))$ operations in F.

IIRREDUCIBLE POLYNOMIALS

Let $f \in F[X]$ be a monic polynomial of degree $\ell > 0$. We develop here an efficient algorithm that determines if f is irreducible.

The theorem says that for every integer $k \geq 1$, the polynomial $X^{q^k} - X$ is the product of all monic irreducibles whose degree divides k. Thus, $\gcd(X^q - X, f)$ is the product of all the distinct linear factors of f. If f has no linear factors, then gcd X^{q^2} $(-X, f)$ is the product of all the distinct quadratic irreducible factors of f. And so on. Now, if f is not irreducible, it must be divisible by some irreducible polynomial of degree at most $\ell/2$, and if g is an irreducible factor of f of minimal degree, say k, then we have $k \leq \ell/2$ and gcd X^{q^k} $(-X, f) \neq 1$.

Conversely, if f is irreducible, then gcd X^{q^k} $(-X, f) = 1$ for all positive integers k up to $\ell/2$. So to test if f is irreducible, it suffices to check if gcd X^{q^k} $(-X, f) = 1$ for all positive integers k up to $\ell/2$ – if so, we may conclude that f is irreducible, and otherwise, we may conclude that f is not irreducible. To carry out the computation efficiently, we note that if $h \equiv X^{q^k} \pmod{f}$, then $\gcd(h - X, f) =$ gcd X^{q^k} $(-X, f)$.

Algorithm IPT

On input f, where $f \in F[X]$ is a monic polynomial of degree $\ell > 0$, determine if f is irreducible as follows:

$h \leftarrow X \bmod f$

for $k \leftarrow 1$ to $\lfloor \ell/2 \rfloor$ do

$h \leftarrow h^q \bmod f$

if $\gcd(h - X, f) \neq 1$ then return false

return true

Theorem: Algorithm RIP uses an expected number of $O(\ell^4 \operatorname{len}(q))$ operations in F, and its output is uniformly distributed over all monic irreducibles of degree ℓ.

Theorem: Algorithm IPT uses $O(\ell^3 \operatorname{len}(q))$ operations in F.

Proof: Consider an execution of a single iteration of the main loop. The cost of the qth-powering step (using a standard repeated-squaring algorithm) is $O(\operatorname{len}(q))$ multiplications modulo f, and so $O(\ell^2 \operatorname{len}(q))$ operations in F. The cost of the gcd computation is $O(\ell^2)$ operations in F. Thus, the cost of a single loop iteration is $O(\ell^2 \operatorname{len}(q))$ operations in F, from which it follows that the cost of the entire algorithm is $O(\ell^3 \operatorname{len}(q))$ operations in F.

Using a standard representation for F, each operation in F takes time $O(\operatorname{len}(q)^2)$ on a RAM, and so the running time of Algorithm IPT on a RAM is $O(\ell^3 \operatorname{len}(q)^3)$, which means that it is a polynomial-time algorithm.

Let us now consider the related problem of constructing an irreducible polynomial of specified degree $\ell > 0$. To do this, we can simply use the result of Theorem, which has the following probabilistic interpretation: if we choose a random, monic polynomial f of degree ℓ over F, then the probability that f is irreducible is at least $1/2\ell$. This suggests the following probabilistic algorithm:

Algorithm RIP

On input ℓ, where ℓ is a positive integer, generate a monic irreducible polynomial $f \in F[X]$ of degree ℓ as follows:

repeat

choose $c_0, \ldots, c_{\ell-1} \in F$ at random

set $f \leftarrow X^\ell + \sum_{i=0}^{\ell-1} c_i X^i$

test if f is irreducible using Algorithm IPT

until f is irreducible

output f

EXERCISE

1. Design and analyze a probabilistic algorithm that, given a monic irreducible polynomial $f \in F[X]$ of degree ℓ as input, generates as output a random monic irreducible polynomial $g \in F[X]$ of degree ℓ (i.e., g should be uniformly distributed over all such polynomials), using an expected number of $O(\ell^{2.5})$ operations in F.
2. This exercise develops an alternative irreducibility test.
   (a) Show that the operation count in part (b) can be reduced to
   $O(\ell^{2.5} \operatorname{len}(\ell) \operatorname{len}(\omega(\ell)) + \ell^2 \operatorname{len}(q))$.

(*b*) Show that a monic polynomial $f \in F[X]$ of degree $\ell > 0$ is irreducible if and only if $X^{q^{\ell}} \equiv X \pmod{f}$ and gcd $X^{q^{\ell/s}}(-X, f) = 1$ for all primes $s \mid \ell$.

(*b*) Using part (a) and the result of the previous exercise, show how to determine if f is irreducible using $O(\ell^{2.5}\operatorname{len}(\ell)\omega(\ell) + \ell^2\operatorname{len}(q))$ operations in F, where $\omega(\ell)$ is the number of distinct prime factors of ℓ.

3. Let $f \in F[X]$ be a monic irreducible polynomial of degree ℓ, let $E := F[X]/(f)$, and let $\xi := [X]_f \in E$. Design and analyze a deterministic algorithm that takes as input the polynomial f defining the extension E, and outputs the values

$$s_j := Tr_{E/F}(\xi^j) \in F \; (j = 0, \ldots, \ell - 1),$$

using $O(\ell^2)$ operations in F. Here, TrE/F is the trace from E to F. Show that given an arbitrary $\alpha \in E$, along with the values $s_0, \ldots, s_{\ell-1}$, one can compute $Tr_{E/F}(\alpha)$ using just $O(\ell)$ operations in F.

4. Design and analyze a deterministic algorithm that takes as input a list of irreducible polynomials $f_1, \ldots, f_r \in F[X]$, where $\ell_i := \deg(f_i)$ for $i = 1, \ldots, r$, and assume that $\{\ell_i\}_{i=1}^{r}$ is pairwise relatively prime. Your algorithm should output an irreducible polynomial $f \in F[X]$ of degree $\ell := \prod_{i=1}^{r} \ell_i$ using $O(\ell^3)$ operations in F.

5. Let $f \in F[X]$ be a monic polynomial of degree $\ell > 0$. Also, let $\xi := [X]_f \in E$, where E is the F-algebra $E := F[X]/(f)$.

(*a*) Given as input $\xi^q \in E$ and a positive integer m, show how to compute the value $\xi^{q^m} \in E$, using $O(\ell^{2.5}\operatorname{len}(m))$ operations in F, and space for $O(\ell^{1.5})$ elements of F.

(*b*) Given as input $\alpha \in E$ and $\xi^{q^m} \in E$ (for some integer $m > 0$), show how to compute the value $\alpha^{q^m} \in E$, using just $O(\ell^{2.5})$ operations in F, and space for $O(\ell^{1.5})$ elements of F.

(*c*) Given as input $\xi^{q^m} \in E$ and $\xi^{q^{m'}} \in E$, where m and m' are positive integers, show how to compute the value $\xi^{q^{m+m'}}$, using $O(\ell^{2.5})$ operations in F, and space for $O(\ell^{1.5})$ elements of F.

DETERMINISTIC FACTORIZATION ALGORITHMS

The algorithms of Cantor and Zassenhaus and of Berlekamp are probabilistic. Deterministic variants of Berlekamp's algorithm, with similar complexity can be developed easily.

This algorithm is only practical for finite fields of small characteristic, and is anyway mainly of theoretical interest, since from a practical perspective, there is nothing wrong with the above probabilistic method. In all of these exercises, we assume that we have access to a basis $\{\varepsilon_i\}_{i=1}^{w}$ for F as a vector space over $\mathbb{Z}_p$. To make the Cantor–Zassenhaus algorithm deterministic, we only need to develop a deterministic variant of Algorithm EDF.

EXERCISE

1. Suppose that $S = \{\lambda_1,..., \lambda_s\}$ is a separating set for f over $\mathbb{Z}_p$, and $\phi_u \in F[X]$ is the minimal polynomial over F of $[\lambda_u]_f \in F[X]/(f)$ for $u = 1,..., s$. Show that each ϕ_u is the product of linear factors over $\mathbb{Z}_p$, and that given S, along with the roots of all the ϕ_u's, we can deterministically factor f using $(|S| + \ell)O(1)$ operations in F.
2. Using the previous exercise, show that the problem of factoring a polynomial over F reduces in deterministic polynomial time to the problem of finding roots of polynomials over $\mathbb{Z}_p$.
3. Let $f = f_1 ... f_r$, where the f_i's are distinct monic irreducible polynomials in $F[X]$. Assume that $r > 1$, and let $\ell := \deg(f)$. For this exercise, the degrees of the f_i's need not be the same. For an intermediate field F', with $\mathbb{Z}_p \subseteq F' \subseteq F$, let us call a set $S = \{\lambda_1,..., \lambda_s\}$, where each $\lambda_u \in F[X]$ with $\deg(\lambda_u) < \ell$, a separating set for f over F' if the following conditions hold:

   (*a*) For $i = 1,..., r$ and $u = 1,..., s$, there exists $c_{ui} \in F'$ such that $\lambda_u \equiv c_{ui} \pmod{f_i}$, and

   (*b*) For every pair of distinct indices i, j, with $1 \leq i < j \leq r$, there exists $u = 1,..., s$ such that $c_{ui} \neq c_{uj}$.

   Show that if S is a separating set for f over $\mathbb{Z}_p$, then the following algorithm completely factors f using $O(p\,|S|\,\ell^2)$ operations in F.

   $H \leftarrow \{f\}$

   for each $\lambda \in S$ do

   $H' \leftarrow \phi$

   for each $a \in \mathbb{Z}_p$ do

   $H' \leftarrow \phi$;

   for each $h \in H$ do

   $d \leftarrow \gcd(\lambda - a, h)$

   if $d = 1$ or $d = h$

   then $H' \leftarrow H' \cup \{h\}$

else $H' \leftarrow H' \cup \{d, h/d\}$

$H \leftarrow H'$

output H

4. Put together all of the above pieces, together with Algorithms SFD and DDF, so as to obtain a deterministic algorithm for factoring polynomials over F that runs in time at most p times a polynomial in the size of the input, and make a careful estimate of the running time of your algorithm.

5. Let f be as in the previous exercise. Show that if S is a sepa rating set for f over F, then the set

$$S' := \left\{ \sum_{i=0}^{w-1} (\varepsilon_j \lambda)^{p^i} \bmod f : 1 \leq w,\ \lambda \in S \right\}$$

is a separating set for f over $\mathbb{Z}_p$. Show how to compute this set using $O(|S|\ell^2 \operatorname{len}(p) w(w-1))$ operations in F.

6. Let f be as in the previous two exercises, but further suppose that each irreducible factor of f is of the same degree, say k. Let $E := F[X]/(f)$ and $\xi := [X]_f \in E$. Define the polynomial $\phi \in E[Y]$ as follows:

$$\prod_{i=0}^{k-1} (Y - \xi^{q^i}).$$

If

$\phi = Y^k + \alpha_{k-1} Y^{k-1} + \dots + \alpha_0,$

with $\alpha_0, \dots, \alpha_{k-1} \in E$, show that the set

$S := \{\operatorname{rep}(\alpha_i) : 0 \leq i \leq k-1\}$

is a separating set for f over F, and can be computed deterministically using $O(k^2 + k \operatorname{len}(q))$ operations in E, and hence $O(k^2 \ell^2 + k \ell^2 \operatorname{len}(q))$ operations in F.

7. It is a fact that when our prime p is odd, then for all integers a, b, with $a \not\equiv b \pmod p$, there exists a non-negative integer $i \leq p^{1/2} \log 2\, p$ such that $(a + i \mid p) \neq (b + i \mid p)$ (here, "(. |.)" is the Legendre symbol). Using this fact, design and analyze a deterministic algorithm for factoring polynomials over F that runs in time at most $p^{1/2}$ times a polynomial in the size of the input.

The following two exercises show that the problem of factoring polynomials over F reduces in deterministic polynomial time to the problem of finding roots of polynomials over $\mathbb{Z}p$.

CANTOR-ZASSENHAUS ALGORITHM

EQUAL DEGREE FACTORIZATION

Each of the factors produced in the distinct degree factorization stage are further factored into their irreducible factors. The algorithm we present for distinct degree factorization is a deterministic, polynomial-time algorithm. The algorithm we present for equal degree factorization is a probabilistic algorithm that runs in expected polynomial time (and whose output is always correct).

DISTINCT DEGREE FACTORIZATION

The input polynomial is decomposed into factors so that each factor is a product of distinct irreducibles of the same degree (and the degree of those irreducibles is also determined).

EQUAL DEGREE FACTORIZATION

The problem, more precisely stated, is this: given a monic polynomial $f \in F[X]$ of degree $\ell > 0$, and an integer $k > 0$, such that f is of the form

$$f = f_1 \cdots f_{rs}$$

for distinct monic irreducible polynomials $f_1, \ldots, f_r$, each of degree k, compute these irreducible factors of f. Note that given f and k, the value of r is easily determined, since $r = \ell / k$.

We begin by discussing the basic mathematical ideas that will allow us to efficiently split f into two non-trivial factors, and then we present a somewhat more elaborate algorithm that completely factors f.

By the Chinese remainder theorem, we have an F-algebra isomorphism

$$\theta: E \to E_1 \times \ldots \times E_r$$

$$[g]_f \mapsto ([g]_{f_1}, \ldots, [g]_{f_r}),$$

where E is the F-algebra $F[X]/(f)$, and for $i = 1, \ldots, r$, E_i is the extension field $F[X]/(f_i)$ of degree k over F.

Recall that $q = p^w$. We have to treat the cases $p = 2$ and $p > 2$ separately. We first treat the case $p = 2$. Let us define the polynomial

$$M_k := \sum_{j=0}^{wk-1} X^{2^j} \in F[X].$$

(The algorithm in the case $p > 2$ will only differ in the definition of M_k.)

For $\alpha \in E$, if $\theta(\alpha) = (\alpha_1, \ldots, \alpha_r)$, then we have

$$(M_k(\alpha_i)) = M_k(\theta(\alpha)) = (M_k(\alpha_1), \ldots, M_k(\alpha_r)).$$

Note that each E_i is an extension of $\mathbb{Z}_2$ of degree wk, and that

$$M_k(\alpha_i) = \sum_{j=0}^{wk-1} \alpha_i^{2^j} = Tr_{E_i/\mathbb{Z}_2}(\alpha_i)$$

where $Tr_{E_i/\mathbb{Z}_2} : E_i \to \mathbb{Z}_2$ is the trace from Ei to $\mathbb{Z}_2$, which is a surjective, $\mathbb{Z}_2$-linear map.

Now, suppose we choose $\alpha \in E$ at random. Then if $\theta(\alpha) = (\alpha_1, \ldots, \alpha_r)$, the family of random variables $\{\alpha_i\}_{i=1}^r$ is mutually independent, with each α_i uniformly distributed over E_i. It follows that the family of random variables $\{M_k(\alpha_i)\}_{i=1}^r$ is mutually independent, with each $M_k(\alpha_i)$ uniformly distributed over $\mathbb{Z}_2$. Thus, if g:= rep $(M_k(\alpha))$ (i.e., $g \in F[X]$ is the polynomial of degree less than ℓ such that $M_k(\alpha) = [g]_f$), then gcd(g, f) will be the product of those factors f_i of f such that $M_k(\alpha_i) = 0$. We will fail to get a non-trivial factorization only if the $M_k(\alpha_i)$ are either all 0 or all 1, which for $r \geq 2$ happens with probability at most 1/2 (the worst case being when $r = 2$).

That is our basic splitting strategy. The algorithm for completely factoring f works as follows. The algorithm proceeds in stages. At any stage, we have a partial factorization $f = \prod_{h \in H} h$, where H is a set of non-constant, monic polynomials. Initially, $H = \{f\}$. With each stage, we attempt to get a finer factorization of f by trying to split each $h \in H$ using the above splitting strategy – if we succeed in splitting h into two non-trivial factors, then we replace h by these two factors. We continue in this way until $|H| = r$.

Algorithm EDF

On input f, k, where $f \in F[X]$ is a monic polynomial of degree $\ell > 0$, and k is a positive integer, such that f is the product of $r := \ell / k$ distinct monic irreducible polynomials, each of degree k, do the following, with M_k:

$H \leftarrow \{f\}$

while $|H| < r$ do

$H' \leftarrow \phi$;

for each $h \in H$ do

choose $\alpha \in F[X]/(h)$ at random

$d \leftarrow$ gcd(rep($M_k(\alpha)$), h)

if $d = 1$ or $d = h$

then $H' \leftarrow H' \cup \{h\}$

else $H' \leftarrow H' \cup \{d, h/d\}$

$H \leftarrow H'$

output H

The correctness of the algorithm is clear from the above discussion. As for its expected running time, we can get a quick-and-dirty upper bound as follows:

- For a given h and $\alpha \in F[X] / (h)$, the value $M_k(\alpha)$ can be computed using $O(k \deg(h)^2 \operatorname{len}(q))$ operations in F, and so the number of operations in F performed in each iteration of the main loop is at most a constant times

$$k\, len(q) \sum_{h \in H} \deg(h)^2 \leq k\, len(q) \left(\sum_{h \in H} \deg(h) \right)^2 = k\ell^2 len(q).$$

- The expected number of iterations of the main loop until we get some nontrivial split is $O(1)$.
- The algorithm finishes after getting $r - 1$ non-trivial splits.
- Therefore, the total expected cost is $O(rk\ell^2 \operatorname{len}(q))$, or $O(\ell^3 \operatorname{len}(q))$, operations in F.

This analysis gives a bit of an over-estimate – it does not take into account the fact that we expect to get fairly "balanced" splits. For the purposes of analyzing the overall running time of the Cantor–Zassenhaus algorithm, this bound suffices; however, the following analysis gives a tight bound on the complexity of Algorithm EDF.

Theorem: In the case $p > 2$, Algorithm EDF uses an expected number of $O(k\ell^2 \operatorname{len}(q))$ operations in F.

Proof: The analysis is essentially the same as in the case $p = 2$, except that now the probability that we fail to split a given pair of irreducible factors is at most 5/9, rather than equal to 1/2. The details are left as an exercise for the reader.

Theorem: In the case $p = 2$, Algorithm EDF uses an expected number of $O(k\ell^2 \operatorname{len}(q))$ operations in F.

Proof: We may assume $r \geq 2$. Let L be the random variable that represents the number of iterations of the main loop of the algorithm. For $n \geq 1$, let H_n be the random variable that represents the value of H at the beginning of the nth loop iteration. For $i, j = 1,..., r$, we define L_{ij} to be the largest value of n (with $1 \leq n \leq L$) such that $f_i \mid h$ and $f_j \mid h$ for some $h \in H_n$.

We first claim that $E[L] = O(\operatorname{len}(r))$. To prove this claim, we make use of the fact that

$$E[L] = \sum_{n \geq 1} P[L \geq n].$$

Now, $L \geq n$ if and only if for some i, j with $1 \leq i < j \leq r$, we have $L_{ij} \geq n$. Moreover, if f_i and f_j have not been separated at the beginning of one loop iteration, then they will be separated at the beginning of the next with probability 1/2. It follows that

$$P[L_{ij} \geq n] = 2^{-(n-1)}.$$

So we have

$$P[L \geq n] \leq \sum_{i \geq j} P[L_{ij} \geq n] \leq r^2 2^{-n}.$$

Therefore,

$$\mathrm{E}[L] = \sum_{n \geq 1} P[L \geq n] = \sum_{n \leq 2\log_2 r} P[L \geq n] + \sum_{n > 2\log_2 r} P[L \geq n]$$

$$\leq 2\log_2 r + \sum_{n > 2\log_2 r} r^2 2^{-n} \leq 2\log_2 r + \sum_{n \geq 0} 2^{-n} = 2\log_2 r + 2,$$

which proves the claim.

As discussed in the paragraph above this theorem, the cost of each iteration of the main loop is $O(k\ell^2 \operatorname{len}(q))$ operations in F. Combining this with the fact that $\mathrm{E}[L] = O(\operatorname{len}(r))$, it follows that the expected number of operations in F for the entire algorithm is $O(\operatorname{len}(r)k\ell^2 \operatorname{len}(q))$. This is significantly better than the above quick-and-dirty estimate, but is not quite the result we are after. For this, we have to work a little harder.

For each polynomial h dividing f, define $\omega(h)$ to be the number of irreducible factors of h. Let us also define the random variable

$$S := \sum_{n=1}^{L} \sum_{h \in H_n} \omega(h)^2.$$

It is easy to see that the total number of operations performed by the algorithm is $O(Sk^3 \operatorname{len}(q))$, and so it will suffice to show that $E[S] = O(r^2)$.

We claim that

$$S := \sum_{i,j} L_{ij},$$

where the sum is over all $i, j = 1,..., r$. To see this, define $\delta_{ij}(h)$ to be 1 if both f_i and f_j divide h, and 0 otherwise. Then we have

$$S := \sum_{n} \sum_{h \in H} \sum_{i,j} \delta_{ij}(h) = \sum_{i,j} \sum_{n} \sum_{h \in H_n} \delta_{ij}(h) = \sum_{i,j} L_{ij},$$

which proves the claim.

We can write

$$S = \sum_{i \neq j} L_{ij} + \sum_{i} L_{ii} = \sum_{i \neq j} L_{ij} + rL.$$

For $i \neq j$, we have

$$E[L_{ij}] = \sum_{n\geq 1} P[L_{ij} \geq n] = \sum_{i\geq 1} 2^{-(n-1)} = 2,$$

and so

$$E[S] = \sum_{i\neq j} E[L_{ij}] + rE[L] = 2r(r-1) + O(r\ \mathrm{len}(r)) = O(r^2).$$

That proves the theorem.

That completes the discussion of Algorithm EDF in the case $p = 2$. Now assume that $p > 2$, so that p, and hence also q, is odd. Algorithm EDF in this case is exactly the same as above, except that in this case, we define the polynomial M_k as

$$M_k := X^{(q^k-1)/2} - 1 \in F[X].$$

Just as before, for $\alpha \in E$ with $\theta(\alpha) = (\alpha_1, \ldots, \alpha_r)$, we have

$$\theta(M_k(\alpha)) = M_k(\theta(\alpha)) = (M_k(\alpha_1), \ldots, M_k(\alpha_r)).$$

Note that each group E_i^* is a cyclic group of order $q^k - 1$, and therefore, the image of the $(q^k - 1)/2$-power map on E_i^* is $\{\pm 1\}$.

Now, suppose we choose $\alpha \in E$ at random. Then if $\theta(\alpha) = (\alpha_1, \ldots, \alpha_r)$, the family of random variables $\{\alpha_i\}_{i=1}^r$ is mutually independent, with each α_i uniformly distributed over E_i. It follows that the family of random variables $\{M_k(\alpha_i)\}_{i=1}^r$ is mutually independent. If $\alpha_i = 0$, which happens with probability $1/q^k$, then $M_k(\alpha_i) = -1$; otherwise, $\alpha_i^{(q^k-1)/2}$ is uniformly distributed over $\{\pm 1\}$, and so $M_k(\alpha_i)$ is uniformly distributed over $\{0, -2\}$. That is to say,

$$M_k(\alpha_i) = \begin{cases} 0 & \text{with probability } (q^k-1)/2q^k, \\ -1 & \text{with probability } 1/q^k, \\ -2 & \text{with probability } (q^k-1)/2q^k. \end{cases}$$

Thus, if $g := \mathrm{rep}(M_k(\alpha))$, then $\gcd(g, f)$ will be the product of those factors f_i of f such that $M_k(\alpha_i) = 0$. We will fail to get a non-trivial factorization only if the $M_k(\alpha_i)$ are either all zero or all non-zero. Assume $r \geq 2$. Consider the worst case, namely, when $r = 2$. In this case, a simple calculation shows that the probability that we fail to split these two factors is

$$\left(\frac{q^k-1}{2q^k}\right)^2 + \left(\frac{q^k+1}{2q^k}\right)^2 = \frac{1}{2}(1 + 1/q^{2k}).$$

The (very) worst case is when $q^k = 3$, in which case the probability of failure is at most 5/9.

DEGREE FACTORIZATION

The problem, more precisely stated, is this: given a monic, square-free polynomial $f \in F[X]$ of degree $\ell > 0$, produce a list of pairs $((g_1, k_1),..., (g_t, k_t))$ where

- Each g_i is the product of monic irreducible polynomials of degree k_i, and
- $f = \prod_{i=1}^{t} g_i$.

This problem can be easily solved using Theorem, using a simple variation of the algorithm for irreducibility testing. The basic idea is this. We can compute $g := \gcd(X^q - X, f)$, so that g is the product of all the linear factors of f. After removing all linear factors from f, we next compute $\gcd_{X^{q^2}}(- X, f)$, which will be the product of all the quadratic irreducibles dividing f, and we can remove these from f – although $X^{q^2} - X$ is the product of all linear and quadratic irreducibles, since we have already removed the linear factors from f, the gcd will give us just the quadratic factors of f. In general, for $k = 1,..., \ell$, having removed all the irreducible factors of degree less than k from f, we compute $\gcd_{X^{q^k}}(-X, f)$ to obtain the product of all the irreducible factors of f of degree k, and then remove these from f.

The above discussion leads to the following algorithm for distinct degree factorization.

Algorithm DDF

On input f, where $f \in F[X]$ is a monic square-free polynomial of degree $\ell > 0$, do the following:

```
initialize an empty list L
        h ← X mod f
        k ← 0
while f ≠ 1 do
        h ← h^q mod f, k ← k + 1
        g ← gcd(h – X, f)
if g ≠ 1 then
append (g, k) to L
        f ← f/g
        h ← h mod f
output L
```

Theorem: Algorithm DDF uses $O(\ell^3 \text{len}(q))$ operations in F.

Proof: Note that the body of the main loop is executed at most ℓ times, since after ℓ iterations, we will have removed all the factors of f. Thus, we perform at most ℓ qth-powering steps, each of which takes $O(\ell^2 \text{len}(q))$ operations in F, and so the total contribution to the running time of these is $O(\ell^3 \text{len}(q))$ operations in F. We also have to take into account the cost of the gcd and division computations. The cost per loop iteration of these is $O(\ell^2)$ operations in F, contributing a term of $O(\ell^3)$ to the total operation count. This term is dominated by the cost of the $q$thpowering steps, and so the total cost of Algorithm DDF is $O(\ell^3 \text{len}(q))$ operations in F.

ANALYSIS OF THE ALGORITHM

Given an arbitrary monic square-free polynomial $f \in F[X]$ of degree $\ell > 0$, the distinct degree factorization step takes $O(\ell^3 \text{len}(q))$ operations in F. This step produces a number of polynomials that must be further subjected to equal degree factorization. If there are t such polynomials, where the ith polynomial has degree ℓ_i, for $i = 1,..., t$, then $\sum_{i=1}^{t} \ell_i = \ell$. Now, the equal degree factorization step for the ith polynomial takes an expected number of $O\,\ell_i^3(\text{len}(q))$ operations in F (actually, our initial, "quick and dirty" estimate is good enough here), and so it follows that the total expected cost of all the equal degree factorization steps is $O \sum_i \ell_i^3 (\text{len}(q))$, which is $O(\ell^3 \text{len}(q))$, operations in F. Putting this all together, we conclude:

Theorem: The Cantor-Zassenhaus factoring algorithm uses an expected number of $O(\ell^3 \text{len}(q))$ operations in F.

This bound is tight, since in the worst case, when the input is irreducible, the algorithm really does do this much work. Also, we have assumed the input to the Cantor–Zassenhaus is a square-free polynomial. However, we may use Algorithm SFD as a preprocessing step to ensure that this is the case.

EXERCISE

1. Let E be an extension field of degree ℓ over F, specified by an irreducible polynomial of degree ℓ over F. Design and analyze an efficient probabilistic algorithm that finds a normal basis for E over F.
2. Show how to modify Algorithm DDF so that the main loop halts as soon as $2k > \deg(f)$.
3. This exercise extends the techniques developed in Exercise 20.1. Let $f \in F[X]$ be a monic polynomial of degree $\ell > 0$, and let $\ell := [X]_f \in E$, where $E := F[X]/(f)$. For each integer $m > 0$, define polynomials

$$T_m := X + X^q + ... + X^{q^{m-1}} \in F[X] \text{ and } N_m : = X. X^q .... X^{q^{m-1}} \in F[X].$$

(*a*) Given as input $\xi^{q^m} \in E$ and $\xi^{q^{m'}}$, where m and m' are positive integers, along with $T_m(\alpha)$ and $T_{m'}(\alpha)$, for some $\alpha \in E$, show how to compute the values $\xi^{q^{m+m'}}$ and $T_{m+m'}(\alpha)$, using $O(\ell^{2.5})$ operations in F, and space for $O(\ell^{1.5})$ elements of F.

(*b*) Given as input ξ^q α E, $\alpha \in E$, and a positive integer m, show how to compute (using part (a)) the value $T_m(\alpha)$, using $O(\ell^{2.5}\ \mathrm{len}(m))$ operations in F, and space for $O(\ell^{1.5})$ elements of F.

(*c*) Repeat parts (a) and (b), except with "*N*" in place of "*T*."

4. Using the result of the previous exercise, show how to implement Algorithm EDF so that it uses an expected number of

$$O(\mathrm{len}(k)\,\ell^{2.5} + \ell^2\,\mathrm{len}(q))$$

operations in F, and space for $O(\ell^{1.5})$ elements of F.

5. Suppose that in Algorithm EDF, we replace the two lines

for each $h \in H$ do

choose $\alpha \in F[X]/(h)$ at random

by the following:

choose $a_0,\ldots, a_{2k-1} \in F$ at random

$g \leftarrow \sum_{j=0}^{2k-1} a_j X^j \in F[X]$

for each $h \in H$ do

$\alpha \leftarrow [g]_h \in F[X]/(h)$

Show that the expected running time bound of Theorem still holds.

Chapter 6

Matrices

In mathematics, plural matrices is a rectangular table of *elements* (or *entries*), which may be numbers or, more generally, any abstract quantities that can be added and multiplied. Matrices are used to describe linear equations, keep track of the coefficients of linear transformations and to record data that depend on multiple parameters. Matrices are described by the field of matrix theory. Matrices can be added, multiplied, and decomposed in various ways, which also makes them a key concept in the field of linear algebra.

The horizontal lines in a matrix are called rows and the vertical lines are called columns. A matrix with m rows and n columns is called an m-by-n matrix (written $m \times n$) and m and n are called its dimensions. The dimensions of a matrix are always given with the number of rows first, then the number of columns. It is commonly said that an m-by-n matrix has an order of m X n ("order" meaning size). Two matrices of the same order whose corresponding entries are equivalent are considered equal.

Almost always capital letters denote matrices with the corresponding lower-case letters with two indices representing the entries. For example, the entry of a matrix A that lies in the i-th row and the j-th column is written as $a_{i,j}$ and called the i,j entry or (i,j)-th entry of A. Alternative notations for that entry are A[i,j] or $A_{i,j}$.

The row is always noted first, then the column. In this example, A (with no subscripts) would symbolize the entire matrix. In addition to using uppercase letters as symbols representing matrices, many authors use a special typographical style, commonly boldface upright (non-italic), to further distinguish matrices from other variables. Following this convention, A is a matrix, distinguished from A, a scalar. An alternate convention is to annotate matrices with their dimensions in small type underneath the symbol, for example, $\underset{r \times c}{A}$ for an r-by-c matrix.

We often write A: $(a_{i,j})_{i=1,m;\, j=1,\ldots n}$ or A: = $(a_{i,j})$ to define an $m \times n$ matrix A. In this case, the entries $a_{i,j}$ are defined separately for all integers $1 \leq i$ £ m and $1 \leq j \leq n$. In some programming languages, the numbering of rows and columns starts at zero. Texts which use any such language extensively frequently follow that convention, so we have $0 \le i \le m-1$ and $0 \le j \le n-1$.

A matrix where one of the dimensions equals one is often called a *vector*, and interpreted as an element of real coordinate space. An $m \times 1$ matrix (one column and m rows) is called a column vector and a $1 \times n$ matrix (one row and n columns) is called a row vector.

MATHEMATICAL DEFINITION

An m X n $(m, n \in \mathbb{N})$ matrix A is a function $A: \{1, 2, ..., m\} \times \{1, 2, ..., n\} \to S$ where S is any non-empty set.

$(\{1, 2, ..., m\} \times \{1, 2, ..., n\}$ is the Cartesian product of sets $\{1, 2, ..., m\}$ and $(\{1, 2, ..., n\})$.

We say that matrix A is a matrix over the set S. Important thing to note is that, if we want to have matrix algebra, the set S must be a ring and matrix A must be a square matrix. Since the set of all square matrices over a ring is also a ring, matrix algebra is usually called matrix ring.

Since this article mainly considers matrices over real numbers, matrices shown here are actually functions $A: (\{1, 2, ..., m\} \times (\{1, 2, ..., n\} \to \mathbb{R}$.

Example. The matrix

$$A = \begin{bmatrix} 9 & 8 & 6 \\ 1 & 2 & 7 \\ 4 & 9 & 2 \\ 6 & 0 & 5 \end{bmatrix} \text{ or } A = \begin{pmatrix} 9 & 8 & 6 \\ 1 & 2 & 7 \\ 4 & 9 & 2 \\ 6 & 0 & 5 \end{pmatrix}$$

is a matrix 4 × 3. The element $a_{2,3}$ or A [2, 3] is 7. In terms of the mathematical definition given above, this matrix is a function $A: \{1, 2, 3, 4\} \times (\{1, 2, 3\} \to \mathbb{R}$ and, for example, A ((2, 3)) = 7 and A ((3, 1)) = 4.

The matrix

$$R = \begin{bmatrix} 1 & 2 & 3 & 4 & 5 & 6 & 7 & 8 & 9 \end{bmatrix}$$

is a 1 × 9 matrix, or 9-element row vector.

BASIC OPERATIONS

Sum

Two or more matrices of identical dimensions m and n can be added. Given m-by-n

matrices A and B, their sum A+B is the m-by-n matrix computed by adding corresponding elements:

For example:

$$\begin{bmatrix}1 & 3 & 1\\ 1 & 0 & 0\\ 1 & 2 & 2\end{bmatrix}+\begin{bmatrix}0 & 0 & 5\\ 7 & 5 & 0\\ 2 & 1 & 1\end{bmatrix}=\begin{bmatrix}1+0 & 3+0 & 1+5\\ 1+7 & 0+5 & 0+0\\ 1+2 & 2+1 & 2+1\end{bmatrix}=\begin{bmatrix}1 & 3 & 6\\ 8 & 5 & 0\\ 3 & 3 & 3\end{bmatrix}$$

Another, much less often used notion of matrix addition is the direct sum.

SCALAR MULTIPLICATION

Given a matrix A and a number c, the scalar multiplication cA is computed by multiplying every element of A by the scalar c (i.e.) $(cA)_{i,j} = c \cdot a_{ij}$). For example:

$$2\cdot\begin{bmatrix}1 & 8 & -3\\ 4 & -2 & 5\end{bmatrix}=\begin{bmatrix}2\cdot 1 & 2\cdot 8 & 2\cdot -3\\ 2\cdot 4 & 2\cdot -2 & 2\cdot 5\end{bmatrix}=\begin{bmatrix}2 & 16 & -6\\ 8 & -4 & 10\end{bmatrix}.$$

Matrix addition and scalar multiplication turn the set $M(m, n, \mathbb{R})$ of all m-by-n matrices with real entries into a real vector space of dimension $m \cdot n$.

MATRIX MULTIPLICATION

Multiplication of two matrices is well-defined only if the number of columns of the left matrix is the same as the number of rows of the right matrix. If A is an m-by-n matrix and B is an n-by-p matrix, then their matrix product AB is the m-by-p matrix given by:

$$(\mathrm{AB})_{ij} = a_{i,1}b_{1,j} + a_{i,2}b_{2,j} + \ldots + a_{i,n}b_{n,j}$$

for each pair (i,j). For example:

$$\begin{bmatrix}1 & 0 & 2\\ -1 & 3 & 1\end{bmatrix}\times\begin{bmatrix}3 & 1\\ 2 & 1\\ 1 & 0\end{bmatrix}=\begin{bmatrix}(1\times 3+0\times 2+2\times 1) & (1\times 1+0\times 1+2\times 0)\\ (-1\times 3+3\times 2+1\times 1) & (-1\times 1+3\times 1+1\times 0)\end{bmatrix}$$

$$=\begin{bmatrix}5 & 1\\ 4 & 2\end{bmatrix}.$$

Matrix multiplication has the following properties:

- (AB)C = A(BC) for all k-by-m matrices A, m-by-n matrices B and n-by-p matrices C ("associativity").
- (A+B)C = AC+BC for all m-by-n matrices A and B and n-by-k matrices C ("right distributivity").

- C(A+B) = CA+CB for all m-by-n matrices A and B and k-by-m matrices C ("left distributivity").

Matrix multiplication is not commutative; that is, given matrices A and B and their product defined, then generally AB ≠ BA. It may also happen that AB is defined but BA is not defined.

Besides the ordinary matrix multiplication just described, there exist other operations on matrices that can be considered forms of multiplication, such as the Hadamard product and the Kronecker product.

LINEAR TRANSFORMATIONS

Matrices can conveniently represent linear transformations because matrix multiplication neatly corresponds to the composition of maps, as will be described next. This same property makes them powerful data structures in high-level programming languages.

Here and in the sequel we identify R^n with the set of "columns" or n-by-1 matrices. For every linear map $f: R^n \rightarrow R^m$ there exists a unique m-by-n matrix A such that $f(x) = Ax$ for all x in R^n. We say that the matrix A "represents" the linear map f. Now if the k-by-m matrix B represents another linear map $g: R^m \rightarrow R^k$, then the linear map $g \circ f$ is represented by BA. This follows from the above-mentioned associativity of matrix multiplication.

More generally, a linear map from an n-dimensional vector space to an m-dimensional vector space is represented by an m-by-n matrix, provided that bases have been chosen for each.

RANKS

The rank of a matrix A is the dimension of the image of the linear map represented by A; this is the same as the dimension of the space generated by the rows of A, and also the same as the dimension of the space generated by the columns of A. It can also be defined without reference to linear algebra as follows: the rank of an m-by-n matrix A is the smallest number k such that A can be written as a product BC where B is an m-by-k matrix and C is a k-by-n matrix (although this is not a practical way to compute the rank).

TRANSPOSE

The transpose of an m-by-n matrix A is the n-by-m matrix A^{tr} (also sometimes written as A^T or ${}^tA$) formed by turning rows into columns and columns into rows, i.e. $A^{tr}[i, j] = A[j, i]$ for all indices i and j. If A describes a linear map with respect to two bases, then the matrix A^{tr} describes the transpose of the linear map with respect to the dual bases.

We have $(A + B)^{tr} = A^{tr} + B^{tr}$ and $(AB)^{tr} = B^{tr} A^{tr}$.

SQUARE ROOT

Given two bounded matrices T and B, B is a square root of T if T = B*B.

SQUARE MATRICES AND RELATED DEFINITIONS

A square matrix is a matrix which has the same number of rows and columns. The set of all square n-by-n matrices, together with matrix addition and matrix multiplication is a ring. Unless $n = 1$, this ring is not commutative.

M(n, R), the ring of real square matrices, is a real unitary associative algebra. M(n, C), the ring of complex square matrices, is a complex associative algebra.

The unit matrix or identity matrix I_n, with elements on the main diagonal set to 1 and all other elements set to 0, satisfies $MI_n = M$ and $I_nN = N$ for any m-by-n matrix M and n-by-k matrix N. For example, if $n = 3$:

$$I_3 = \begin{bmatrix} 1 & 0 & 0 \\ 0 & 1 & 0 \\ 0 & 0 & 1 \end{bmatrix}.$$

The identity matrix is the identity element in the ring of square matrices.

Invertible elements in this ring are called invertible matrices or non-singular matrices. An n by n matrix A is invertible if and only if there exists a matrix B such that

$$AB = I_n \ (= BA).$$

In this case, B is the inverse matrix of A, denoted by A^{-1}. The set of all invertible n-by-n matrices forms a group (specifically a Lie group) under matrix multiplication, the general linear group.

If λ is a number and v is a non-zero vector such that $Av = \lambda v$, then we call v an eigenvector of A and λ the associated eigenvalue. (*Eigen* means "own" in German and in Dutch.) The number λ is an eigenvalue of A if and only if $A-\lambda I_n$ is not invertible, which happens if and only if $p_A(\lambda) = 0$. Here $p_A(x)$ is the characteristic polynomial of A.

This is a polynomial of degree n and has therefore n complex roots (counting multiple roots according to their multiplicity). In this sense, every square matrix has n complex eigenvalues.

The determinant of a square matrix A is the product of its n eigenvalues, but it can also be defined by the *Leibniz formula*. Invertible matrices are precisely those matrices with a nonzero determinant.

The Gaussian elimination algorithm is of central importance: it can be used to compute determinants, ranks and inverses of matrices and to solve systems of linear equations.

SPECIAL TYPES OF MATRICES

In many areas in mathematics, matrices with certain structure arise. A few important examples are

- Symmetric matrices are such that elements symmetric about the *main diagonal* (from the upper left to the lower right) are equal, that is, $a_{i,j} = a_{j,i} \Leftrightarrow A^T = A$.
- Skew-symmetric matrices are such that elements symmetric about the *main diagonal* are the negative of each other, that is $a_{i,j} = a_{j,i} \Leftrightarrow A^T = -A$. In a skew-symmetric matrix, all diagonal elements are zero, that is, $a_{i,i} = -a_{i,i} = a_{i,i} \Rightarrow = 0$.
- Hermitian (or self-adjoint) matrices are such that elements symmetric about the diagonal are each others complex conjugates, that is, $a_{i,j} = \overline{a}_{j,i} \Leftrightarrow A^H = A$, , where $\overline{z}$ signifies the complex conjugate of a complex number z and A^H the conjugate transpose of A.
- Toeplitz matrices have common elements on their diagonals, that is, $a_{i,j} = a_{j+1\ j+1}$.
- Stochastic matrices are square matrices whose rows are probability vectors; they are used to define Markov chains.
- A square matrix A is called idempotent if $A^2 = AA = A$.

BASIC PROPERTIES

Throughout this section, R denotes a ring.

For positive integers m and n, an m X n matrix A over a ring R is a rectangular array

$$A = \begin{pmatrix} a_{11} & a_{12} & \cdots & a_{1n} \\ a_{21} & a_{22} & \cdots & a_{2n} \\ \vdots & \vdots & & \vdots \\ a_{m1} & a_{m2} & \cdots & a_{mn} \end{pmatrix}$$

where each entry a_{ij} in the array is an element of R; the element a_{ij} is called the (i, j) entry of A, which we denote by $A(i, j)$. For $i = 1, ..., m$, the ith row of A is

$$(a_{i1}, ..., a_{in}),$$

which we denote by $\text{Row}_i(A)$, and for $j = 1, ..., n$, the jth column of A is

$$\begin{pmatrix} a_{1j} \\ a_{2j} \\ \vdots \\ a_{mj} \end{pmatrix}$$

which we denote by $\text{Col}_j(A)$. We regard a row of A as a $1 \times n$ matrix, and a column of A as an $m \times 1$ matrix.

The set of all $m \times n$ matrices over R is denoted by $R^{m \times n}$. Elements of $R^{1 \times n}$ are called row vectors (of dimension n) and elements of $R^{m \times 1}$ are called column vectors (of dimension m). Elements of $R^{n \times n}$ are called square matrices (of dimension n). We do not make a distinction between $R^{1 \times n}$ and $R^{\times n}$; that is, we view standard n-tuples as row vectors.

We can define the familiar operations of matrix addition and scalar multiplication:

- If $A, B \in R^{m \times n}$, then $A + B$ is the $m \times n$ matrix whose (i, j) entry is $A(i, j) + B(i, j)$.
- If $c \in R$ and $A \in R^{m \times n}$, then cA is the $m \times n$ matrix whose (i, j) entry is $cA(i, j)$.

In addition, we define the $m \times n$ zero matrix to be the $m \times n$ matrix, all of whose entries are 0_R; we denote this matrix by $0_R^{m \times n}$.

Theorem:

(*i*) Matrix multiplication is associative; that is, $A(BC) = (AB)C$ for all $A \in R^{m \times n}$, $B \in R^{n \times p}$, and $C \in R^{p \times q}$.

(*ii*) Matrix multiplication distributes over matrix addition; that is, $A(C + D) = AC + AD$ and $(A + B)C = AC + BC$ for all $A, B \in R^{m \times n}$ and $C, D \in R^{n \times p}$.

(*iii*) The $n \times n$ identity matrix $I \in R^{n \times n}$ acts as a multiplicative identity; that is, $AI = A$ and $IB = B$ for all $A \in R^{m \times n}$ and $B \in R^{n \times m}$; in particular, $CI = C = IC$ for all $C \in R^{n \times n}$.

(*iv*) Scalar multiplication and matrix multiplication associate; that is, $c(AB) = (cA)B = A(cB)$ for all $c \in R$, $A \in R^{m \times n}$, and $B \in R^{n \times p}$.

Proof: All of these are trivial, except for (i), which requires just a bit of computation to show that the (i, ℓ) entry of both $A(BC)$ and $(AB)C$ is equal to (as the reader may verify)

$$\sum_{j=1}^{n}\sum_{k=1}^{p} A(i,j)B(j,k)C(k,\ell).$$

Some simple but useful facts to keep in mind are the following:

- If $A \in R^{m \times n}$ and $B \in R^{n \times p}$, then the kth column of AB is equal to Av, where $v = \text{Col}_k(B)$; also, the ith row of AB is equal to wB, where $w = \text{Row}_i(A)$.
- If $A \in R^{m \times n}$ and $u = (c_1, \ldots, c_m) \in R^{1 \times m}$, then

$$uA = \sum_{i=1}^{m} c_i Row_i(A).$$

In words: uA is a linear combination of the rows of A, with coefficients taken from the corresponding entries of u.

- If $A \in R^{m \times n}$ and

$$v = \begin{pmatrix} d_1 \\ \vdots \\ d_n \end{pmatrix} \in R^{n \times 1},$$

then

$$Av = \sum_{j=1}^{n} d_j \ \text{Col}_j(A).$$

In words: Av is a linear combination of the columns of A, with coefficients taken from the corresponding entries of v.

If $A \in R^{mXn}$, the transpose of A, denoted by A^T, is defined to be the n X m matrix whose (j, i) entry is $A(i, j)$.

Theorem: If A, $B \in R^{mXn}$, $C \in R^{nXp}$, and $c \in R$, then

(*i*) $(A + B)^T = A^T + B^T$,

(*ii*) $(cA)^T = cA^T$,

(*iii*) $(A^T)^T = A$,

(*iv*) $(AC)^T = C^T A^T$.

If A_i is an n_i X n_{i+1} matrix, for $i = 1, ..., k$, then by associativity of matrix multiplication, we may write the product matrix $A_1 \ldots A_k$, which is an n_1 X n_{k+1} matrix, without any ambiguity.

For an $n X n$ matrix A, and a positive integer k, we write A^k to denote the product $A \ldots A$, where there are k terms in the product. Note that $A^1 = A$. We may extend this notation to $k = 0$, defining A^0 to be the $n X n$ identity matrix. One may readily verify the usual rules of exponent arithmetic: for all non-negative integers k, ℓ, we have

$$(A^{\ell})_k = A^{k\ell} = (A^k)^{\ell} \text{ and } A^k A^{\ell} = A^{k+\ell}.$$

It is easy also to see that part (iv) of previous theorem implies that for all non-negative integers k, we have

$$(A^k)^T = (A^T)^k.$$

Theorem: With addition and scalar multiplication as defined above, R^{mXn} is an R-module. The matrix $0_R^{m \times n}$ is the additive identity, and the additive inverse of a matrix $A \in R^{mXn}$ is the $m \times n$ matrix whose (i, j) entry is $-A(i, j)$.

Proof: To prove this, one first verifies that matrix addition is associative and commutative, which follows from the associativity and commutativity of addition in R. The claims made about the additive identity and additive inverses are also easily verified. These observations establish that R^{mXn} is an abelian group. We can also define the familiar operation of matrix multiplication:

- If $A \in R^{m \times n}$ and $B \in R^{n \times p}$, then AB is the $m \times p$ matrix whose (i, k) entry is

$$\sum_{j=1}^{n} A(i,j)B(j,k).$$

In addition, we define the n X n identity matrix to be the matrix $I \in R^{nXn}$, where $I(i, i) := 1_R$ and $I(i, j) := 0_R$ for $i \neq j$.

ALGORITHM USED IN MATRICES

For computational purposes, matrices are represented in the obvious way as arrays of elements of R. As remarked at the beginning of this chapter, we shall treat R as an "abstract data type," and not worry about how elements of R are actually represented; in discussing the complexity of algorithms, we shall simply count "operations in R," by which we mean additions, subtractions, and multiplications; we shall sometimes also include equality testing and computing multiplicative inverses as "operations in R."

In any real implementation, there will be other costs, such as incrementing counters, and so on, which we may safely ignore, as long as their number is at most proportional to the number of operations in R.

The following statements are easy to verify:

- We can multiply an $m \times n$ matrix by a scalar using mn operations in R.
- We can add two $m \times n$ matrices using mn operations in R.
- We can multiply an $m \times n$ matrix and an n p matrix using $O(mnp)$ operations in R.

It is also easy to see that given an $n X n$ matrix A, and a non-negative integer e, we can adapt the repeated squaring algorithm so as to compute A^e using $O(\text{len}(e))$ multiplications of $n X n$ matrices, and hence $O(\text{len}(e)n^3)$ operations in R.

EXERCISE

Let $A \in R^{mXn}$. Show that if vA is the 1 X n zero matrix for all $v \in R^{1Xm}$, then A is the $m \times n$ zero matrix.

THE INVERSE OF A MATRIX

Let R be a ring. For a square matrix $A \in R^{nXn}$, we call a matrix $B \in R^{nXn}$ an inverse of A if $BA = AB = I$, where I is the n X n identity matrix. It is easy to see that if A has an inverse, then the inverse is unique: if B and C are inverses of A, then

$$B = BI = B(AC) = (BA)C = IC = C.$$

Because the inverse of A is uniquely determined, we denote it by A^{-1}. If A has an inverse, we say that A is invertible, or non-singular. If A is not invertible, it is sometimes called singular. We will use the terms "invertible" and "not invertible."

Observe that A is the inverse of A^{-1}; that is, $(A^{-1})^{-1} = A$.

If A and B are invertible $n \times n$ matrices, then so is their product: in fact, it is easily verified that $(AB)^{-1} = B^{-1}A^{-1}$. It follows that if A is an invertible matrix, and k is a non-negative integer, then A^k is invertible with inverse $(A^{-1})^k$, which we also denote by A^{-k}.

It is also easy to see that A is invertible if and only if the transposed matrix A^T is invertible, in which case $(A^T)^{-1} = (A^{-1})^T$. Indeed, $AB = I = BA$ holds if and only if $B^TA^T = I = A^TB^T$.

We now develop a connection between invertible matrices and R-module isomorphisms. Recall from the previous section the R-module isomorphism

$$\Lambda : R^{n\times n} \to \mathrm{Hom}_R (R^{1\times n}, R^{1\times n})$$
$$A \mapsto \lambda_A,$$

where for each $A \in R^{nXn}$, A is the corresponding R-linear map

$$\lambda_A: R^{1\times n} \to R^{1\times n}$$
$$v \mapsto vA.$$

Evidently, λ_I is the identity map.

Theorem: Let $A \in R^{nXn}$, and let $\lambda_A: R^{1Xn} \to R^{1Xn}$ be the corresponding R-linear map. Then A is invertible if and only if λ_A is bijective, in which case $\lambda_{A^{-1}} = \lambda_A^{-1}$.

Proof: Suppose A is invertible, and that B is its inverse. We have $AB = BA = I$, and hence $\lambda_{AB} = \lambda_{BA} = \lambda_I$, from which it follows that $\lambda_B \circ \lambda_A = \lambda_A \circ \lambda_B = \lambda_I$. Since λ_I is the identity map, this implies λ_A is bijective.

Suppose λ_A is bijective. We know that the inverse map λ_A^{-1} is also an R-linear map, and since the mapping Λ above is surjective, we have $\lambda_A^{-1} = \lambda_B$ for some $B \in R^{nXn}$. Therefore, we have $\lambda_B \circ \lambda_A = \lambda_A \circ \lambda_B = \lambda_I$, and hence $\lambda_{AB} = \lambda_{BA} = \lambda_I$. Since the mapping Λ is injective, it follows that $AB = BA = I$. This implies A is invertible, with $A^{-1} = B$.

Theorem: Let $A \in R^{nXn}$. The following are equivalent:

(*i*) A is invertible;

(*ii*) $\{\mathrm{Row}_i (A)\}_{i=1}^n$ is a basis for R^{1Xn};

(*iii*) $\{\mathrm{Col}_j(A)\}_{j=1}^n$ is a basis for R^{nX1}.

Proof: We first prove the equivalence of (i) and (ii). By the previous theorem, A is invertible if and only if λ_A is bijective. Also, in the previous section, we observed that A is surjective if and only if $\{Row_i(A)\}_{i=1}^{n}$ spans R^{1Xn}, and that λ_A is injective if and only if $\{Row_i(A)\}_{i=1}^{n}$ is linearly independent.

The equivalence of (i) and (iii) follows by considering the transpose of A.

EXERCISE

1. Let R be a ring, and let A be a square matrix over R. Let us call B a left inverse of A if $BA = I$, and let us call C a right inverse of A if $AC = I$.
   (*a*) Show that if A has both a left inverse B and a right inverse C, then $B = C$ and hence A is invertible.
   (*b*) Assume that R is a field. Show that if A has either a left inverse or a right inverse, then A is invertible.
2. Show that if A and B are two square matrices over a field such that their product AB is invertible, then both A and B themselves must be invertible.
3. Show that if A is a square matrix over an arbitrary ring, and A^k is invertible for some $k > 0$, then A is invertible.

GAUSSIAN ELIMINATION

Throughout this section, F denotes a field.

A matrix $B \in F^{mXn}$ is said to be in reduced row echelon form if there exists a sequence of integers $(p_1, ..., p_r)$, with $0 \leq r \leq m$ and $1 \leq p_1 < p_2 < ... < p_r \leq n$, such that the following holds:

- For $i = 1, ..., r$, all of the entries in row i of B to the left of entry (i, p_i) are zero; that is, $B(i, j) = 0$ for $j = 1, ..., p_i - 1$;
- For $i = 1, ..., r$, all of the entries in column p_i of B above entry (i, p_i) are zero; that is, $B(i', p_i) = 0$ for $i' = 1, ..., i - 1$;
- For $i = 1, ..., r$, we have $B(i, p_i) = 1$;
- All entries in rows $r + 1, ..., m$ of B are zero; that is, $B(i, j) = 0$ for $i = r + 1, ..., m$ and $j = 1, ..., n$.

It is easy to see that if B is in reduced row echelon form, then the sequence $(p_1, ..., p_r)$ above is uniquely determined, and we call it the pivot sequence of B.

Several further remarks are in order:

- All of the entries of B are completely determined by the pivot sequence, except for the entries (i, j) with $1 \leq i \leq r$ and $j > p_i$ with $j \notin \{p_{i+1}, ..., p_r\}$, which may be arbitrary.

- If B is an $n \times n$ matrix in reduced row echelon form whose pivot sequence is of length n, then B must be the $n \times n$ identity matrix.
- We allow for an empty pivot sequence (i.e., $r = 0$), which will be the case precisely when $B = 0^{m \times n}$.

Example 14.3. The following 46 matrix B over the rational numbers is in reduced row echelon form:

$$B = \begin{pmatrix} 0 & 1 & -2 & 0 & 0 & 3 \\ 0 & 0 & 0 & 1 & 0 & 2 \\ 0 & 0 & 0 & 0 & 1 & -4 \\ 0 & 0 & 0 & 0 & 0 & 0 \end{pmatrix}$$

The pivot sequence of B is (2, 4, 5). Notice that the first three rows of B form a linearly independent family of vectors, that columns 2, 4, and 5 form a linearly independent family of vectors, and that all of other columns of B are linear combinations of columns 2, 4, and 5. Indeed, if we truncate the pivot columns to their first three rows, we get the 3 X 3 identity matrix.

Generalizing the previous example, if a matrix is in reduced row echelon form, it is easy to deduce the following properties, which turn out to be quite useful:

Theorem: If B is a matrix in reduced row echelon form with pivot sequence $(p_1, ..., p_r)$, then

(*i*) Rows 1, 2, ..., r of B form a linearly independent family of vectors;

(*ii*) Columns p_1, ..., p_r of B form a linearly independent family of vectors, and all other columns of B can be expressed as linear combinations of columns p_1, ..., p_r.

Proof: Exercise– just look at the matrix!

Gaussian elimination is an algorithm that transforms a given matrix $A \in F^{mXn}$ into a matrix $B \in F^{mXn}$, where B is in reduced row echelon form, and is obtained from A by a sequence of elementary row operations. There are three types of elementary row operations:

Type I: Swap two rows,

Type II: Multiply a row by a non-zero scalar,

Type III: Add a scalar multiple of one row to a different row.

The application of any specific elementary row operation to an m X n matrix C can be affected by multiplying C on the left by a suitable m X m matrix X. Indeed, the matrix X corresponding to a particular elementary row operation is simply the matrix obtained by applying the same elementary row operation to the m X m identity matrix. It is easy to see that for every elementary row operation, the corresponding matrix X is invertible.

We now describe the basic version of Gaussian elimination. The algorithm works as follows. First, it makes a copy B of A (this is not necessary if the original matrix A is not needed afterwards). The algorithm proceeds column by column, starting with the left-most column, so that after processing column j, the first j columns of B are in reduced row echelon form, and the current value of r represents the length of the pivot sequence.

To process column j, in steps 3–6 the algorithm first searches for a non-zero element among $B(r+1, j), \ldots, B(m, j)$; if none is found, then the first $j + 1$ columns of B are already in reduced row echelon form.

Otherwise, one of these non-zero elements is selected as the pivot element (the choice is arbitrary), which is then used in steps 8–13 to bring column j into the required form.

After incrementing r, the pivot element is brought into position (r, j), using a Type I operation in step 9. Then the entry (r, j) is set to 1, using a Type II operation in step 10. Finally, all the entries above and below entry (r, j) are set to 0, using Type III operations in steps 11–13. Note that because columns $1, \ldots, j-1$ of B were already in reduced row echelon form, none of these operations changes any values in these columns.

1. $B \leftarrow A, r \leftarrow 0$
2. For $j \leftarrow 1$ to n do
3. $\ell \leftarrow 0, i \leftarrow r$
4. While $\ell = 0$ and $i \leq m$ do
5. $i \leftarrow i + 1$
6. If $B(i, j) \neq 0$ then $\ell \leftarrow i$
7. If $\ell \neq 0$ then
8. $r \leftarrow r + 1$
9. Swap rows r and ℓ of B
10. $\text{Row}_r(B) \leftarrow B(r, j)^{-1}\, \text{Row}_r(B)$
11. For $i \leftarrow 1$ to m do
12. If $i \neq r$ then
13. $\text{Row}_i(B) \leftarrow \text{Row}_i(B) - B(i, j)\, \text{Row}_r(B)$
14. output B

As for the complexity of the algorithm, it is easy to see that it performs $O(mn)$ elementary row operations, each of which takes $O(n)$ operations in F, so a total of $O(mn^2)$ operations in F.

Example. Consider the execution of the Gaussian elimination algorithm on input

$$A = \begin{pmatrix} [0] & [1] & [1] \\ [2] & [1] & [2] \\ [2] & [2] & [0] \end{pmatrix} \in \mathbb{Z}_3^{3\times 3}$$

After copying A into B, the algorithm transforms B as follows:

$$\begin{pmatrix} [0] & [1] & [1] \\ [2] & [1] & [2] \\ [2] & [2] & [0] \end{pmatrix} \xrightarrow{Row_1 \leftrightarrow Row_2} \begin{pmatrix} [2] & [1] & [2] \\ [0] & [1] & [1] \\ [2] & [2] & [0] \end{pmatrix} \xrightarrow{Row_1 \leftrightarrow [2]Row_1} \begin{pmatrix} [1] & [2] & [1] \\ [0] & [1] & [1] \\ [2] & [2] & [0] \end{pmatrix}$$

$$\xrightarrow{Row_3 \leftarrow Row_3 - [2]Row_1} \begin{pmatrix} [1] & [2] & [1] \\ [0] & [1] & [1] \\ [0] & [1] & [1] \end{pmatrix} \xrightarrow{Row_1 \leftarrow Row_1 - [2]Row_2} \begin{pmatrix} [1] & [0] & [2] \\ [0] & [1] & [1] \\ [0] & [1] & [1] \end{pmatrix}$$

$$\xrightarrow{Row_3 \leftarrow Row_3 - Row_2} \begin{pmatrix} [1] & [0] & [2] \\ [0] & [1] & [1] \\ [0] & [0] & [0] \end{pmatrix}.$$

Suppose the Gaussian elimination algorithm performs a total of t elementary row operations. Then as discussed above, the application of the eth elementary row operation, for $e = 1, ..., t$, amounts to multiplying the current value of the matrix B on the left by a particular invertible m X m matrix X_e. Therefore, the final output value of B satisfies the equation

$$B = XA \text{ where } X = X_t X_{t-1} \text{ .... } X_1.$$

Since the product of invertible matrices is also invertible, we see that X itself is invertible.

Although the algorithm as presented does not compute the matrix X, it can be easily modified to do so. The resulting algorithm, which we call extended Gaussian elimination, is the same as plain Gaussian elimination, except that we initialize the matrix X to be the m m identity matrix, and we add the following steps:

- Just before step 9: swap rows r and ℓ of X.
- Just before step 10: $\text{Row}_r(X) \leftarrow B(r, j)^{-1}\,\text{Row}_r(X)$.
- Just before step 13: $\text{Row}_i(X) \leftarrow \text{Row}_i(X) - B(i, j)\,\text{Row}_r(X)$.

At the end of the algorithm we output X in addition to B.

So we simply perform the same elementary row operations on X that we perform on B. The reader may verify that the above algorithm is correct, and that it uses $O(mn(m + n))$ operations in F.

Example. The execution of the extended Gaussian elimination algorithm initializes X to the identity matrix, and then transforms X as follows:

$$\begin{pmatrix} [1] & [0] & [0] \\ [0] & [1] & [0] \\ [0] & [0] & [1] \end{pmatrix} \xrightarrow{Row_1 \leftrightarrow Row_2} \begin{pmatrix} [0] & [1] & [0] \\ [1] & [0] & [0] \\ [0] & [0] & [1] \end{pmatrix} \xrightarrow{Row_1 \leftarrow [2] Row_1} \begin{pmatrix} [0] & [2] & [0] \\ [1] & [0] & [0] \\ [0] & [0] & [1] \end{pmatrix}$$

$$\xrightarrow{Row_3 \leftarrow Row_3 - [2] Row_1} \begin{pmatrix} [0] & [2] & [0] \\ [1] & [0] & [0] \\ [0] & [2] & [1] \end{pmatrix} \xrightarrow{Row_1 \leftarrow Row_1 - [2] Row_2} \begin{pmatrix} [1] & [2] & [0] \\ [1] & [0] & [0] \\ [0] & [2] & [1] \end{pmatrix}$$

$$\xrightarrow{Row_3 \leftarrow Row_3 - Row_2} \begin{pmatrix} [1] & [2] & [0] \\ [1] & [0] & [0] \\ [2] & [2] & [1] \end{pmatrix}.$$

EXERCISE

1. Given a matrix $B \in F^{mXn}$ in reduced row echelon form, show how to compute its pivot sequence using $O(n)$ operations in F.
2. For each type of elementary row operation, describe the matrix X which corresponds to it, as well as X^{-1}.

APPLICATIONS OF GAUSSIAN ELIMINATION

Throughout this section, A is an arbitrary m X n matrix over a field F, and $XA = B$, where X is an invertible m X m matrix, and B is an m X n matrix in reduced row echelon form with pivot sequence $(p_1, ..., p_r)$. This is precisely the information produced by the extended Gaussian elimination algorithm, given A as input.

Let

$$\lambda_A: F^{1\times m} \to F^{1\times n}$$
$$v \mapsto vA$$

be the linear map corresponding to A.

MATRICES AND LINEAR MAPS

Let R be a ring. For positive integers m and n, consider the R-modules R^{1Xm} and R^{1Xn}. If A is an m X n matrix over R, then the map

$$\lambda_A: R^{1\times m} \to R^{1\times n}$$
$$v \mapsto vA$$

is easily seen to be an R-linear map – this follows immediately from parts (ii) and (iv) of previous theorem. We call λA the linear map corresponding to A.

If $v = (c_1, ..., c_m) \in R^{1Xm}$, then

$$\lambda_A(v) = vA = \sum_{i=1}^{m} c_i \ \mathrm{Row}_i(A).$$

From this, it is clear that

- The image of λ_A is the submodule of $R^{1\times n}$ spanned by $\{Row_i(A)\}_{i=1}^{m}$; in particular, A is surjective if and only if $\{Row_i(A)\}_{i=1}^{m}$ spans $R^{1\times n}$;
- λ_A is injective if and only if $\{Row_i(A)\}_{i=1}^{m}$ is linearly independent.

There is a close connection between matrix multiplication and composition of corresponding linear maps. Specifically, let $A \in R^{mXn}$ and $B \in R^{nXp}$, and consider the corresponding linear maps $\lambda_A: R^{1Xm} \to R^{1Xn}$ and $\lambda_B: R^{1Xn} \to R^{1Xp}$. Then we have

$$\lambda_B \circ \lambda_A = \lambda_{AB}.$$

This follows immediately from the associativity of matrix multiplication.

We have seen how vector/matrix multiplication defines a linear map. Conversely, we shall now see that the action of any R-linear map can be viewed as a vector/ matrix multiplication, provided the R-modules involved have bases (which will always be the case for finite dimensional vector spaces).

Let M be an R-module, and suppose that $S = \{\alpha_i\}_{i=1}^{m}$ is a basis for M, where $m > 0$. As we know, every element $\alpha \in M$ can be written uniquely as $c_1\alpha_1 + ... + c_m\alpha_m$, where the c_i's are in R. Let us define

$$\mathrm{Vec}_S(\alpha) := (c_1, ..., c_m) \in R^{1\times m}.$$

We call $\mathrm{Vec}_S(\alpha)$ the coordinate vector of α relative to S. The function

$$\mathrm{Vec}_S: M \to R^{1\times m}$$

is an R-module isomorphism.

Let N be another R-module, and suppose that $T = \{\beta_j\}_{j=1}^{n}$ is a basis for N, where $n > 0$. Just as in the previous paragraph, every element $\beta \in N$ has a unique coordinate vector $\mathrm{Vec}_T(\beta) \in R^{1Xn}$ relative to T.

Now let $\rho: M \to N$ be an arbitrary R-linear map. Our goal is to define a matrix $A \in R^{mXn}$ with the following property:

$$\mathrm{Vec}_T(\rho(\alpha)) = \mathrm{Vec}_S(\alpha)A \text{ for all } \alpha \in M.$$

In words: if we multiply the coordinate vector of on the right by A, we get the coordinate vector of $\rho(\alpha)$.

Constructing such a matrix A is easy: we define A to be the matrix whose ith row, for $i = 1, \ldots, m$, is the coordinate vector of $\rho(\alpha_i)$ relative to T. That is,

$$\text{Row}_i(A) = \text{Vec}_T(\rho(\alpha_i)) \text{ for } i = 1, \ldots, m.$$

Then for an arbitrary $\alpha \in M$, if $(c_1, \ldots, c_m)$ is the coordinate vector of relative to S, we have

$$\rho(\alpha) = \rho\left(\sum_i c_i \alpha_i\right) = \sum_i c_i \rho(\alpha_i)$$

and so

$$Vec_T(\rho(\alpha)) = \sum_i c_i \, \text{Vec}_T(\rho(\alpha_i)) = \sum_i c_i \text{Row}_i(A) = Vec_S(\alpha)A.$$

then subtracting, we obtain

$$\text{Vec}_S(\alpha)(A - A') = 0_R^{1\times n}$$

for all $\alpha \in M$. Since the map $\text{Vec}_S: M \to R^{1Xm}$ is surjective, this means that $v(A - A')$ is zero for all $v \in R^{1Xm}$, and from this, it is clear that $A - A'$ is the zero matrix, and so $A = A'$.

We call the unique matrix A the matrix of ρ relative to S and T, and denote it by $\text{Mat}_{S,T}(\rho)$.

Recall that $\text{Hom}_R(M, N)$ is the R-module consisting of all R-linear maps from M to N. We can view $\text{Mat}_{S,T}$ as a function mapping elements of $\text{Hom}_R(M, N)$ to elements of R^{mXn}.

Theorem: The function $\text{Mat}_{S,T}: \text{Hom}_R(M, N) \to R^{mXn}$ is an R-module isomorphism. In particular, for every $A \in R^{mXn}$, the pre-image of A under $\text{Mat}_{S,T}$ is $Vec_T^{-1} \circ \lambda_A \circ \text{Vec}_S$, where $\lambda_A: R^{1Xm} \to R^{1Xn}$ is the linear map corresponding to A.

Proof: To show that $\text{Mat}_{S,T}$ is an R-linear map, let $\rho, \rho' \in \text{Hom}_R(M, N)$, and let $c \in R$. Also, let $A := \text{Mat}_{S,T}(\rho)$ and $A' := \text{Mat}_{S,T}(\rho')$. Then for all $\lambda \in M$, we have

$$\text{Vec}_T((\rho + \lambda')(\alpha)) = \text{Vec}_T(\rho(\alpha) + \rho'(\alpha)) = \text{Vec}_T(\rho(\alpha)) + \text{Vec}_T(\rho'(\alpha))$$
$$= \text{Vec}_S(\alpha)A + \text{Vec}_S(\alpha)A' = \text{Vec}_S(\alpha)(A + A').$$

As this holds for all $\alpha \in M$, and since the matrix of a linear map is uniquely determined, we must have $\text{Mat}_{S,T}(\rho + \rho') = A + A'$. A similar argument shows that $\text{Mat}_{S,T}(c) = cA$. This shows that $\text{Mat}_{S,T}$ is an R-linear map.

To show that the map $\text{Mat}_{S,T}$ is injective, it suffices to show that its kernel is trivial. If is in the kernel of this map, then setting $A := 0_R^{m\times n}$, we see that $\text{Vec}_T(\rho(\alpha))$ is zero for all $\alpha \in M$. But since the map $\text{Vec}_T: N \to R^{1Xn}$ is injective, this implies $\rho(\alpha)$ is zero for all $\alpha \in M$. Thus, ρ must be the zero map.

To show surjectivity, we show that every $A \in R^{m \times n}$ has a pre-image under $\mathrm{Mat}_{S,T}$ as described in the statement of the theorem. So let A be an $m \times n$ matrix, and let $\rho := Vec_T^{-1} \circ \lambda_A \circ \mathrm{Vec}_S$. Again, since the matrix of a linear map is uniquely determined, it suffices to show that holds for this particular A and ρ. For every $\alpha \in M$, we have

$$\mathrm{Vec}_T(\rho(\alpha)) = \mathrm{Vec}_T(Vec_T^{-1}(\lambda_A(\mathrm{Vec}_S(\alpha)))) = \lambda_A(\mathrm{Vec}_S(\alpha))$$
$$= \mathrm{Vec}_S(\alpha)A.$$

That proves the theorem.

As a special case of the above, suppose that $M = R^{1 \times m}$ and $N = R^{1 \times n}$, and S and T are the standard bases for M and N. In this case, the functions Vec_S and Vec_T are the identity maps, and the previous theorem implies that the function

$$\Lambda : R^{m \times n} \to \mathrm{Hom}_R(R^{1 \times m}, R^{1 \times n})$$
$$A \mapsto \lambda_A$$

is the inverse of the function $\mathrm{Mat}_{S,T}: \mathrm{Hom}_R(R^{1 \times m}, R^{1 \times n}) \to R^{m \times n}$. Thus, the function Λ is also an R-module isomorphism.

We see that an R-linear map ρ from M to N, together with particular bases for M and N, uniquely determine a matrix A such that the action of multiplication on the right by A implements the action of ρ with respect to the given bases.

There may be many bases for M and N to choose from, and different choices will in general lead to different matrices.

Also, note that in general, a basis may be indexed by an arbitrary finite set; however, in defining coordinate vectors and matrices of linear maps, the index set must be ordered in some way.

In any case, from a computational perspective, the matrix A gives us an efficient way to compute the map ρ, assuming elements of M and N are represented as coordinate vectors with respect to the given bases.

We have taken a "row-centric" point of view. Of course, if one prefers, by simply transposing everything, one can equally well take a "column-centric" point of view, where the action of ρ corresponds to multiplication of a column vector on the left by a matrix.

Example. Consider the quotient ring $E = R[X]/(f)$, where $f \in R[X]$ with $\deg(f) = \ell > 0$ and $\mathrm{lc}(f) \in R$. Let $\xi := [X]_f \in E$. As an R-module, E has a basis $S := \{\xi^{i-1}\}_{i=1}^{\ell}$. Let $\rho: E \to E$ be the ξ-multiplication map, which sends $\alpha \in E$ to $\xi\alpha \in E$. This is an R-linear map. If $f = c_0 + c_1X + \ldots + c_{\ell-1}X^{\ell-1} + c_{\ell}X^{\ell}$, then the matrix of ρ relative to S is the $\ell \times \ell$ matrix

$$A = \begin{pmatrix} 0 & 1 & 0 & \cdots & 0 \\ 0 & 0 & 1 & \cdots & 0 \\ & & & \ddots & \\ 0 & 0 & 0 & \cdots & 1 \\ -c_0/c_\ell & -c_1/c_\ell & -c_2/c_\ell & \cdots & -c_{\ell-1}/c_\ell \end{pmatrix}$$

where for $i = 1, \ldots, \ell - 1$, the ith row of A contains a 1 in position $i + 1$, and is zero everywhere else. The matrix A is called the companion matrix of f.

Example. Let $x_1, \ldots, x_k \in R$. Let $R[X]_{<k}$ be the set of polynomials $g \in R[X]$ with deg(g) $< k$, which is an R-module with a basis S:= $\{X^{i-1}\}_{i=1}^{k}$. The multi-point evaluation map

$$\rho: R[X]_{<k} \to R^{1\times k}$$

$$g \mapsto (g(x_1), \ldots, g(x_k))$$

is an R-linear map. Let T be the standard basis for R^{1Xk}. Then the matrix of ρ relative to S and T is the k X k matrix

$$A = \begin{pmatrix} 1 & 1 & \cdots & 1 \\ x_1 & x_2 & \cdots & x_k \\ x_1^2 & x_2^2 & \cdots & x_k^2 \\ \vdots & \vdots & & \vdots \\ x_1^{k-1} & x_2^{k-1} & \cdots & x_k^{k-1} \end{pmatrix}$$

The matrix A is called a Vandermonde matrix.

EXERCISE

1. Let F be a finite field, and let A be a non-zero m X n matrix over F. Suppose one chooses a vector $v \in F^{1Xm}$ at random. Show that the probability that vA is the zero vector is at most $1 / |F|$.
2. Let $\sigma: M \to N$ and $\tau: N \to P$ be R-linear maps, and suppose that M, N, and P have bases S, T, and U, respectively. Show that

   $$\text{Mat}_{S,U}(\tau \circ \sigma) = \text{Mat}_{S,T}(\sigma) \,.\, \text{Mat}_{T,U}(\tau).$$
3. Design and analyze a probabilistic algorithm that takes as input matrices A, B, C $\in \mathbb{Z}_p^{m\times m}$, where p is a prime. The algorithm should run in time $O(m^2 \text{ len}(p)^2)$ and should output either "yes" or "no" so that the following holds:

- If $C = AB$, then the algorithm should always output "yes";
- If $C \neq AB$, then the algorithm should output "no" with probability at least 0.999.

4. Let V be a vector space over a field F with basis $S = \{\alpha_i\}_{i=1}^{m}$. Suppose that U is a subspace of V of dimension $\ell < m$. Show that there exists a matrix $A \in F^{m\times(m-\ell)}$ such that for all $\alpha \in V$, we have $\alpha \in U$ if and only if $\mathrm{Vec}_S(\alpha)A$ is zero. Such a matrix A is called a parity check matrix for U.

IMAGE

Consider first the row space of A, by which we mean the subspace of $F^{1\times n}$ spanned by $\{Row_i(A)\}_{i=1}^{m}$, or equivalently, the image of λ_A.

We claim that the row space of A is the same as the row space of B. To see this, note that since $B = XA$, for every $v \in F^{1\times m}$, we have $vB = v(XA) = (vX)A$, and so the row space of B is contained in the row space of A. For the other containment, note that since X is invertible, we can write $A = X^{-1}B$, and apply the same argument.

Further, note that row space of B, and hence that of A, clearly has dimension r. Rows 1, ..., r of B form a basis for the row space of B.

Consider next the kernel K of λ_A, or what we might call the row null space of A. We claim that $\{Row_i(X)\}_{i=r+1}^{m}$ is a basis for K. Clearly, just from the fact that $XA = B$ and the fact that rows $r + 1$, ..., m of B are zero, it follows that rows $r + 1$, ..., m of X are contained in K. Furthermore, as X is invertible, $\{Row_i(X)\}_{i=1}^{m}$ is a basis for $F^{1\times m}$. Thus, the family of vectors $\{Row_i(X)\}_{i=r+1}^{m}$ is linearly independent and spans a subspace K' of K. It suffices to show that $K' = K$. Suppose to the contrary that $K' \subseteq K$, and let $v \in K \setminus K'$.

As $\{Row_i(X)\}_{i=1}^{m}$ spans $F^{1\times m}$, we may write $v = \sum_{i=1}^{m} c_i\ Row_i(X)$; moreover, as $v \notin K'$, we must have $c_i \neq 0$ for some $i = 1, ..., r$. Setting $\tilde{v} := (c_1, ..., c_m)$, we see that $v = \tilde{v}X$, and so

$$\lambda_A(v) = vA = (\tilde{v}X)A = \tilde{v}(XA) = \tilde{v}B.$$

Furthermore, since $\{Row_i(B)\}_{i=1}^{r}$ is linearly independent, rows $r + 1$, ..., m of B are zero, and $\tilde{v}$ has a non-zero entry in one of its first r positions, we see that $\tilde{v}B$ is not the zero vector. We have derived a contradiction, and hence may conclude that $K' = K$.

Finally, note that if $m = n$, then A is invertible if and only if its row space has dimension m, which holds if and only if $r = m$, and in the latter case, B is the identity matrix, and hence X is the inverse of A.

Let us summarize the above discussion:

- The first r rows of B form a basis for the row space of A (i.e., the image of λ_A).
- The last $m - r$ rows of X form a basis for the row null space of A (i.e., the kernel of λ_A).

- If $m = n$, then A is invertible (i.e., λ_A is an isomorphism) if and only if $r = m$, in which case X is the inverse of A (i.e., the matrix of λ_A^{-1} A relative to the standard basis).

So we see that from the output of the extended Gaussian elimination algorithm, we can simply "read off" bases for both the image and the kernel, as well as the inverse (if it exists), of a linear map represented as a matrix with respect to given bases.

SYSTEMS OF LINEAR EQUATIONS

Suppose that in addition to the matrix A, we are given $w \in F^{1Xn}$, and want to find a solution $v \in F^{1Xm}$ (or perhaps describe all solutions), to the equation

$$vA = w.$$

Equivalently, we can phrase the problem as finding an element (or describing all elements) of the set $\lambda_A^{-1}(w)$.

Now, if there exists a solution at all, say $v \in F^{1Xm}$, then $\lambda_A(v) = \lambda_A(v')$ if and only if $v \equiv v'$ (mod K), where K is the kernel of λ_A. It follows that the set of all solutions to is $v + K = \{v + v' : v_0 \in K\}$.

As we have discussed above, the last $m - r$ rows of X form a basis for K, so it suffices to determine if $w \in \text{Im } \lambda_A$, and if so, determine a single pre-image v of w. Also as we discussed, Im λ_A, that is, the row space of A, is equal to the row space of B, and because of the special form of B, we can quickly and easily determine if the given w is in the row space of B, as follows. By definition, w is in the row space of B if and only if there exists a vector $\bar{v} \in F^{1Xm}$ such that $\bar{v}B = w$. We may as well assume that all but the first r entries of $\bar{v}$ are zero. Moreover, $\bar{v}B = w$ implies that for $i = 1, ..., r$, the ith entry of $\bar{v}$ is equal to the p_ith entry of w. Thus, the vector $\bar{v}$, if it exists, is completely determined by the entries of w at positions $p_1, ..., p_r$. We can construct $\bar{v}$ satisfying these conditions, and then test if $\bar{v}B = w$. If not, then we may conclude that it has no solutions; otherwise, setting $v := \bar{v}X$, we see that $vA = (\bar{v}X)A = \bar{v}(XA) = \bar{v}B = w$, and so v is a solution.

One easily verifies that if we implement the above procedure as an algorithm, the work done in addition to running the extended Gaussian elimination algorithm amounts to $O(m(n + m))$ operations in F.

A special case of the above procedure is when $m = n$ and A is invertible, in which case has a unique solution, namely, $v := wX$, since in this case, $X = A^{-1}$.

The Orthogonal Complement of a Subspace

So as to give equal treatment to rows and columns, one can also define the column null space of A to be the kernel of the linear map defined by multiplication on the left by A; that is, the column null space of A is $\{z \in F^{nX1} : Az = 0\}$. By applying the results above to the transpose of A, we see that the column null space of A has dimension $n - r$, where r is the rank of A.

Let $U \subseteq F^{1Xn}$ be the row space of A, and let $U^{\perp} \subseteq F^{1Xn}$ denote the set of all vectors $\bar{u} \in F^{1Xn}$ whose transpose $\bar{u}^{T}$ belongs to the column null space of A. Now, U is a subspace of F^{1Xn} of dimension r and $U^{\perp}$ is a subspace of F^{1Xn} of dimension $n - r$. The space $U^{\perp}$ consists precisely of all vectors $\bar{u} \in F^{1Xn}$ that are "orthogonal" to all vectors $u \in U$, in the sense that the "inner product" $u\bar{u}^{-T}$ is zero. For this reason, $U^{\perp}$ is sometimes called the "orthogonal complement of U."

Clearly, $U^{\perp}$ is determined by the subspace U itself, and does not depend on the particular choice of matrix A. It is also easy to see that the orthogonal complement of $U^{\perp}$ is U; that is, $(U^{\perp}) = U$. This follows immediately from the fact that $U(U^{\perp})^{\perp}$ and $\dim_F((U^{\perp})^{\perp}) = n - \dim_F(U^{\perp}) = \dim_F(U)$. Now suppose that $U \cap U^{\perp} = \{0\}$. Then by Theorem, we have an isomorphism of $U\, U^{\perp}$ with $U + U^{\perp}$, and since $U \times U^{\perp}$ has dimension n, it must be the case that $U + U^{\perp} = F^{1Xn}$. It follows that every element of F^{1Xn} can be expressed uniquely as $u + \bar{u}$, where $u \in U$ and $\bar{u} \in U^{\perp}$.

We emphasize that the observations in the previous paragraph hinged on the assumption that $U \cap U^{\perp} = \{0\}$, which itself holds provided U contains no nonzero "self-orthogonal vectors" u such that uu^T is zero. If F is the field of real numbers, then of course there are no non-zero self-orthogonal vectors, since uu^T is the sum of the squares of the entries of u. However, for other fields, there may very well be non-zero self-orthogonal vectors.

As an example, if $F = \mathbb{Z}_2$, then any vector u with an even number of 1-entries is self orthogonal. So we see that while much of the theory of vector spaces and matrices carries over without change from familiar ground fields, like the real numbers, to arbitrary ground fields F, not everything does. In particular, the usual decomposition of a vector space into a subspace and its orthogonal complement breaks down, as does any other procedure that relies on properties specific to "inner product spaces."

For the following three exercises, as above, A is an arbitrary $m \times n$ matrix over a field F, and $XA = B$, where X is an invertible $m \times m$ matrix, and B is in reduced row echelon form.

THE RANK OF A MATRIX

We define the row rank of A to be the dimension of its row space, which is equal to $\dim_F(\text{Im } \lambda_A)$. The column space of A is defined as the subspace of F^{mX1} spanned by $\{Col_j(A)\}_{j=1}^n$; that is, the column space of A is $\{Az: z \in F^{nX1}\}$. The column rank of A is the dimension of its column space.

Now, the column space of A may not be the same as the column space of B, but from the identity $B = XA$, and the fact that X is invertible, it easily follows that these two subspaces are isomorphic (via the map that sends $y \in F^{mX1}$ to Xy), and hence have the same dimension. Moreover, by Theorem, the column rank of B is r, which is the same as the row rank of A.

So we may conclude: The column rank and row rank of A are the same. Because of this, we may define the rank of a matrix to be the common value of its row and column rank.

EXERCISE

1. Show that the matrix B is uniquely determined by A; more precisely, show that if $X'A = B'$, where X' is an invertible $m\,X\,m$ matrix, and B' is in reduced row echelon form, then $B' = B$.

   In the following two exercises, the theory of determinants could be used; however, they can all be solved directly, without too much difficulty, using just the ideas developed so far in the text.

2. The purpose of this exercise is to use linear algebra to prove that any pairwise independent family of hash functions must contain a large number of hash functions. More precisely, let $\{\Phi_r\}_{r\in R}$ be a pairwise independent family of hash functions from S to T, with $|T| \geq 2$. Our goal is to show that $|R| \geq |S|$. Let $n := |S|$, and $m := |T|$, and $\ell := |R|$. Write $R = \{r_1, ..., r_\ell\}$ and $S = \{s_1, ..., s_n\}$. Without loss of generality, we may assume that T is a set of non-zero real numbers that sum to zero (e.g., $T = \{1, ..., m-1, -m(m-1)/2\}$). Now define the matrix $A \in \mathbb{R}^{n\times\ell}$ with $A(i, j) := \Phi_{r_j}(s_i)$. Show that $\{Row_i(A)\}_{i=1}^n$ is a pairwise orthogonal family of non-zero vectors (see previous exercise). From this, deduce that $\ell \geq n$.

3. Let p be a prime. A matrix $A \in \mathbb{Z}^{m\times m}$ is called invertible modulo p if and only if there exists a matrix $B \in \mathbb{Z}^{mXm}$ such that $AB \equiv BA \equiv I \pmod{p}$, where I is the $m\,X\,m$ integer identity matrix. Here, two matrices are considered congruent with respect to a given modulus if and only if their corresponding entries are congruent. Show that A is invertible modulo p if and only if A is invertible over $\mathbb{Q}$, and the entries of A^{-1} lie in $\mathbb{Q}^{(p)}$.

4. Show that the column null space of A is the same as the column null space of B.

5. Let R be the ring $\mathbb{Z}_{p^e}$, where p is prime and $e > 1$. Let $\pi := [p] \in R$. The goal of this exercise is to develop an efficient algorithm for the following problem: given a

matrix $A \in R^{mXn}$, with $m > n$, find a vector $v \in R^{1Xm}$ such that $vA = 0^{1Xn}$ but $v \notin \pi R^{1Xm}$.

(*a*) Show how to modify the extended Gaussian elimination algorithm to solve the following problem: given a matrix $A \in R^{m\times n}$, compute $X \in R^{m\times m}$ and $B \in R^{m\times n}$, such that $XA = B$, X is invertible, and B is in row echelon form. Your algorithm should run in time $O(mn(m + n)e^2 \operatorname{len}(p)^2)$. Assume that the input includes the values p and e.

(*b*) Using the fact that the matrix X computed in part (a) is invertible, argue that none of its rows belong to $\pi R^{1\times m}$.

(*c*) Argue that if $m > n$ and the matrix B computed in part (a) has pivot sequence $(p_1, ..., p_r)$, then $m - r > 0$ and if v is any one of the last $m - r$ rows of X, then $vA = 0^{1\times n}$.

(*d*) Give an example that shows that $\{Row_i(B)\}_{i=1}^{r}$ need not be linearly independent, and that $\{Row_i(X)\}_{i=r+1}^{m}$ need not span the kernel of the linear map λ_A corresponding to A.

6. Let R be the ring $\mathbb{Z}_{\ell}$, where $\ell > 1$ is an integer. You are given a matrix $A \in R^{mXn}$. Show how to efficiently compute $X \in R^{mXm}$ and $B \in R^{mXn}$ such that $XA = B$, X is invertible, and B is in row echelon form. Your algorithm should run in time $O(mn(m + n) \operatorname{len}(\ell)^2)$.

7. Show how to compute a basis for the column null space of A using $O(r(n - r))$ operations in F, given A and B.

8. Show that $A \in F^{mX(m-\ell)}$ is a parity check matrix for U if and only if $\{Col_j(A)^T\}_{i=1}^{m-\ell}$ is a basis for the orthogonal complement of $\mathrm{Vec}_S(U) \subseteq F^{1Xm}$.

9. Let $\{v_i\}_{i=1}^{n}$ be a family of vectors, where $v_i \in \mathbb{R}^{1\times\ell}$ for each $i = 1, ..., n$. We say that $\{v_i\}_{i=1}^{n}$ is pairwise orthogonal if $v_i v_j^T$ is zero for all $i \neq j$. Show that every pairwise orthogonal family of non-zero vectors over $\mathbb{R}$ is linearly independent.

Chapter 7

Probability Fundamental

Probability is the likelihood or chance that something is the case or will happen. Probability theory is used extensively in areas such as statistics, mathematics, science and philosophy to draw conclusions about the likelihood of potential events and the underlying mechanics of complex systems.

INTERPRETATIONS

The word *probability* does not have a consistent direct definition. In fact, there are two broad categories of probability interpretations:

1. Frequentists talk about probabilities only when dealing with well defined random experiments. The probability of a random event denotes the *relative frequency of occurrence* of an experiment's outcome, when repeating the experiment.
2. Bayesians, however, assign probabilities to any statement whatsoever, even when no random process is involved. Probability, for a Bayesian, is a way to represent an individual's *degree of belief* in a statement, given the evidence.

MATHEMATICAL TREATMENT

In mathematics a probability of an event, A is represented by a real number in the range from 0 to 1 and written as $P(A)$, $p(A)$ or $Pr(A)$. An impossible event has a probability of 0, and a certain event has a probability of 1. However, the converses are not always true: probability 0 events are not always impossible, nor probability 1 events certain. The rather subtle distinction between "certain" and "probability 1" is treated at greater length in the article on "almost surely".

The *opposite* or *complement* of an event A is the event [not A] (that is, the event of A not occurring); its probability is given by $P(\text{not } A) = 1 - P(A)$. As an example, the chance of not rolling a six on a six-sided die is 1 – (chance of rolling a six) $= 1 - \frac{1}{6} = \frac{5}{6}$.

If two events, A and B are independent then the joint probability is

$$P(A \text{ and } B) = P(A \cap B) = P(A)\,P(B),$$

for example if two coins are flipped the chance of both being heads is $\frac{1}{2} \times \frac{1}{2} = \frac{1}{4}$.

If two events are mutually exclusive then the probability of either occurring is

$$P(A \text{ or } B) = P(A \cup B) = P(A) + P(B).$$

For example, the chance of rolling a 1 or 2 on a six-sided die is $P(1 \text{ or } 2) = P(1) + P(2) = \frac{1}{6} + \frac{1}{6} = \frac{1}{3}$.

If the events are not mutually exclusive then

$$P(A \text{ or } B) = P(A) + P(B) - P(A \text{ and } B).$$

For example, when drawing a single card at random from a regular deck of cards, the chance of getting a heart or a face card (J,Q,K) (or one that is both) is $\frac{13}{52} + \frac{12}{52} - \frac{3}{52} = \frac{11}{26}$, because of the 52 cards of a deck 13 are hearts, 12 are face cards, and 3 are both: here the possibilities included in the "3 that are both" are included in each of the "13 hearts" and the "12 face cards" but should only be counted once.

Conditional probability is the probability of some event A, given the occurrence of some other event B. Conditional probability is written $P(A|B)$, and is read "the probability of A, given B". It is defined by

$$P(A \mid B) = \frac{P(A \cap B)}{P(B)}.$$

If $P(B) = 0$ then $P(A \mid B)$ is undefined.

Summary of Probabilities

Event	**Probability**
A	$P(A) \in [0, 1]$
not A	$P(A') = 1 - P(A)$
A or B	$P(A \cap B) = P(A) + P(B) - P(P(A \cap B)$ $= P(A) + P(B)$ if A and B are mutually exclusive
A and B	$P(A \cap B) = P(A/B)\,P(B)$ $= P(A)\,P(B)$ if A and B are independent
A given B	$P(A|B)$

THEORY

Like other theories, the theory of probability is a representation of probabilistic concepts in formal terms-that is, in terms that can be considered separately from their meaning. These formal terms are manipulated by the rules of mathematics and logic, and any results are then interpreted or translated back into the problem domain.

There have been at least two successful attempts to formalize probability, namely the Kolmogorov formulation and the Cox formulation.

In Kolmogorov's formulation, sets are interpreted as events and probability itself as a measure on a class of sets. In Cox's theorem, probability is taken as a primitive (that is, not further analyzed) and the emphasis is on constructing a consistent assignment of probability values to propositions. In both cases, the laws of probability are the same, except for technical details.

There are other methods for quantifying uncertainty, such as the Dempster-Shafer theory and possibility theory, but those are essentially different and not compatible with the laws of probability as they are usually understood.

APPLICATIONS

Two major applications of probability theory in everyday life are in risk assessment and in trade on commodity markets. Governments typically apply probabilistic methods in environmental regulation where it is called "pathway analysis", often measuring well-being using methods that are stochastic in nature, and choosing projects to undertake based on statistical analyses of their probable effect on the population as a whole.

It is not correct to say that statistics are involved in the modelling itself, as typically the assessments of risk are one-time and thus require more fundamental probability models, e.g. "the probability of another 9/11". A law of small numbers tends to apply to all such choices and perception of the effect of such choices, which makes probability measures a political matter.

A good example is the effect of the perceived probability of any widespread Middle East conflict on oil prices - which have ripple effects in the economy as a whole. An assessment by a commodity trader that a war is more likely vs. less likely sends prices up or down, and signals other traders of that opinion.

Accordingly, the probabilities are not assessed independently nor necessarily very rationally. The theory of behavioral finance emerged to describe the effect of such groupthink on pricing, on policy, and on peace and conflict.

It can reasonably be said that the discovery of rigorous methods to assess and combine probability assessments has had a profound effect on modern society. Accordingly, it may be of some importance to most citizens to understand how odds and probability assessments are made, and how they contribute to reputations and to decisions, especially in a democracy. Another significant application of probability theory in everyday life is reliability. Many consumer products, such as automobiles and consumer electronics, utilize reliability theory in the design of the product in order to reduce the probability of failure. The probability of failure is also closely associated with the product's warranty.

RELATION TO RANDOMNESS

In a deterministic universe, based on Newtonian concepts, there is no probability if all conditions are known. In the case of a roulette wheel, if the force of the hand and the period of that force are known, then the number on which the ball will stop would be a certainty.

Of course, this also assumes knowledge of inertia and friction of the wheel, weight, smoothness and roundness of the ball, variations in hand speed during the turning and so forth. A probabilistic description can thus be more useful than Newtonian mechanics for analysing the pattern of outcomes of repeated rolls of roulette wheel.

Physicists face the same situation in kinetic theory of gases, where the system, while deterministic *in principle*, is so complex (with the number of molecules typically the order of magnitude of Avogadro constant ($6 \cdot 10^{23}$) that only statistical description of its properties is feasible.

A revolutionary discovery of 20th century physics was the random character of all physical processes that occur at microscopic scales and are governed by the laws of quantum mechanics. The wave function itself evolves deterministically as long as no observation is made, but, according to the prevailing Copenhagen interpretation, the randomness caused by the wave function collapsing when an observation is made, is fundamental.

BASICS OF PROBABILITY

Formally speaking, we will add a new type of instruction to our random access machine:

Random bit This type of instruction is of the form $\gamma \leftarrow$ RAND, where takes the same form as in arithmetic instructions. Execution of this type of instruction assigns to γ a value sampled from the uniform distribution on $\{0, 1\}$, independently from the execution of all other random-bit instructions.

Algorithms that use random-bit instructions are called probabilistic (or randomized), while those that do not are called deterministic.

In describing probabilistic algorithms at a high level, we shall write "$y \xleftarrow{\$} \{0, 1\}$" to denote the assignment of a random bit to the variable y, and "$y \xleftarrow{\$} \{0, 1\}^{X\ell}$" to denote the assignment of a random bit string of length ℓ to the variable y.

To analyze the behavior of a probabilistic algorithm, we first need a probability distribution that appropriately models its execution. Once we have done this, we shall define the running time and output to be random variables associated with this distribution.

DEFINING THE DISTRIBUTION

It would be desirable to define a probability distribution that could be used for all algorithms and all inputs. While this can be done in principle, it would require notions from the theory of probability more advanced than those we developed in the previous chapter. Instead, for a given probabilistic algorithm A and input x, we shall define a discrete probability distribution that models A's execution on input x. Thus, every algorithm/input pair yields a different distribution.

We could view the sample space in that example to be the set of all bit strings consisting of zero or more 0 bits, followed by a single 1 bit, and to each such bit string ω of this special form, we assign the probability $2^{-|\omega|}$, where $|\omega|$ denotes the length of ω. The "random experiment" we have in mind is to generate random bits one at a time until one of these special "halting" strings is generated.

In developing the definition of the probability distribution for a probabilistic algorithm, we simply consider more general sets of "halting" strings, as determined by the algorithm and its input.

So consider a fixed algorithm A and input x. Let λ be a finite bit string of length, say, ℓ. We can use λ to "drive" the execution of A on input x for up to ℓ execution steps, as follows: for each step $i = 1, ..., \ell$, if the ith instruction executed by A is $\gamma \leftarrow$ RAND, the ith bit of λ is assigned to γ. In this context, we shall refer to λ as an execution path. The reader may wish to visualize λ as a finite path in an infinite binary tree, where we start at the root, branching to the left if the next bit in λ is a 0 bit, and branching to the right if the next bit in λ is a 1 bit.

After using λ to drive A on input x for up to ℓ steps, we might find that the algorithm executed a halt instruction at some point during the execution, in which case we call λ a complete execution path; moreover, if this halt instruction was the ℓth instruction executed by A, then we call λ an exact execution path.

Our intent is to define the probability distribution associated with A on input x to be $P: \Omega \rightarrow [0, 1]$, where the sample space Ω is the set of all exact execution paths, and $P(\omega) := 2^{-|\omega|}$ for each $\omega \in \Omega$. However, for this to work, all the probabilities must sum to 1. The next theorem at least guarantees that these probabilities sum to at most 1. The only property of Ω that really matters in the proof of this theorem is that it is prefix free, which means that no exact execution path is a proper prefix of any other.

Theorem: Let Ω be the set of all exact execution paths for A on input x. Then $\sum_{\omega\in\Omega} 2^{-|\omega|} \leq 1$.

Proof: Let k be a non-negative integer. Let $\Omega_k \subseteq \Omega$ be the set of all exact execution paths of length at most k, and let $k{:} = \sum_{\omega\in\Omega_k} 2^{-|\omega|}$. We shall show below that

$$\alpha_k \leq 1.$$

From this, it will follow that

$$\sum_{\omega \in \Omega} 2^{-|\omega|} = \lim_{k \to \infty} \alpha_k \leq 1.$$

To prove the inequality (9.1), consider the set C_k of all complete execution paths of length equal to k. We claim that

$$\alpha_k = 2^{-k} | C_k |,$$

from which follows, since clearly, $| C_k | \leq 2^k$. Observe that by definition, each $\lambda \in C_k$ extends some $\omega \in \Omega_k$; that is, ω is a prefix of λ; moreover, ω is uniquely determined by λ, since no exact execution path is a proper prefix of any other exact execution path. Also observe that for each $\omega \in \Omega_k$, if $C_k(\omega)$ is the set of execution paths $\lambda \in C_k$ that extend ω, then $| C_k(\omega) | = 2^{k-|\omega|}$, and by the previous observation, $\{C_k(\omega)\}_{\omega \in \Omega_k}$ is a partition of C_k. Thus, we have

$$\alpha_k = \sum_{\omega \in \Omega_k} 2^{-|\omega|} = \sum_{\lambda \in C_k(\omega)} 2^{-k+|\omega|} = 2^{-k} \sum_{\omega \in \Omega_k} \sum_{\lambda \in C_k(\omega)} 1 = 2^{-k} | C_k |,$$

which proves.

From the above theorem, if Ω is the set of all exact execution paths for A on input x, then

$$\alpha := \sum_{\omega \in \Omega} 2^{-|\omega|} \leq 1,$$

and we say that A halts with probability α on input x. If = 1, we define the distribution P: $\Omega \to [0, 1]$ associated with A on input x, where $P(\omega) := 2^{-|\omega|}$ for each $\omega \in \Omega$.

We shall mainly be interested in algorithms that halt with probability 1 on all inputs. The following four examples provide some simple criteria that guarantee this.

Example. Suppose A and B are probabilistic algorithms that both halt with probability 1 on all inputs. Using A and B as subroutines, we can form their serial composition; that is, we can construct the algorithm

C(*x*): output *B*(*A*(*x*)),

which on input x, first runs A on input x, obtaining a value y, then runs B on input y, obtaining a value z, and finally, outputs z. We claim that C halts with probability 1 on all inputs.

For simplicity, we may assume that A places its output y in a location in memory where B expects to find its input, and that B places its output in a location in memory where C's output should go. With these assumptions, the program for C is obtained by simply concatenating the programs for A and B, making the following adjustments: every halt instruction in A's program is translated into an instruction that branches to the first instruction of B's program, and every target in a branch instruction in B's program is increased by the length of A's program.

Let Ω be the sample space representing A's execution on an input x. Each $\omega \in \Omega$ determines an output y, and a corresponding sample space Ω'_ω representing B's execution on input y. The sample space representing C's execution on input x is

$$\Omega'' = \{\omega\omega' : \omega \in \Omega, \omega' \in \Omega'_\omega\},$$

where $\omega\omega'$ is the concatenation of ω' and ω', We have

$$\sum_{\omega\omega' \in \Omega''} 2^{-|\omega\omega'|} = \sum_{\omega \in \Omega'} 2^{-|\omega|} \sum_{\omega' \in \Omega'_\omega} 2^{-|\omega'|} = \sum_{\omega \in \Omega} 2^{-|\omega|}.1 = 1,$$

which shows that C halts with probability 1 on input x.

Example. Suppose A, B, and C are probabilistic algorithms that halt with probability 1 on all inputs, and that A always outputs either true or false. Then we can form the conditional construct

$D(x)$: if $A(x)$ then output $B(x)$ else output $C(x)$.

By a calculation similar to that in the previous example, it is easy to see that D halts with probability 1 on all inputs.

Example. Suppose A and B are probabilistic algorithms that halt with probability 1 on all inputs, and that A always outputs either true or false. We can form the iterative construct

$C(x)$: while $A(x)$ do $x \leftarrow B(x)$

output x.

Algorithm C may or may not halt with probability 1. To analyze C, we define an infinite sequence of algorithms $\{C_n\}_{n=0}^{\infty}$; namely, we define C_0 as

$C_0(x)$: halt,

and for $n > 0$, we define C_n as

$C_n(x)$: if $A(x)$ then $C_{n-1}(B(x))$.

Essentially, C_n drives C for up to n loop iterations before halting, if necessary, in C_0. By the previous three examples, it follows by induction on n that each C_n halts with probability 1 on all inputs. Therefore, we have a well-defined probability distribution for each C_n and each input x.

Consider a fixed input x. For each $n \geq 0$, let β_n be the probability that on input x, C_n terminates by executing algorithm C_0. Intuitively, β_n is the probability that C executes at least n loop iterations; however, this probability is defined with respect to the probability distribution associated with algorithm C_n on input x. It is not hard to see that the sequence $\{\beta_n\}_{n=0}^{\infty}$ is non-increasing, and so the limit $\beta := \lim_{n\to\infty} \beta_n$ exists; moreover, C halts with probability $1 - \beta$ on input x.

On the one hand, if the loop in algorithm C is guaranteed to terminate after a finite number of iterations (as in a "for loop"), then C certainly halts with probability 1. Indeed, if on input x, there is a bound ℓ (depending on x) such that the number of loop iterations is always at most ℓ, then $\beta_{\ell+1} = \beta_{\ell+2} = \ldots = 0$. On the other hand, if on input x, C enters into a good, old-fashioned infinite loop, then C certainly does not halt with probability 1, as $\beta_0 = \beta_1 = \ldots = 1$. Of course, there may be in-between cases, which require further analysis.

We now illustrate the above criteria with a couple of some simple, concrete examples.

Example. Consider the following algorithm:

$i \leftarrow 0$

repeat

$i \leftarrow i + 1$

$\sigma \xleftarrow{\ell} \{0, 1\}^{\times i}$

until $\sigma = 0^{\times i}$

For each positive integer n, let β_n be the probability that the algorithm executes at least n loop iterations, in the sense of Example. It is not hard to see that

$$\beta_n = \prod_{i=1}^{n-1}(1-2^{-i}) \geq \prod_{i=1}^{n-1} e^{-2^{-i+1}} = e^{-\sum_{i=0}^{n-2} 2^{-i}} \geq e^{-2},$$

Therefore,

$$\lim_{n\to\infty} \beta_n \geq e^{-1} > 0,$$

and so the algorithm does not halt with probability 1, even though it never falls into an infinite loop.

Example. Suppose that on input x, A always halts within a finite number of steps, regardless of its random choices. More precisely, this means that there is a bound ℓ (depending on A and x), such that all execution paths of length ℓ are complete. In this case, we say that A's running time on input x is strictly bounded by ℓ, and it is clear that A halts with probability 1 on input x. Moreover, one can much more simply model A's computation on input x by working with the uniform distribution on execution paths of length ℓ.

Example. Consider the following algorithm, which models an experiment in which we toss a fair coin repeatedly until it comes up heads:

repeat

$b \xleftarrow{\$} \{0, 1\}$

until $b = 1$

For each positive integer n, let β_n be the probability that the algorithm executes at least n loop iterations, in the sense of previous Example. It is not hard to see that $\beta_n = 2^{-n+1}$, and since $\beta_n \to 0$ as $n \to \infty$, the algorithm halts with probability 1, even though the loop is not guaranteed to terminate after any particular, finite number of steps.

DEFINING THE RUNNING TIME AND OUTPUT

Let A be a probabilistic algorithm that halts with probability 1 on a fixed input x. We may define the random variable Z that represents A's running time on input x, and the random variable Y that represents A's output on input x.

Formally, Z and Y are defined using the probability distribution on the sample space Ω. The sample space Ω consists of all exact execution paths for A on input x. For each $\omega \in \Omega$, $Z(\omega) := |\omega|$, and $Y(\Omega)$ is the output produced by A on input x, using ω to drive its execution.

The expected running time of A on input x is defined to be $E[Z]$. Note that in defining the expected running time, we view the input as fixed, rather than drawn from some probability distribution. Also note that the expected running time may be infinite.

We say that A runs in expected polynomial time if there exist constants a, b, and c, such that for all n, and for all inputs x of size n, the expected running time of A on input x is at most $an^b + c$. We say that A runs in strict polynomial time if there exist constants a, b, and c, such that for all n, and for all inputs x of size n, A's running time on input x is strictly bounded by $an^b + c$.

Example. Consider the following probabilistic algorithm that takes as input a positive integer m. It models an experiment in which we toss a fair coin repeatedly until it comes up heads m times.

$k \leftarrow 0$

repeat

$b \xleftarrow{\$} \{0, 1\}$

if $b = 1$ then $k \leftarrow k + 1$

until $k = m$

Let L be the random variable that represents the number of loop iterations executed the algorithm on a fixed input m. We claim that $E[L] = 2m$. To see this, define random variables $L_1, ..., L_m$, where L_1 is the number of loop iterations needed to get $b = 1$ for the first time, L_2 is the number of additional loop iterations needed to get $b = 1$ for the second time, and so on. Clearly, we have $L = L_1 + ... + L_m$, and moreover, $E[L_i] = 2$ for $i = 1, ..., m$; therefore, by linearity of expectation, we have $E[L] = E[L_1] + ... + E[L_m] = 2m$. It follows that the expected running time of this algorithm on input m is $O(m)$.

Example. Consider again the algorithm in Example. Let L be the random variable that represents the number of loop iterations executed by the algorithm. The distribution of L is a geometric distribution, with associated success probability 1/2. Therefore, $E[L] = 2$. Let Z be the random variable that represents the running time of the algorithm. We have $Z \leq cL$, for some implementation-dependent constant c. Therefore, $E[Z] \leq c\, E[L] = 2c$.

Example. Consider the following algorithm:

$n \leftarrow 0$

repeat $n \leftarrow n + 1$, $b \xleftarrow{\not{c}} \{0, 1\}$ until $b = 1$

repeat $\sigma \xleftarrow{\not{c}} \{0, 1\}^{\times n}$ until $= 0^{\times n}$

The expected running time is infinite (even though it does halt with probability 1). To see this, define random variables L_1 and L_2, where L_1 is the number of iterations of the first loop, and L_2 is the number of iterations of the second. As in Example, the distribution of L_1 is a geometric distribution with associated success probability 1/2, and $E[L_1] = 2$. For each $k \geq 1$, the conditional distribution of L_2 given $L_1 = k$ is a geometric distribution with associated success probability $1/2_k$, and so $E[L_2 \mid L_1 = k] = 2^k$. Therefore,

$$E[L_2] = \sum_{k\geq 1} E[L_2 \mid L_1 = k] P[L_2 = k] = \sum_{k\geq 1} 2^k . 2^{-k} = \sum_{k\geq 1} 1 = \infty.$$

We have presented a fairly rigorous definitional framework for probabilistic algorithms, but from now on, we shall generally reason about such algorithms at a higher, and more intuitive, level. Nevertheless, all of our arguments can be translated into this rigorous framework, the details of which we leave to the interested reader. Moreover, all of the algorithms we shall present halt with probability 1 on all inputs, but we shall not go into the details of proving this (but the criteria in Examples 9.1 – 9.4 can be used to easily verify this).

EXERCISE

1. Let A be a probabilistic algorithm that on a given input x, halts with probability 1, and produces an output in the set T. Let P be the corresponding probability distribution, and let Y and Z be random variables representing the output and running time, respectively. For each $k \geq 0$, let P_k be the uniform distribution on all execution paths λ of length k. We define random variables Y_k and Z_k, associated with P_k, as follows: if λ is complete, we define $Y_k(\lambda)$ to be the output produced by A, and

$Z_k(\lambda)$ to be the actual number of steps executed by A; otherwise, we define $Y_k(\lambda)$ to be the special value "$\perp$" and $Z_k(\lambda)$ to be k. For each $t \in T$, let p_{tk} be the probability (relative to P_k) that $Y_k = t$, and let μ_k be the expected value (relative to P_k) of Z_k. Show that:

(*a*) For each $t \in T$, $P[Y = t] = \lim_{k\to\infty} p_{tk}$;

(*b*) $E[Z] = \lim_{k\to\infty} \mu_k$.

2. Let A_1 and A_2 be probabilistic algorithms. Let B be any probabilistic algorithm that always outputs 0 or 1. For $i = 1, 2$, let A_i' be the algorithm that on input x computes and outputs $B(A_i(x))$. Fix an input x, and let Y_1 and Y_2 be random variables representing the outputs of A_1 and A_2, respectively, on input x, and let Y_1' and Y_2' be random variables representing the outputs of A_1' and A_2', respectively, on input x. Assume that the images of Y_1 and Y_2 are finite, and let $\delta := \Delta\,[Y_1; Y_2]$ be their statistical distance. Show that $|\,P[Y_1' = 1] - P[Y_2' = 1]| \leq \delta$.
3. Suppose A is a probabilistic algorithm that halts with probability 1 on input x, and let $P: \Omega \to [0, 1]$ be the corresponding probability distribution. Let be an execution path of length ℓ, and assume that no proper prefix of λ is exact. Let $\varepsilon_\lambda := \{\omega \in \Omega:$ ω extends $\omega\}$. Show that $P[\varepsilon_\lambda] = 2^{-\ell}$.

GENERATE AND TEST PARADIGM

Let us suppose we have two probabilistic algorithms, A and B, and we combine them to form a new algorithm

$C(x)$: repeat $y \leftarrow A(x)$ until $B(x, y)$

output y.

Here, we assume that $B(x, y)$ always outputs either true or false.

Our goal is to answer the following questions about C for a fixed input x:

1. Does C halt with probability 1?
2. What is the expected running time of C?
3. What is the output distribution of C?

The answer to the first question is "yes," provided (i) A halts with probability 1 on input x, (ii) for all possible outputs y of $A(x)$, B halts with probability 1 on input (x, y), and (iii) for some possible output y of $A(x)$, $B(x, y)$ outputs true with non-zero probability. We shall assume this from now on. To address the second and third questions, let us define random variables L, Z, and Y, where L is the total number of loop iterations of C, Z is the total running time of C, and Y is the output of C. We can reduce the study of L, Z, and Y to the study of a single iteration of the main loop.

Instead of working with a new probability distribution that directly models a single iteration of the loop, it is more convenient to simply study the first iteration of the loop in C.

To this end, we define random variables Z_1 and Y_1, where Z_1 is the running time of the first loop iteration of C, and Y_1 is the value assigned to y in the first loop iteration of C. Also, let H_1 be the event that the algorithm halts in the first loop iteration, and let T be the set of possible outputs of $A(x)$. Note that by the assumption in the previous paragraph, $P[H_1] > 0$.

Theorem: Under the assumptions above,

(*i*) L has a geometric distribution with associated success probability $P[H_1]$, and in particular, $E[L] = 1/P[H_1]$;

(*ii*) $E[Z] = E[Z_1]\ E[L] = E[Z_1]/P[H_1]$;

(*iii*) for every $t \in T$, $P[Y = t] = P[Y_1 = t \mid H_1]$.

Proof: (i) is clear.

To prove (ii), for $i \geq 1$, let Z_i be the time spent by the algorithm in the ith loop iteration, so that $Z = \sum_{i \geq 1} Z_i$. Now, the conditional distribution of Z_i given $L \geq i$ is the same as the distribution of Z_1; moreover, $Z_i = 0$ when $L < i$. Therefore, by the law of total expectation, for each $i \geq 1$, we have

$$E[Z_i] = E[Z_i \mid L \geq i]\ P[L \geq i] + E[Z_i \mid L < i]\ P[L < i] = E[Z_1]\ P[L \geq i].$$

We may assume that $E[Z_1]$ is finite, as otherwise (ii) is trivially true. By Theorem and the infinite version of Theorem, we have

$$E[Z] = \sum_{i \geq 1} E[Z_i] = \sum_{i \geq 1} E[Z_1] P[L \geq i]$$

$$= E[Z_1] \sum_{i \geq 1} P[L \geq i] = E[Z_1] E[L].$$

To prove (iii), for $i \geq 1$, let Y_i be the value assigned to y in loop iteration i, with $Y_i := \perp$ if $L < i$, and let H_i be the event that the algorithm halts in loop iteration i (i.e., H_i is the event that $L = i$). By a calculation similar to that made in the proof of Theorem, for each $t \in T$, we have

$$P[Y = t] = \sum_{i \geq 1} P[(Y = t) \cap H_i] = \sum_{i \geq 1} P[(Y_i = t) \cap H_i \mid L \geq i] P[L \geq i]$$

$$= P[(Y_1 = t) \cap H_1] \sum_{i \geq 1} P[L \geq i] = P[(Y_1 = t) \cap H_1].E[L]$$

$$= P[Y_1 = t) \cap H_1] / P[H_1] = P[Y_1 = t \mid H_1].$$

Example. Let us analyze the following algorithm:

repeat

$y \xleftarrow{\not{c}} \{1, 2, 3, 4\}$

$z \xleftarrow{\not{c}} \{1, ..., y\}$

until $z = 1$

output y

With each loop iteration, the algorithm chooses y uniformly at random, and then decides to halt with probability $1/y$. Let L denote the number of loop iterations, and Y the output. Also, let Y_1 be the value of y in the first iteration, and H_1 the event that the algorithm halts in the first iteration. Y_1 is uniformly distributed over $\{1, ..., 4\}$, and for $t = 1, ..., 4$, $P[H_1 \mid Y_1 = t] = 1/t$. Therefore,

$$P[H_1] = \sum_{t=1}^{4} P[H_1 \mid Y_1 = t] P[Y_1 = t] = \sum_{t=1}^{4} (1/t)(1/4) = 25/48.$$

Thus, $E[L] = 48/25$. For the output distribution, for $t = 1, ..., 4$, we have

$$P[Y = t] = P[Y_1 = t \mid H_1] = P[(Y_1 = t) \cap H_1] / P[H_1]$$

$$= P[H_1 \mid Y_1 = t]\ P[Y_1 = t] / P[H_1] = (1/t)(1/4)(48/25) = \frac{12}{25t}.$$

This example illustrates how a probabilistic test can be used to create a biased output distribution.

Example. Suppose T is a finite set, and T' is a non-empty, finite subset of T. Consider the following generalization of Algorithm RN:

repeat

$y \xleftarrow{\not{c}} T$

until $y \in T'$

output y

Here, we assume that we have an algorithm to generate a random element of T (i.e., uniformly distributed over T), and an efficient algorithm to test for membership in T'. Let L denote the number of loop iterations, and Y the output. Also, let Y_1 be the value of y in the first iteration, and H_1 the event that the algorithm halts in the first iteration. Since Y_1 is uniformly distributed over T, and H_1 is the event that $Y_1 \in T'$, we have $P[H_1] = | T' | / |$

T |. It follows that $E[L] = |T| / |T'|$. As for the output, for every $t \in T$, we have

$$P[Y = t] = P[Y_1 = t \mid H_1] = P[Y_1 = t \mid Y_1 \in T'],$$

which is 0 if $t \notin T'$ and is $1 / |T'|$ if $t \in T'$. It follows that Y is uniformly distributed over T'.

EXERCISE

1. Now consider the following modification to the algorithm in the previous exercise:

   $j \leftarrow 2$

   repeat

   $j \leftarrow j + 1, n \xleftarrow{\not c} \{0, ..., j - 1\}$

   until $n = 0$ or $n = 1$

   Show that the expected running time of this algorithm is finite.

2. Design and analyze an efficient probabilistic algorithm that takes as input an integer $n \geq 2$, and outputs a random element of $\mathbb{Z}_n^*$.

3. Consider the following probabilistic algorithm that takes as input a positive integer m:

   $S \leftarrow \varnothing;$

   repeat

   $n \xleftarrow{\not c} \{1, ..., m\}, S \leftarrow \cup \{n\}$

   until $|S| = m$

   Show that the expected number of iterations of the main loop is $\sim m \log m$.

4. Consider the following algorithm (which takes no input):

$j \leftarrow 1$

repeat

$j \leftarrow j + 1, n \xleftarrow{\not c} \{0, ..., j - 1\}$

until $n = 0$

Show that the expected running time of this algorithm is infinite (even though it does halt with probability 1).

GENERATING A RANDOM NUMBER

Suppose we want to generate a number, uniformly at random from the interval $\{0, ..., m - 1\}$, for a given positive integer m.

If m is a power of 2, say $m = 2^{\ell}$, then we can do this directly as follows: generate a random ℓ-bit string σ, and convert σ to the integer $I(\sigma)$ whose base-2 representation is σ; that is, if $\sigma = b_{\ell-1} b_{\ell-2} \dots b_0$, where the b_i's are bits, then

$$I(\sigma) := \sum_{i=0}^{\ell-1} b_i 2^i.$$

In the general case, we do not have a direct way to do this, since we can only directly generate random bits. But the following algorithm does the job:

Algorithm RN. On input m, where m is a positive integer, do the following, where $\ell := \lceil \log_2 m \rceil$:

repeat

$\sigma \xleftarrow{\not{\ell}} \{0, 1\}^{\times \ell}$

$y \leftarrow I(\sigma)$

until $y < m$

output y

Theorem: The expected running time of Algorithm RN is $O(\text{len}(m))$, and its output is uniformly distributed over $\{0, \dots, m-1\}$.

Proof: Note that $m \leq 2^{\ell} < 2m$. Let L denote the number of loop iterations of this algorithm, and Z its running time. With every loop iteration, the algorithm halts with probability $m/2^{\ell}$, and so the distribution of L is a geometric distribution with associated success probability $m/2^{\ell} > 1/2$. Therefore, $E[L] = 2^{\ell}/m < 2$. Since $Z \leq c\, \text{len}(m).L$ for some constant c, it follows that $E[Z] = O(\text{len}(m))$.

Next, we analyze the output distribution. Let Y denote the output of the algorithm. We want to show that Y is uniformly distributed over $\{0, \dots, m-1\}$. This is perhaps intuitively obvious, but let us give a rigorous justification of this claim.

To do this, for $i = 1, 2, \dots$, let Y_i denote the value of y in the ith loop iteration; for completeness, if the ith loop iteration is not executed, then we define $Y_i := \perp$. Also, for $i = 1, 2, \dots$, let H_i be the event that the algorithm halts in the ith loop iteration (i.e., H_i is the event that $L = i$). Let $t \in \{0, \dots, m-1\}$ be fixed.

First, by total probability, we have

$$P[Y = t] = \sum_{i \geq 1} P[(Y = t) \cap H_i] = \sum_{i \geq 1} P[(Y_i = t) \cap H_i].$$

Next, observe that as each loop iteration works the same as any other, it follows that for each $i \geq 1$, we have

$$P[(Y_i = t) \cap \mathrm{H}_i \mid L \geq i] = \mathrm{P}[(Y_1 = t) \cap \mathrm{H}_1] = \mathrm{P}[Y_1 = t] = 2^{-\ell}.$$

Moreover, since H*i* implies *L i*, we have

$$P[(Yi = t) \cap H_i] = P[(Y_i = t) \cap H_i \cap (L \geq i)]$$

$$= P[(Y_i = t) \cap H_i \mid L \geq i]\, P[L \geq i] = 2^{-\ell} P[L \geq i],$$

and so using and the infinite version of Theorem, we have

$$P[Y = t] = \sum_{i \geq 1} P[(Y_i = t) \cap H_i] = \sum_{i \geq 1} 2^{-\ell} P[L \geq i] = 2^{-\ell} \sum_{i \geq 1} P[L \geq i]$$

$$= 2^{-\ell} . E[L] = 2^{-\ell} . 2^{\ell}/m = 1/m.$$

This shows that Y is uniformly distributed over $\{0, ..., m-1\}$.

Of course, by adding an appropriate value to the output of Algorithm RN, we can generate random numbers uniformly in the interval $\{m_1, ..., m_2\}$, for any given m_1 and m_2. In what follows, we shall denote the execution of this algorithm as

$$y \xleftarrow{\not{c}} \{m_1, ..., m_2\}.$$

More generally, if T is any finite, non-empty set for which we have an efficient algorithm whose output is uniformly distributed over T, we shall denote the execution of this algorithm as

$$y \xleftarrow{\not{c}} T.$$

For example, we may write

$$y \xleftarrow{\not{c}} \mathbb{Z}_m$$

to denote assignment to y of a randomly chosen element of $\mathbb{Z}_m$. Of course, this is done by running Algorithm RN on input m, and viewing its output as a residue class modulo m.

We also mention the following alternative algorithm for generating an almostrandom number from an interval.

Algorithm *RN'*. On input m, k, where both m and k are positive integers, do the following, where $\ell := \lceil \log_2 m \rceil$:

$\sigma \xleftarrow{\not{c}} \{0, 1\}^{\times(\ell + k)}$

$y \leftarrow I(\sigma) \bmod m$

output y

Compared with Algorithm RN, Algorithm RN' has the advantage that there are no loops – it always halts in a bounded number of steps; however, it has the disadvantage that its output is not uniformly distributed over the interval $\{0, ..., m - 1\}$. Nevertheless, the statistical distance between its output distribution and the uniform distribution on $\{0, ..., m - 1\}$ is at most 2^{-k}. Thus, by choosing k suitably large, we can make the output distribution "as good as uniform" for most practical purposes.

EXERCISE

1. You are to design and analyze an efficient probabilistic algorithm B that takes as input two integers n and y, with $n > 0$ and $0 \leq y \leq n$, and always outputs 0 or 1. Your algorithm should satisfy the following property. Suppose A is a probabilistic algorithm that takes two inputs, n and x, and always outputs an integer between 0 and n. Let Y be a random variable representing A's output on input n, x. Then for all inputs n, x, we should have P[$B(n, A(n, x))$ outputs 1] = E[Y]/n.
2. Prove that if m is not a power of 2, there is no probabilistic algorithm whose running time is strictly bounded and whose output distribution is uniform on $\{0, ..., m - 1\}$.

GENERATING RANDOM SEQUENCE

RANDOM SEQUENCE ALGORITHM

On input m, where m is an integer ≥ 2, do the following:

$n_0 \leftarrow m$

$k \rightarrow 0$

repeat

$k \leftarrow k + 1$

nk $\xleftarrow{\not c}$ $\{1, ..., n_{k-1}\}$

until $n_k = 1$

output $(n_1, ..., n_k)$

We analyze first the output distribution, and then the running time.

RUNNING TIME ANALYSIS

Let $\ell :=$ len(m), and let K be the random variable that represents the number of loop iterations performed by the algorithm. With the random variables $O_2, ..., O_m$ defined as above, we can write $K = 1 + \sum_{j=2}^{m} O_j$. Moreover, for each j, $O_j + 1$ has a geometric distribution with associated success probability $1 - 1/j$, and hence

$$E[O_j] = \frac{1}{1-1/j} - 1 = \frac{1}{j-1}.$$

Thus,

$$E[K] = 1 + \sum_{j=2}^{m} E[O_j] = 1 + \sum_{j=1}^{m-1} \frac{1}{j} \leq 2 + \int_1^m \frac{dy}{y} = \log m + 2,$$

where we have estimated the sum by an integra.

Intuitively, this is roughly as we would expect, since with probability 1/2, each successive *ni* is at most one half as large as its predecessor, and so after $O(\ell)$ steps, we expect to reach 1.

Let Z be the total running time of the algorithm. We may bound E[Z] using essentially the same argument that was used in the proof of Theorem. First, write $Z = \sum_{i \geq 1} Z_i$, where Z_i is the time spent in the *i*th loop iteration. Each loop iteration, if executed at all, runs in expected time $O(\ell)$. That is, there exists a constant c, such that for each $i \geq 1$,

$$E[Z_i \mid K \geq i] \leq c\ell \text{ and } E[Z_i \mid K < i] = 0.$$

Thus,

$$E[Z_i] = E[Z_i \mid K \geq i]\, P[K \geq i] + E[Z_i \mid K < i]\, P[K < i] \leq c\ell\, P[K \geq i],$$

and so

$$E[Z] = \sum_{i \geq 1} E[Z_i] \leq c\ell \sum_{i \geq 1} P[K \geq i] = c\ell E[K] = O(\ell^2).$$

In summary, we have shown:

the expected running time of Algorithm RS on '-bit inputs is $O(\ell^2)$.

EXERCISE. Show that when Algorithm RS runs on input m, the expected number of (not necessarily distinct) primes in the output sequence is ~ log log m.

OUTPUT DISTRIBUTION ANALYSIS[H]

Let $N_1, N_2, \ldots$ be random variables denoting the choices of $n_1, n_2, \ldots$ (for completeness, define $N_i := 1$ if loop i is never entered).

A particular output of the algorithm is a non-increasing sequence $(j_1, \ldots, j_h)$, where $j_1 \geq j_2 \cdots \geq j_{h-1} > j_h = 1$. For any such sequence, we have

$$P\left[\bigcap_{v=1}^{h} (N_v = j_v)\right] = P[N_1 = 1] . \prod_{v=2}^{h} P\left[N_v = j_v \mid \bigcap_{w<v} (N_w = j_w)\right]$$

$$= \frac{1}{m} . \frac{1}{j_1} \cdots \frac{1}{j_{h-1}}.$$

This completely describes the output distribution, in the sense that we have determined the probability with which each non-increasing sequence appears as an output. However, there is another way to characterize the output distribution that is significantly more useful.

For $j = 2, ..., m$, define the random variable O_j to be the number of occurrences of the integer j in the output sequence. The O_j's determine the N_i's, and vice versa. Indeed, $O_m = e_m, ..., O_2 = e_2$ if and only if the output of the algorithm is the sequence

$$(\underbrace{m,...,m,}_{e_m\text{times}})\underbrace{m-1,...,m-1,}_{e_{m-1}\text{times}}...,\underbrace{2,...,2,}_{e_2\text{times}}1).$$

From, we can therefore directly compute

$$P\left[\bigcap_{j=2}^{m}(O_j = e_j)\right] = \frac{1}{m}\prod_{j=2}^{m}\frac{1}{j^{e_j}}.$$

Moreover, we can write $1/m$ as a telescoping product,

$$\frac{1}{m} = \frac{m-1}{m}.\frac{m-2}{m-1}.....\frac{2}{m}.\frac{1}{2} = \prod_{j=2}^{m}(1-1/j),$$

and so re-write (9.5) as

$$P\left[\prod_{j=2}^{m}(O_j = e_j)\right] = \prod_{j=2}^{m} j^{-e_j}(1-1/j).$$

Notice that for $j = 2, ..., m$,

$$\sum_{e_j \geq 0} j^{-e_j}(1-1/j) = 1,$$

and so by (a discrete version of) Theorem, the family of random variables $\{O_j\}_{j=2}^{m}$ is mutually independent, and for each $j = 2, ..., m$ and each integer $e_j \geq 0$, we have

$$P[O_j = e_j] = j^{-e_j}\ (1 - 1/j).$$

In summary, we have shown:

That the family $\{O_j\}_{j=2}^{m}$ is mutually independent, where for each $j = 2, ..., m$, the variable $O_j + 1$ has a geometric distribution with an associated success probability of $1 - 1/j$.

Another, perhaps more intuitive, analysis of the distribution of the O_j's runs as follows. Conditioning on the event $O_m = e_m, ..., O_{j+1} = e_{j+1}$, one sees that the value of O_j is the number of times the value j appears in the sequence $N_i, N_{i+1}, \ldots$, where $i = e_m + \ldots + e_{j+1} + 1$; moreover, in this conditional probability distribution, it is not too hard to convince oneself that N_i is uniformly distributed over $[1, \ldots, j\}$.

Hence the probability that $O_j = e_j$ in this conditional probability distribution is the probability of getting a run of exactly e_j copies of the value j in an experiment in which we successively choose numbers between 1 and j at random, and this latter probability is clearly j^{-e_j} $(1 - 1/j)$.

GENERATING A RANDOM FACTORED NUMBER

We now present an efficient algorithm that generates a random factored number. That is, on input $m \geq 2$, the algorithm generates a number y uniformly distributed over the interval $\{1, \ldots, m\}$, but instead of the usual output format for such a number y, the output consists of the prime factorization of y.

As far as anyone knows, there are no efficient algorithms for factoring large numbers, despite years of active research in search of such an algorithm. So our algorithm to generate a random factored number will not work by generating a random number and then factoring it.

We shall assume the existence of an efficient, deterministic primality test IsPrime. In the analysis of the algorithm, we shall make use of Mertens' theorem.

Algorithm RFN. On input m, where m is an integer ≥ 2, do the following:

repeat

run Algorithm RS on input m, obtaining $(n_1, \ldots, n_k)$

(*) let $(p_1, \ldots, p_r)$ be the subsequence of primes in $(n_1, \ldots, n_k)$

(**) $y \leftarrow p_1 \cdots p_r$

if $y \leq m$ then

$x \xleftarrow{\not{c}} \{1, \ldots, m\}$

if $x \leq y$ then output $(p_1, \ldots, p_r)$ and halt

forever

Notes:

(*) For $i = 1, \ldots, k - 1$, the number n_i is tested for primality using algorithm Is Prime. The sequence $(n_1, \ldots, n_k)$ may contain duplicates, and if these are prime, they are appear in $(p_1, \ldots, p_r)$ with the same multiplicity.

(**) We assume that the product is computed by a simple iterative procedure that halts as soon as the partial product exceeds m. This ensures that the time spent forming the product is always $O(\text{len}(m)^2)$, which simplifies the analysis.

We now analyze the running time and output distribution of Algorithm RFN on input m, using the generate-and-test paradigm discussed earlier; here, the "generate" part consists of the first two lines of the loop body, which generates the sequence $(p_1, ..., p_r)$, while the "test" part consists of the last four lines of the loop body.

Let $\ell := \text{len}(m)$. We assume that each call to IsPrime takes time at most $\tau(\ell)$, and for simplicity, we assume $\tau(\ell) = \Omega(\ell)$.

Let K_1 be the value of k in the first loop iteration, Z_1 be the running time of the first loop iteration, Y_1 be the value of y in the first loop iteration, and H_1 be the event that the algorithm halts in the first loop iteration. Also, let Z be the total running time of the algorithm, and let Y be the value of y in the last loop iteration (i.e., the number whose factorization is output).

We begin with three preliminary calculations.

First, let $t = 1, ..., m$ be a fixed integer, and let us calculate the probability that $Y_1 = t$. Suppose $t = \prod_{p \le m} p^{e_p}$ is the prime factorization of t. Let $O_2, ..., O_m$ be random variables as defined in §9.5, so that O_j represents the number of occurrences of j in the output sequence of the first invocation of Algorithm RS. Then $Y_1 = t$ if and only if $O_p = e_p$ for all primes $p \le m$, and so by the analysis in §9.5, we have

$$P[Y_1 = t] = \prod_{p \le m} p^{-e_p}(1 - 1/p) = \frac{g(m)}{t},$$

where

$$g(m) := \prod_{p \le m} (1 - 1/p).$$

Second, we calculate $P[H_1]$. Observe that for $t = 1, ..., m$, we have

$$P[H_1 \mid Y_1 = t] = t/m,$$

and so

$$P[H_1] = \sum_{t=1}^{m} P[H_1 \mid Y_1 = t] P[Y_1 = t] = \sum_{t=1}^{m} \frac{t}{m} \frac{g(m)}{t} = g(m).$$

Third, let $t = 1, ..., m$ be a fixed integer, and let us calculate the conditional probability that $Y_1 = t$ given H_1. We have

$$P[Y_1 = t \mid H_1] = \frac{P[(Y_1 = t) \cap H_1]}{P[H_1]} = \frac{P[H_1 \mid Y_1 = t] P[Y_1 = t]}{P[H_1]}$$

$$= \frac{(t/m)(g(m)/t)}{g(m)} = \frac{1}{m}.$$

We may now easily analyze the output distribution of Algorithm RFN. By Theorem, for each $t = 1, ..., m$, we have

$$P[Y = t] = P[Y_1 = t \mid H_1] = \frac{1}{m},$$

which shows that the output is indeed uniformly distributed over all integers in [1, ..., m}, represented in factored form.

Finally, we analyze the expected running time of Algorithm RFN. It is easy to see that $E[Z_1] = O(E[K_1]\tau(\ell) + \ell^2)$, we know that $E[K_1] = O(\ell)$, and hence $E[Z_1] = O(\ell\tau(\ell))$. By Theorem, we have

$$E[Z] = E[Z_1]/ P[H_1] = E[Z_1]g(m)^{-1}.$$

By Mertens' theorem, $g(m)^{-1} = O(\ell)$. We conclude that

$$E[Z] = O(\ell^2\tau(\ell)).$$

That is, the expected running time of Algorithm RFN is $O(\ell^2\tau(\ell))$.

PROBABILISTIC PRIMALITY TEST

We can analyze the behavior of Algorithm RFN under the assumption that IsPrime is a probabilistic algorithm which may erroneously indicate that a composite number is prime with probability at most ε. Let $\ell :=$ len(m), let $\mp(\ell)$ be a bound on the expected running time of IsPrime for all inputs of bit length at most ℓ (and again, assume $\mp(\ell) = \Omega(\ell)$).

The random variables K_1, Z_1, Y_1, Z, Y and the event H_1 are defined as above. Let us also define F_1 to be the event that the primality test makes a mistake in the first loop iteration, and F to be the event that the output of the algorithm is not a list of primes. Let $\delta := P[F_1]$.

Again, we begin with three preliminary calculations.

First, let $t = 1, ..., m$ be fixed and let us calculate $P[(Y_1 = t) \cap F_1]$. To do this, define the random variable Y_1' to be the product of the actual primes among the output of the first invocation of Algorithm RS (because the primality test may err, Y_1 may contain additional factors). Evidently, the events $(Y_1 = t) \cap \overline{F}_1$ and $Y_1'(= t) \cap \overline{F}_1$ are the same.

Moreover, we claim that the events $Y_1' = t$ and $\overline{F}_1$ are independent. To see this, recall that the family $\{O_j\}_{j=2}^{m}$ is mutually independent, and also observe that the event $Y01 = t$ depends only on the random variables O_j, where j is prime, while the event $\overline{F}_1$ depends only on the random variables O_j, where j is composite, along with the execution paths of IsPrime on corresponding inputs. Thus, by a calculation analogous to one we made above,

$$P[(Y_1 = t) \cap \overline{F}_1] = P[Y_1' = t]\ P[\overline{F}_1] = \frac{g(m)}{t}(1-\delta).$$

Second, we calculate $P[H_1 \cap \overline{F}_1]$. Observe that for $t = 1, \ldots, m$, we have

$$P[H_1 \mid (Y_1 = t) \cap \overline{F}_1] = t/m,$$

and so

$$P[H_1 \cap \overline{F}_1] = \sum_{t=1}^{m} P[H_1 \cap (Y_1 = t) \cap \overline{F}_1]$$

$$= \sum_{t=1}^{m} P[H_1 \mid (Y_1 = t) \cap \overline{F}_1] P[Y_1 = t \cap \overline{F}_1]$$

$$= \sum_{t=1}^{m} \frac{t}{m} \frac{g(m)}{t}(1-\delta) = g(m)(1-\delta)$$

Third, let $t = 1, \ldots, m$ be a fixed integer, and let us calculate the conditional probability that $(Y_1 = t) \cap \overline{F}_1$ given H_1. We have

$$P[(Y_1 = t) \cap \overline{F}_1 \mid H_1] = \frac{P[(Y_1 = t) \cap \overline{F}_1 \cap H_1]}{P[H_1]}$$

$$= \frac{P[H_1 \mid (Y_1 = t) \cap \overline{F}_1] P[(Y_1 = t) \cap \overline{F}_1]}{P[H_1]}$$

$$= \frac{(t/m)((1-\delta)g(m)/t)}{P[H_1]} = \frac{g(m)(1-\delta)}{mP[H_1]}.$$

We may now easily analyze the output distribution of Algorithm RFN. By Theorem, for each $t = 1, \ldots, m$, we have

$$P[(Y = t) \cap \overline{F}] = P[(Y_1 = t) \cap \overline{F}_1 \mid H_1] = \frac{g(m)(1-\delta)}{mP[H_1]}$$

Thus, every integer between 1 and m is equally likely to be output by Algorithm RFN in correct factored form.

Let us also calculate an upper bound on the probability P[F] that Algorithm RFN outputs an integer that is not in correct factored form. We have

$$P[F_1 \mid H_1] = \frac{P[F_1 \cap H_1]}{P[H_1]} \le \frac{P[F_1]}{P[F_1 \cup H_1]}.$$

Moreover,

$$P[F_1 \cup H_1] = P[F_1] + P[H_1 \cap \bar{F}_1] = \delta + g(m)(1 - \delta)$$
$$\geq g(m)\delta + g(m)(1 - \delta) = g(m).$$

By Theorem, it follows that

$$P[F] = P[F_1 \mid H_1] \leq \delta/g(m).$$

Now, the reader may verify that

$$\delta \leq \varepsilon\,(E[K_1] - 1),$$

and by our calculations, $E[K_1] \leq \log m + 2$. Thus,

$$\delta \leq \varepsilon\,.\,(\log m + 1),$$

and so by Mertens' theorem,

$$P[F] = O(\ell^2 \varepsilon).$$

We may also analyze the statistical distance Δ between the output distribution of Algorithm RFN and the uniform distribution on $\{1, ..., m\}$ (in factored form). It follows from Theorem that $\Delta = P[F] \leq \delta/g(m) = O(\ell^2 \varepsilon)$.

Finally, we analyze the expected running time of Algorithm RFN. We have

$$P[H_1] \geq P[H_1 \cap \bar{F}_1] = g(m)(1 - \delta).$$

We leave it to the reader to verify that $E[Z_1] = O(\mp\ell(\ell))$, from which it follows by Theorem that

$$E[Z] = E[Z_1] = P[H_1] = O(\ell^2 \mp(\ell)/(1 - \delta)).$$

If ε is moderately small, so that $\varepsilon(\log m + 1) \leq 1/2$, and hence $\delta \leq 1/2$, then

$$E[Z] = O(\ell^2 \mp(\ell)).$$

GENERATING A RANDOM PRIME

Suppose we are given an integer $m \geq 2$, and want to generate a random prime between 2 and m. One way to proceed is simply to generate random numbers until we get a prime. This idea will work, assuming the existence of an efficient, deterministic algorithm

Is Prime that determines whether or not a given integer is prime. For the moment, we shall just assume we have such an algorithm, and use it as a "black box." Let us assume that on inputs of bit length at most ℓ, IsPrime runs in time at most $\tau(\ell)$.

Let us also assume (quite reasonably) that $\tau(\ell) = \Omega(\ell)$.

Algorithm RP. On input m, where m is an integer ≥ 2, do the following:

repeat

$n \xleftarrow{\not\ell} \{2, ..., m\}$

until Is Prime (n)

output n

We now wish to analyze the running time and output distribution of Algorithm RP on an input m, where $\ell := \text{len}(m)$.

This is easily done, and more specifically. The expected number of loop iterations performed by Algorithm RP is $(m - 1)/\pi(m)$, where $\pi(m)$ is the number of primes up to m. By Chebyshev's theorem, $\pi(m) = \Theta(m/\ell)$. It follows that the expected number of loop iterations is $\Theta(\ell)$.

Furthermore, the expected running time of any one loop iteration is $O(\tau(\ell))$ (the expected running time for generating n is $O(\ell)$, and this is where we use the assumption that $\tau(\ell) = \Omega(\ell)$). It follows that the expected total running time is $O(\tau\,\ell\,(\ell))$. As for the output, it is clear that it is uniformly distributed over the set of primes up to m.

GENERATING ℓ-BIT PRIME

Instead of generating a random prime between 2 and m, we may instead want to generate a random ℓ-bit prime, that is, a prime between $2^{\ell-1}$ and $2^{\ell} - 1$. Bertrand's postulate tells us that there exist such primes for every $\ell \geq 2$, and that in fact, there are $\Omega(2^{\ell}/\ell)$ such primes. Because of this, we can modify Algorithm RP, so that each candidate n is chosen at random from the interval

$\{2^{\ell-1}, ..., 2^{\ell} - 1]$, and all of the results for that algorithm carry over essentially without change. In particular, the expected number of trials until the algorithm halts is $O(\ell)$, and if a probabilistic primality test is used, with an error probability of ℓ, the probability that the output is not prime is $O(\ell\varepsilon)$.

EXERCISE

1. Suppose Algorithm RP is implemented using an imperfect random number generator, so that the statistical distance between the output distribution of the random number generator and the uniform distribution on $\{2, ..., m\}$. Assume that $2\delta < \pi(m)/(m - 1)$. Also, let Δ denote the expected number of iterations of the main loop of Algorithm RP, let Δ denote the statistical distance between its output distribution and the uniform distribution on the primes up to m, and let $\ell := \text{len}(m)$.
   (*a*) Assuming the primality test is probabilistic, with one-sided error ε, show that $\mu = O(\ell)$ and $\Delta = O((\delta + \varepsilon)\ell)$.
   (*b*) Assuming the primality test is deterministic, show that $\mu = O(\ell)$ and $\Delta = O(\delta\ell)$.

PRIMALITY TEST OF PROBABILITY

There are in fact simpler and far more efficient primality tests that are probabilistic. We shall discuss such an algorithm in detail in the next chapter. This algorithm (like several other probabilistic primality tests) has one-sided error, in the following sense: if the input n is prime, then the algorithm always outputs true; otherwise, if n is composite, the output may be true or false, but the probability that the output is true is at most ε, where ε is a very small number (the algorithm may be easily tuned to make ε quite small, e.g., 2^{-100}).

Let us analyze the behavior of Algorithm RP under the assumption that Is Prime is implemented by a probabilistic algorithm with an error probability for composite inputs bounded by ε, as discussed in the previous paragraph. Let $\bar{T}(\ell)$ be a bound on the expected running time of this algorithm for all inputs of bit length at most ℓ. Again, we assume that $\bar{T}(\ell) = \Omega(\ell)$.

Let us consider a fixed input m, and let $\ell := \text{len}(m)$. Let L, Z, and N be random variables representing, respectively, the number of loop iterations, the total running time, and output of Algorithm RP on input m. Also, let Z_1 be the random variable representing the running time of the first loop iteration, and let N_1 be the random variable representing the value assigned to n in the first loop iteration. Let H_1 be the event that the algorithm halts in the first loop iteration, and let C_1 be the event that N_1 is composite.

Clearly, N_1 is uniformly distributed over $\{2, ..., m\}$. Also, by our assumptions about Is Prime, we have

$$E[Z_1] = O(\bar{T}(\ell)),$$

and moreover, for each $j \in \{2, ..., m\}$, we have

$$P[H_1 \mid N_1 = j] \leq \varepsilon \text{ if } j \text{ is composite,}$$

and

$$P[H_1 \mid N_1 = j] = 1 \text{ if } j \text{ is prime.}$$

In particular,

$$P[H_1 \mid C_1] \leq \varepsilon \text{ and } P[H_1 \mid C_1] = 1.$$

It follows that

$$P[H_1] = P[H_1 \mid C_1]\,P[C_1] + P[H_1 \mid \overline{C_1}]\,P[\overline{C_1}] \geq P[H_1 \mid \overline{C_1}]\,P[\overline{C_1}]$$
$$= \pi(m)/(m - 1).$$

Therefore,

$$E[L] \leq (m - 1)/\pi(m) = O(\ell)$$

and

$$E[Z] = E[L]\, E[Z_1] = O(\ell \,\bar{\tau}(\ell)).$$

That takes care of the running time. Now consider the output. For every $j \in \{2, ..., m\}$, we have

$$P[N = j] = P[N_1 = j \mid H_1].$$

If j is prime, then

$$P[N = j] = P[N_1 = j \mid H_1] = \frac{P[(N_1 = j) \cap H_1]}{P[H_1]}$$

$$= \frac{P[H_1 \mid N_1 = j \mid P[N_1 = j]}{P[H_1]} = \frac{1}{(m-1)P[H_1]}.$$

Thus, every prime is output with equal probability; however, the algorithm may also output a number that is not prime. Let us bound the probability of this event. One might be tempted to say that this happens with probability at most ε; however, in drawing such a conclusion, one would be committing the fallacy of previous Example—to correctly analyze the probability that Algorithm RP mistakenly outputs a composite, one must take into account the rate of incidence of the "primality disease," as well as the error rate of the test for this disease. Indeed, if C is the event that N is composite, then we have

$$P[C] = P[C_1 \mid H_1] = \frac{P[C_1 \cap H_1]}{P[H_1]} = \frac{P[H_1 \mid C_1]P[C_1]}{P[H_1]}$$

$$\leq \frac{\varepsilon}{P[H_1]} \leq \frac{\varepsilon}{\pi(m)/(m-1)} = O(\ell\varepsilon).$$

Another way of analyzing the output distribution of Algorithm RP is to consider its statistical distance Δ from the uniform distribution on the set of primes between 2 and m. As we have already argued, every prime between 2 and m is equally likely to be output, and in particular, any fixed prime is output with probability at most $1{=}\pi(m)$. It follows from earlier read Theorem that $\Delta = P[C] = O(\ell\, \varepsilon)$.

THEOREM OF COMPLEX

We close this chapter with a few observations about probabilistic algorithms from a more "complexity theoretic" point of view.

Suppose f is a function mapping bit strings to bit strings. We may have an algorithm A that approximately computes f in the following sense: there exists a constant ε, with $0 \leq \varepsilon < 1/2$, such that for all inputs x, $A(x)$ outputs $f(x)$ with probability at least $1 - \varepsilon$. The value ε is a bound on the error probability, which is defined as the probability that $A(x)$ does not output $f(x)$.

LANGUAGE RECOGNITION

An important special case of approximately computing a function is when the output of the function f is either 0 or 1 (or equivalently, false or true). In this case, f may be viewed as the characteristic function of the language $L := \{x: f(x) = 1\}$. (It is the tradition of computational complexity theory to call sets of bit strings "languages.") There are several "flavors" of probabilistic algorithms for approximately computing the characteristic function f of a language L that are traditionally considered—for the purposes of these definitions, we may restrict ourselves to algorithms that output either 0 or 1:

- We call a probabilistic, expected polynomial-time algorithm an Atlantic City algorithm for recognizing L if it approximately computes f with error probability bounded by a constant $\varepsilon < 1/2$.
- We call a probabilistic, expected polynomial-time algorithm A a Monte Carlo algorithm for recognizing L if for some constant $\delta > 0$, we have:

  – $P[A(x) \text{ outputs } 1] \geq \delta$ for all $x \in L$;

  – $P[A(x) \text{ outputs } 1] = 0$ for all $x \notin L$.
- We call a probabilistic, expected polynomial-time algorithm a Las Vegas algorithm for recognizing L if it computes f correctly on all inputs x.

STRICT POLYNOMIAL TIME

If we have an algorithm A that runs in expected polynomial time, and which approximately computes a function f, then we can easily turn it into a new algorithm A' that runs in strict polynomial time, and also approximates f, as follows.

Suppose that $\varepsilon < 1/2$ is a bound on the error probability, and $Q(n)$ is a polynomial bound on the expected running time for inputs of size n. Then A' simply runs A for at most $kQ(n)$ steps, where k is any constant chosen so that $\varepsilon + 1/k < 1/2$ — if A does not halt within this time bound, then A' simply halts with an arbitrary output.

The probability that A' errs is at most the probability that A errs plus the probability that A runs for more than $kQ(n)$ steps. By Markov's inequality, the latter probability is at most $1/k$, and hence A' approximates f as well, but with an error probability bounded by $\varepsilon + 1/k$.

REDUCING THE ERROR PROBABILITY

There is a standard "trick" by which one can make the error probability very small; namely, run A on input x some number, say k, times, and take the majority output as the answer. Suppose $\varepsilon < 1/2$ is a bound on the error probability. Using the Cherno–bound, the error probability for the iterated version of A is bounded by

$$\exp[-(1/2 - \varepsilon)^2 k/2],$$

and so the error probability decreases exponentially with the number of iterations. This bound is derived as follows. For $i = 1, ..., k$, let X_i be the indicator variable for the event that the *i*th iteration of $A(x)$ does not output $f(x)$. The expected value of the sample mean $\overline{X} := \frac{1}{k}\sum_{i=1}^{k} X_i$ is at most ε, and if the majority output of the iterated algorithm is wrong (or indeed, if there is no majority), then $\overline{X}$ exceeds its expectation by at least 1/2 – ε.

EXERCISE

1. Show that a language is recognized by a Las Vegas algorithm if and only if the language and its complement are recognized by Monte Carlo algorithms.
2. Show that if *L* is recognized by a Las Vegas algorithm that runs in strict polynomial time, then *L* may be recognized in deterministic polynomial time.
3. Show that if *L* is recognized by an Atlantic City algorithm that runs in expected polynomial time, then it is recognized by an Atlantic City algorithm that runs in strict polynomial time, and whose error probability is at most 2_{-n} on inputs of size *n*.
4. Show that every language recognized by a Las Vegas algorithm is also recognized by a Monte Carlo algorithm, and that every language recognized by a Monte Carlo algorithm is also recognized by an Atlantic City algorithm.
5. Show that if *L* is recognized by a Monte Carlo algorithm that runs in expected polynomial time, then it is recognized by a Monte Carlo algorithm that runs in strict polynomial time, and whose error probability is at most 2^{-n} on inputs of size *n*.

TESTING WITH PROBABILITY

TRIAL DIVISION

Suppose we are given an integer $n > 1$, and we want to determine whether *n* is prime or composite. The simplest algorithm to describe and to program is trial division. We simply divide *n* by 2, 3, and so on, testing if any of these numbers evenly divide *n*. Of course, we only need to divide by primes up to $\sqrt{n}$, since if *n* is composite, it must have a prime factor no greater than $\sqrt{n}$.

Not only does this algorithm determine whether *n* is prime or composite, it also produces a non-trivial factor of *n* in case *n* is composite. Of course, the drawback of this algorithm is that it is terribly inefficient: it requires Q($\sqrt{n}$) arithmetic operations, which is exponential in the bit length of *n*. Thus, for practical purposes, this algorithm is limited to quite small *n*. Suppose, for example, that *n* has 100 decimal digits, and that a computer can perform 1 billion divisions per second (this is much faster than any computer existing today).

Then it would take on the order of 1033 years to perform $\sqrt{n}$ divisions. In this chapter, we discuss a much faster primality test that allows 100-decimaldigit numbers to be tested for primality in less than a second. Unlike the above test, however, this test does not find a factor of n when n is composite. Moreover, the algorithm is probabilistic, and may in fact make a mistake. However, the probability that it makes a mistake can be made so small as to be irrelevant for all practical purposes. Indeed, we can easily make the probability of error as small as 2^{-100} – should one really care about an event that happens with such a miniscule probability?

EULER'S PHI FUNCTION

In this section, we use some of the ideas developed to analyze the Miller–Rabin test to prove that the problem of factoring n and the problem of computing $\varphi(n)$ are equivalent. By equivalent, we mean that given an efficient algorithm to solve one problem, we can efficiently solve the other, and vice versa.

Clearly, one direction is easy: if we can factor n into primes, so

$$n = p_1^{e_1} \ldots p_r^{e_r},$$

then we can simply compute $\varphi(n)$ using the formula

$$\varphi(n) = p_1^{e_1-1}(p_1 - 1) \ldots p_r^{e_r-1}(p_r - 1)$$

For the other direction, first consider the special case where $n = pq$, for distinct primes p and q. Suppose we are given n and $\varphi(n)$, so that we have two equations in the unknowns p and q:

$$n = pq \text{ and } \varphi(n) = (p - 1)(q - 1).$$

Substituting n/p for q in the second equation, and simplifying, we obtain

$$p^2 + (\varphi(n) - n - 1)p + n = 0,$$

which can be solved using the quadratic formula.

For the general case, it is just as easy to prove a stronger result: given any nonzero multiple of the exponent of $\mathbb{Z}_n^*$, we can efficiently factor n. In particular, this will show that we can efficiently factor Carmichael numbers.

Before stating the algorithm in its full generality, we can convey the main idea by considering the special case where $n = pq$, where p and q are distinct primes, with $p \equiv q \equiv 3 \pmod 4$. Suppose we are given such an n, along with a non-zero multiple f of the exponent of $\mathbb{Z}_n^*$. Now, $\mathbb{Z}_n^* \cong \mathbb{Z}_p^* \times \mathbb{Z}_q^*$, and since $\mathbb{Z}_p^*$ is a cyclic group of order $p - 1$ and $\mathbb{Z}_q^*$ is a cyclic group of order $q - 1$, this means that f is a non-zero common multiple of $p - 1$ and $q - 1$. Let $f = t2^h$, where t is odd, and consider the following probabilistic algorithm:

$\alpha \xleftarrow{\not{c}} \mathbb{Z}_n^+$

$d \leftarrow \gcd(\text{rep}(\alpha), n)$

if $d \neq 1$ then output d and halt

$\beta \leftarrow \alpha^t$

$d' \leftarrow \gcd(\text{rep}(\beta) + 1, n)$

if $d' \notin \{1, n\}$ then output d' and halt

output "failure"

Recall that rep(α) denotes the canonical representative of α, that is, the unique integer a such that $[a]_n =$ and $0 \leq a < n$. We shall prove that this algorithm outputs a non-trivial divisor of n with probability at least 1/2.

Let ρ be the t-power map on $\mathbb{Z}_n^*$, and let $G := \rho^{-1}(\{\pm 1\})$. We shall show that

- $G \subseteq \mathbb{Z}_n^*$, and
- If the algorithm chooses a $\notin G$, then it splits n.

Since G is a subgroup of $\mathbb{Z}_n^*$, it follows that $|G|/|\mathbb{Z}_n^+| \leq |G|/|\mathbb{Z}_n^*| \leq 1/2$, and this implies the algorithm succeeds with probability at least 1/2.

Let θ: $\mathbb{Z}_n \to \mathbb{Z}_p \times \mathbb{Z}_q$ be the ring isomorphism from the Chinese remainder theorem. The assumption that $p \equiv 3 \pmod 4$ means that $(p - 1)/2$ is an odd integer, and since f is a multiple of $p - 1$, it follows that $\gcd(t, p - 1) = (p - 1)/2$, and hence the image of $\mathbb{Z}_p^*$ under the t-power map is the subgroup of $\mathbb{Z}_p^*$ of order 2, which is $\{\pm 1\}$. Likewise, the image of $\mathbb{Z}_q^*$ under the t-power map is $\{\pm 1\}$. Thus

$$\theta(\text{Im } \rho) = \theta((\mathbb{Z}_n^*)^t) = (\theta(\mathbb{Z}_n^*))^t = (\mathbb{Z}_p^*)^t \times (\mathbb{Z}_q^*)^t = \{\pm 1\} \times \{\pm 1\},$$

and so Im ρ consists of the four elements:

$$1 = \theta^{-1}(1, 1), -1 = \theta^{-1}(-1, -1), \theta_{-1}(-1, 1), \theta^{-1}(1, -1).$$

By the observations in the previous paragraph, not all elements of $\mathbb{Z}_n^*$ map to ± 1 under ρ, which means that $G \subseteq \mathbb{Z}_n^*$. Suppose that the algorithm chooses $\alpha \in \mathbb{Z}_n^+ \setminus G$. We want to show that n gets split. If $\alpha \notin \mathbb{Z}_n^*$, then gcd(rep($\alpha$), n) is a non-trivial divisor of n, and the algorithm splits n. So let us assume that $\alpha \in \mathbb{Z}_n^* \setminus G$. Consider the value $\beta = \alpha^t = \rho(\alpha)$ computed by the algorithm. Since $\alpha \notin G$, we have $\beta \neq \pm 1$, and by the observations in the previous paragraph, we have $\theta(\beta) = (-1, 1)$ or $\theta(\beta) = (1, -1)$. In the first case, $\theta(\beta + 1) = (0, 2)$, and so gcd(rep($\beta$) + 1, n) = p, while in the second case, $\theta(\beta + 1) = (2, 0)$, and so gcd(rep($\beta$) + 1, n) = q. In either case, the algorithm splits n.

We now consider the general case, where n is an arbitrary positive integer. Let (n) denote the exponent of $\mathbb{Z}_n^*$. If the prime factorization of n is as in, then by the Chinese remainder theorem, we have

$$\lambda(n) = lcm(\lambda(p_1^{e_1}),...,\lambda(p_r^{e_r})).$$

Moreover, for every prime power p^e, by Theorem, we have

$$\lambda(p^e) = \begin{cases} p^{e-1}(p-1) & \text{if } p \neq 2 \text{ or } e \leq 2 \\ 2^{e-2} & \text{if } p = 2 \text{ and } e \geq 3. \end{cases}$$

In particular, if $d \mid n$, then (d) | $\lambda(n)$.

Now, assume we are given n, along with a non-zero multiple f of $\lambda(n)$. We would like to calculate the complete prime factorization of n.

We may proceed recursively: first, if $n = 1$, we may obviously halt; otherwise, we test if n is prime, using an efficient primality test, and if so, halt (if we are using the Miller–Rabin test, then we may erroneously halt even when n is composite, but we can ensure that this happens with negligible probability); otherwise, we split n as $n = d_1 d_2$, using an algorithm to be described below, and then recursively factor both d_1 and d_2; since $\lambda(d_1) \mid f$ and $\lambda(d_2) \mid f$, we may use the same value f in the recursion.

So let us assume that $n > 1$ and n is not prime, and our goal now is to use f to obtain a non-trivial factorization of n. If n is even, then we can certainly do this. Moreover, if n is a perfect power, that is, if $n = a^b$ for some integers $a > 1$ and $b > 1$, we can also obtain a non-trivial factorization of n.

So let us assume not only that $n > 1$ and n is not prime, but also that n is odd, and n is not a perfect power. Let $f = t2^h$, where t is odd. Consider the following probabilistic algorithm:

$\alpha \xleftarrow{\not{c}} \mathbb{Z}_n^+$

$d \leftarrow$ gcd(rep(α), n)

if $d \neq 1$ then output d and halt

$\beta \leftarrow \alpha^t$

for $j \leftarrow 0$ to $h - 1$ do

$d' \leftarrow \gcd(\text{rep}(\beta) + 1, n)$

if $d' \notin \{1, n\}$ then output d' and halt

$\beta \leftarrow \beta^2$

output "failure"

We want to show that this algorithm outputs a non-trivial factor of n with probability at least 1/2. To do this, suppose the prime factorization of n. Then by our assumptions about n, we have $r \geq 2$ and each p_i is odd. Let $\lambda(p_i^{e_i}) = t_i 2^{h_i}$, where t_i is odd, for $i = 1, ..., r$, and let $g := \max\{h_1, ..., h_r\}$.

Note that since $\lambda(n) \mid f$, we have $1 \leq g \leq h$.

Let ρ be the $(t2^{g-1})$-power map on $\mathbb{Z}_n^*$, and let $G := \rho^{-1}(\{\pm 1\})$. As above, we shall show that

- G N $\mathbb{Z}_n^*$, and
- If the algorithm chooses $\alpha \notin G$, then it splits n,

which will prove that the algorithm splits n with probability at least 1/2.

Let

$$\theta : \mathbb{Z}_n \to \mathbb{Z}_{p_1^{e_1}} \times ... \times \mathbb{Z}_{p_r^{e_r}}$$

be the ring isomorphism of the Chinese remainder theorem. We have

$$\rho(\text{Im}\ \rho) = G_1 \times ... \times G_r,$$

where

$$G_i := \left(\mathbb{Z}^*_{p_i^{e_i}}\right)^{t2^{g-1}} \quad \text{for } i = 1, ..., r.$$

Let us assume the pi's are ordered so that $h_i = g$ for $i = 1, ..., r'$, and $h_i < g$ for $i = r' + 1, ..., r$, where we have $1 \leq r' \leq r$. Then we have $G_i = \{\pm 1\}$ for $i = 1, ..., r'$, and $G_i = \{\pm 1\}$ for $i = r' + 1, ..., r$.

By the observations in the previous paragraph, and the fact that $r \geq 2$, the image of ρ contains elements other than ± 1; for example, $\theta^{-1}(-1, 1, ..., 1)$ is such an element. This means that $G \subseteq \mathbb{Z}_n^*$. Suppose the algorithm chooses $\alpha \in \mathbb{Z}_n^+ \backslash G$. We want to show that n gets split. If $\alpha \notin \mathbb{Z}_n^*$, then $\gcd(\text{rep}(\alpha), n)$ is a non-trivial divisor of n, and so the algorithm certainly splits n. So assume $\alpha \in \mathbb{Z}_n^* \backslash G$. In loop iteration $j = g - 1$, the value of β is equal to $\rho(\alpha)$, and writing $\theta(\beta) = (\beta_1, ..., \beta_r)$, we have $i = 1$ for $i = 1, ..., r$. Let S be the set of indices i such that $\beta_i = -1$.

As $\alpha \notin G$, we know that $\beta \neq \pm 1$, and so $\varnothing \subseteq S \subseteq \{1, ..., r\}$. Thus,

$$\gcd(\text{rep}(\beta) + 1, n) = \prod_{i \in S} p_i^{e_i}$$

is a non-trivial factor of n. This means that the algorithm splits n in loop iteration $j = g - 1$ (if not in some earlier loop iteration).

So we have shown that the above algorithm splits n with probability 1/2. If we iterate the algorithm until n gets split, the expected number of loop iterations required will be at most 2. Combining this with the above recursive algorithm, we get an algorithm that completely factors an arbitrary n in expected polynomial time.

EXERCISE

1. Suppose there is a probabilistic algorithm A that takes as input an integer n of the form $n = pq$, where p and q are distinct, ℓ-bit primes, with $p = 2p' + 1$ and $q = 2q' + 1$, where p' and q' are prime. The algorithm also takes as input $\alpha, \beta \in (\mathbb{Z}_n^*)^2$. It outputs either "failure," or integers x, y, not both zero, such that $\alpha^x\beta^y = 1$. Furthermore, assume that A runs in expected polynomial time, and that for all n of the above form, and for randomly chosen $\alpha, \beta \in (\mathbb{Z}_n^*)^2$, A succeeds in finding x, y as above with probability $\varepsilon(n)$. Here, the probability is taken over the random choice of α and β, as well as the random choices made during the execution of A on input (n, α, β). Show how to use A to construct another probabilistic algorithm A' that takes as input n as above, runs in expected polynomial time, and that satisfies the following property:
   if $\varepsilon(n) \geq 0.001$, then A' factors n with probability at least 0.999.
2. Suppose you are given an integer n of the form $n = pq$, where p and q are distinct, ℓ-bit primes, with $p = 2p' + 1$ and $q = 2q' + 1$, where p' and q' are themselves prime. Suppose that you are also given an integer t such that $\gcd(t, p'q') \neq 1$. Show how to efficiently factor n.

MILLER–RABIN TEST

We describe in this section a fast (polynomial time) test for primality, known as the Miller–Rabin test. As discussed above, the algorithm is probabilistic, and may (with small probability) make a mistake.

We assume for the remainder of this section that the number n we are testing for primality is an odd integer greater than 1.

We recall some basic algebraic facts that will play a critical role in this section. Suppose $n = p_1^{e_1} \ldots p_r^{e_r}$ is the prime factorization of n (since n is odd, each p_i is odd). The Chinese remainder theorem gives us a ring isomorphism

$$\theta : \mathbb{Z}_n \to \mathbb{Z}_{p_1^{e_1}} \times \ldots \times \mathbb{Z}_{p_r^{e_r}}$$

$$[a]_n \mapsto ([a]_{p_1^{e_1}}, \ldots, [a]_{p_r^{e_r}}),$$

and restricting θ to $\mathbb{Z}_n^*$ yields a group isomorphism

$$\mathbb{Z}_n^* \cong \mathbb{Z}_{p_1^{e_1}}^* \times \ldots \times \mathbb{Z}_{p_r^{e_r}}^*.$$

Moreover, Theorem says that each $\mathbb{Z}_{p_i^{e_i}}^*$ is a cyclic group, whose order, of course, is $\varphi(p_i^{e_i}) = p_i^{e_i - 1}(p_i - 1)$, where φ is Euler's phi function.

Several probabilistic primality tests, including the Miller–Rabin test, have the following general structure. Define $\mathbb{Z}_n^+$ to be the set of non-zero elements of $\mathbb{Z}_n$; thus, $|\mathbb{Z}_n^+| = n - 1$, and if n is prime, $\mathbb{Z}_n^+ = \mathbb{Z}_n^*$. Suppose also that we define a set $L_n \subseteq \mathbb{Z}_n^+$ such that:

- There is an efficient algorithm that on input n and $\alpha \in \mathbb{Z}_n^+$, determines if $\alpha \in L_n$;
- If n is prime, then $L_n = \mathbb{Z}_n^*$;
- If n is composite, $|L_n| \leq c(n - 1)$ for some constant $c < 1$.

To test n for primality, we set a "repetition parameter" k, and choose random elements $\alpha_1, \ldots, \alpha_k \in \mathbb{Z}_n^+$. If $\alpha_i \in L_n$ for all $i = 1, \ldots, k$, then we output true; otherwise, we output false.

It is easy to see that if n is prime, this algorithm always outputs true, and if n is composite this algorithm outputs true with probability at most c^k. If $c = 1/2$ and k is chosen large enough, say $k = 100$, then the probability that the output is wrong is so small that for all practical purposes, it is "just as good as zero."

We now make a first attempt at defining a suitable set L_n. Let us define

$$L_n := \{\alpha \in \mathbb{Z}_n^+ : \alpha^{n-1} = 1\}.$$

Note that $L_n \subseteq \mathbb{Z}_n^+$, since if $\alpha^{n-1} = 1$, then α has a multiplicative inverse, namely, α^{n-2}. We can test if $\alpha \in L_n$ in time $O(\text{len}(n)^3)$, using a repeated-squaring algorithm.

Theorem: Every Carmichael number n is of the form $n = p_1 \ldots p_r$, where the p_i's are distinct primes, $r \geq 3$, and $(p_i - 1) \mid (n - 1)$ for $i = 1, \ldots, r$.

Proof: Let $n = p_1^{e_1} \ldots p_r^{e_r}$ be a Carmichael number. By the Chinese remainder theorem, we have an isomorphism of $\mathbb{Z}_n^*$ with the group

$$\mathbb{Z}_{p_1^{e_1}}^* \times \ldots \times \mathbb{Z}_{p_r^{e_r}}^*,$$

and we know that each group $\mathbb{Z}_{p_i^{e_i}}^*$ is cyclic of order $p_i^{e_i-1}(p_i - 1)$. Thus, the power $n - 1$ kills the group $\mathbb{Z}_n^*$ if and only if it kills all the groups $\mathbb{Z}_{p_i^{e_i}}^*$, which happens if and only if $p_i^{e_i-1}(p_i - 1) \mid (n - 1)$. Now, on the one hand, $n \equiv 0 \pmod{pi}$. On the other hand, if $e_i > 1$, we would have $n \equiv 1 \pmod{p_i}$, which is clearly impossible. Thus, we must have $e_i = 1$.

It remains to show that $r \geq 3$. Suppose $r = 2$, so that $n = p_1 p_2$. We have

$$n - 1 = p_1 p_2 - 1 = (p_1 - 1)p_2 + (p_2 - 1).$$

Since $(p_1 - 1) \mid (n - 1)$, we must have $(p_1 - 1) \mid (p_2 - 1)$. By a symmetric argument, $(p_2 - 1) \mid (p_1 - 1)$. Hence, $p_1 = p_2$, a contradiction.

To obtain a good primality test, we need to define a different set L_n', which we do as follows. Let $n - 1 = t2^h$, where t is odd (and $h \geq 1$ since n is assumed odd), and define

$$L_n' := \{\alpha \in \mathbb{Z}_n^+ : \alpha^{t2^h} = 1 \text{ and}$$

$$\alpha^{t2^{j+1}} = \alpha^{t2^j} = \pm 1 \text{ for } j = 0, \ldots, h - 1\}.$$

The Miller–Rabin test uses this set L_n', in place of the set L_n defined above. It is clear from the definition that $L_n' \subseteq L_n$.

Testing whether a given $\alpha \in \mathbb{Z}_n^+$ belongs to L_n' can be done using the following procedure:

```
β ← α^t
if β = 1 then return true
for j ← 0 to h − 1 do
if β = −1 then return true
if β = +1 then return false
β ← β^2
return false
```

It is clear that using a repeated-squaring algorithm, this procedure runs in time

$O(\text{len}(n)^3)$. We leave it to the reader to verify that this procedure correctly determines membership in L'_n.

Theorem: If n is prime, then $L_n = \mathbb{Z}^*_n$. If n is composite and $L_n \not\subset \mathbb{Z}^*_n$, then $|L_n| \leq (n-1)\,2$.

Proof: Note that L_n is the kernel of the $(n-1)$-power map on $\mathbb{Z}^*_n$, and hence is a subgroup of $\mathbb{Z}^*_n$.

If n is prime, then we know that $\mathbb{Z}^*_n$ is a group of order $n-1$. Since the order of a group element divides the order of the group, we have $\alpha^{n-1} = 1$ for all $\alpha \in \mathbb{Z}^*_n$. That is, $L_n = \mathbb{Z}^*_n$.

Suppose that n is composite and $L_n \not\subseteq \mathbb{Z}^*_n$. Since the order of a subgroup divides the order of the group, we have $|\mathbb{Z}^*_n| = t\,|L_n|$ for some integer $t > 1$. From this, we conclude that

$$|L_n| = \frac{1}{t}|\mathbb{Z}^*_n| \leq \frac{1}{2}|\mathbb{Z}^*_n| \leq \frac{n-1}{2}.$$

Unfortunately, there are odd composite numbers n such that $L_n = \mathbb{Z}^*_n$. Such numbers are called Carmichael numbers. The smallest Carmichael number is

$$561 = 3\,.\,11\,.\,17.$$

Carmichael numbers are extremely rare, but it is known that there are infinitely many of them, so we cannot ignore them. The following theorem puts some constraints on Carmichael numbers.

Theorem: If n is prime, then $L'_n = \mathbb{Z}^*_n$. If n is composite, then $|L'_n| \leq (n-1)/4$.

Proof: Let $n - 1 = t2^h$, where t is odd.

Case 1: N is prime. Let $\alpha \in \mathbb{Z}^*_n$. Since $\mathbb{Z}^*_n$ is a group of order $n-1$, and the order of a group element divides the order of the group, we know that $\alpha^{t2^h} = \alpha^{n-1} = 1$. Now consider any index $j = 0, \ldots, h-1$ such that $\alpha^{t2^{j+1}} = 1$, and consider the value $\beta = \alpha^{t2^j}$. Then since $\beta^2 = \alpha^{t22^{j+1}} = 1$, the only possible choices for β are ± 1 – this is because $\mathbb{Z}^*_n$ is cyclic of even order and so there are exactly two elements of $\mathbb{Z}^*_n$ whose multiplicative order divides 2, namely ± 1. So we have shown that $\alpha \in L'_n$.

Case 2: $n = p^e$, where p is prime and $e > 1$. Certainly, L'_n is contained in the kernel K

of the $(n-1)$-power map on $\mathbb{Z}_n^*$. By Theorem, $|K| = \gcd(\varphi(n), n-1)$. Since $n = p^e$, we have $\varphi(n) = p^{e-1}(p-1)$, and so

$$|L_n'| \leq |K| = \gcd(p^{e-1}(p-1), p^e-1) = p-1 = \frac{p^e-1}{p^{e-1}+...+1} \leq \frac{n-1}{4}.$$

Case 3: $n = p_1^{e_1}...p_r^{e_r}$ is the prime factorization of n, and $r > 1$. Let

$$\theta : \mathbb{Z}_n \to \mathbb{Z}_{p_1^{e_1}} \times ... \times \mathbb{Z}_{p_r^{e_r}}$$

be the ring isomorphism provided by the Chinese remainder theorem. Also, let $\varphi(p_i^{e_i}) = t_i 2^{h_i}$, with t_i odd, for $i = 1, ..., r$, and let $g := \min\{h, h_1, ..., h_r\}$. Note that $g \geq 1$, and that each $\mathbb{Z}^*_{p_i^{e_i}}$ is a cyclic group of order $t_i 2^{h_i}$.

We first claim that for every $\alpha \in L_n'$, we have $\alpha^{t2^g} = 1$. To prove this, first note that if $g = h$, then by definition, $\alpha^{t2^g} = 1$, so suppose that $g < h$. By way of contradiction, suppose that $\alpha^{t2^g} \neq 1$, and let j be the smallest index in the range $g, ..., h-1$ such that $\alpha^{t2^{j+1}} = 1$.

By the definition of L_n', we must have $\alpha^{t2^j} = -1$. Since $g < h$, we must have $g = h_i$ for some particular index $i = 1, ..., r$. Writing $\theta(\alpha) = (\alpha_1, ..., \alpha_r)$, we have $\alpha_i^{t2^j} = -1$. This implies that the multiplicative order of α_i^t is equal to 2^{j+1}. However, since $j \geq g = h_i$, this contradicts the fact that the order of a group element (in this case, α_i^t) must divide the order of the group (in this case, $\mathbb{Z}^*_{p_i^{e_i}}$).

For $j = 0, ..., h$, let us define j to be the $(t2^j)$-power map on $\mathbb{Z}_n^*$. From the claim in the previous paragraph, and the definition of L_n', it follows that each $\alpha \in L_n'$ satisfies $\alpha^{t2^{g-1}} = \pm 1$. In other words, $L_n' \subseteq \rho_{g-1}^{-1}(\{\pm 1\})$, and hence

$$|L_n'| \leq 2\,|\text{Ker}\,\rho_{g-1}|.$$

From the group isomorphism $\mathbb{Z}_n^* \cong \mathbb{Z}^*_{p_1^{e_1}} \times ... \times \mathbb{Z}^*_{p_r^{e_r}}$, and Theorem, we have

$$| Ker\ \rho_j | = \prod_{i=1}^{r} \gcd(t_i 2^{h_i}, t2^j)$$

for each $j = 0, ..., h$. Since $g \leq h$, and $g \leq h_i$ for $i = 1, ..., r$, it follows immediately from that

$$2^r | \text{Ker } \rho_{g-1}| = | \text{Ker } \rho_g | \leq | \text{Ker } \rho_h |.$$

we obtain

$$| L_n^{'} | \leq 2^{-r+1} | \text{Ker } \rho_h |.$$

If $r \geq 3$, then directly implies that $| L_n^{'} | \leq |\mathbb{Z}_n^*| /4 \leq (n - 1)/4$, and we are done. So suppose that $r = 2$. In this case, Theorem implies that n is not a Carmichael number, which implies that $| \text{Ker } \rho_h | \leq |\mathbb{Z}_n^*| /2$, and so again, implies $| L_n^{'} | \leq |\mathbb{Z}_n^*| /4 \leq (n - 1)/4$.

EXERCISE

1. Let p be a prime. Show that $n := 2p + 1$ is a prime if and only if $2^{n-1} \equiv 1 \pmod{n}$.
2. Show that an integer $n > 1$ is prime if and only if there exists an element in $\mathbb{Z}_n^*$ of multiplicative order $n-1$.
3. Here is another primality test that takes as input an odd integer $n > 1$, and a positive integer parameter k. The algorithm chooses $\alpha_1, ..., \alpha_k \in \mathbb{Z}_n^+$ at random, and computes

   $$\beta_i := \alpha_i^{(n-1)/2} \ (i = 1, ..., k).$$

   If $(\beta_1, ..., \beta_k)$ is of the form $(\pm 1, \pm 1, ..., \pm 1)$, but is not equal to $(1, 1, ..., 1)$, the algorithm outputs true; otherwise, the algorithm outputs false. Show that if n is prime, then the algorithm outputs false with probability at most 2^{-k}, and if n is composite, the algorithm outputs true with probability at most 2^{-k}.
4. Show that Carmichael numbers satisfy Fermat's little theorem; that is, if n is a Carmichael number, then $\alpha^n =$ for all $\alpha \in \mathbb{Z}_n$.

The algorithm in the above exercise is an example of an "Atlantic City" algorithm for the language of prime numbers (or equivalently, the language of composite numbers), while the Miller–Rabin test is an example of a "Monte Carlo" algorithm for the language of composite numbers.

GENERATING RANDOM PRIMES

The Miller–Rabin test is the most practical algorithm known for testing primality, and because of this, it is widely used in many applications, especially cryptographic applications where one needs to generate large, random primes. In this section, we discuss how one uses the Miller–Rabin test in several practically relevant scenarios where one must generate large primes.

GENERATING A RANDOM PRIME BETWEEN 2 AND M

Suppose we are given an integer $m \geq 2$, and want to generate a random prime between 2 and m. We can do this by simply picking numbers at random until one of them passes a primality test. In this section, we discuss aspects of this problem that are specific to the situation where the Miller–Rabin test is used to implement IsPrime. To be more precise, let us define the following algorithm:

Algorithm MR. On input n, k, where n and k are integers with $n > 1$ and $k \geq 1$, do the following:

if $n = 2$ then return true

if n is even then return false

repeat k times

$\alpha \xleftarrow{\ell} \mathbb{Z}_n^+$

if $a \notin L_n'$ return false

return true

So we shall implement IsPrime(.) as MR(., k), where k is an auxiliary parameter. By Theorem, if n is prime, the output of MR(n, k) is always true, while if n is composite, the output is true with probability at most 4^{-k}. Thus, this implementation of IsPrime satisfies the assumptions, with $\varepsilon = 4^{-k}$.

Let $\gamma(m, k)$ be the probability that the output of Algorithm – using this implementation of IsPrime – is composite. Then as we discussed,

$$\gamma(m,k) \leq 4^{-k}.\frac{m-1}{\pi(m)} = O(4^{-k}\ell),$$

where $\ell := \text{len}(m)$, and $\pi(m)$ is the number of primes up to m. Furthermore, if the output of Algorithm RP is prime, then every prime is equally likely; that is, the conditional distribution of the output, given that the output is prime, is the uniform distribution on the set of primes up to m.

Let us now consider the expected running time of Algorithm RP. The expected number of iterations of the main loop in Algorithm RP is $O(\ell)$. Clearly, the expected running time of a single loop iteration is $O(k\ell^3)$, since MR(n, k) executes at most k iterations of the Miller–Rabin test, and each such test takes time $O(\ell^3)$.

This leads to a bound on the expected total running time of Algorithm RP of $O(k\ell^4)$. However, this estimate is overly pessimistic, because when n is composite, we expect to perform very few Miller–Rabin tests—only when n is prime do we actually perform all k of them. To make a rigorous argument, let us define random variables measuring various quantities during the first iteration of the main loop in Algorithm RP: N_1 is the value of n; K_1 is the number of Miller–Rabin tests actually performed; Z_1 is the running time.

Of course, N_1 is uniformly distributed over $\{2, ..., m\}$. Let C_1 be the event that $N1$ is composite. Consider the conditional distribution of K_1 given C_1. This is not exactly a geometric distribution, since K_1 never takes on values greater than k; nevertheless, using Theorem, we can easily calculate

$$E[K_1 \mid C_1] = \sum_{i\geq 1} P[K_1 \geq i \mid C_1] \leq \sum_{i\geq 1} (1/4)^{i-1} = 4/3.$$

Using the law of total expectation, it follows that

$$E[K_1] = E[K_1 \mid C_1]\, P[C_1] + E[K_1 \mid \bar{C}_1]\, P[\bar{C}_1]$$

$$\leq 4/3 + k\pi(m)/(m-1).$$

Thus, $E[K_1] \leq 4/3 + O(k/\ell)$, and hence $E[Z_1] = O(\ell^3 E[K_1]) = O(\ell^3 + k\ell^2)$.

Therefore, if Z is the total running time of Algorithm *RP*, then $E[Z] = O\ell(E[Z_1])$, and so

$$E[Z] = O(\ell^4 + k\ell^3).$$

Note that the above estimate for $\gamma(m, k)$ is actually quite pessimistic. This is because the error probability 4^{-k} is a worst-case estimate; in fact, for "most" composite integers n, the probability that MR(n, k) outputs true is much smaller than this. In fact, $\gamma(m, 1)$ is very small for large m.

Theorem: We have

$$\gamma(m, 1) \leq \exp[-(1 + o(1)) \log(m) \log(\log(\log(m)))/ \log(\log(m))].$$

Proof: The bound in the above theorem goes to zero quite quickly: faster than $(\log m)^{-c}$ for every positive constant c. While the above theorem is asymptotically very good, in practice, one needs explicit bounds. For example, the following lower bounds for $-\log_2(\gamma(2^{\ell}, 1))$ are known:

ℓ	200	300	400	500	600
	3	19	37	55	74

Given an upper bound on $\gamma(m, 1)$, we can bound (m, k) for $k \geq 2$ using the following inequality:

$$\gamma(m,k) \le \frac{\gamma(m,1)}{1-\gamma(m,1)} 4^{-k+1}.$$

To prove, it is not hard to see that on input m, the output distribution of Algorithm RP is the same as that of the following algorithm:

repeat

repeat

$n' \xleftarrow{\not{c}} \{2, ..., m\}$

until MR (n', k – 1)

$n \leftarrow n'$

until MR(n, k – 1)

output n

Let N_1 be the random variable representing the value of n in the first iteration of the main loop in this algorithm, let C_1 be the event that N_1 is composite, and let H_1 be the event that this algorithm halts at the end of the first iteration of the main loop. Using Theorem, we see that

$$\gamma(m, k) = P[C_1 \mid H_1] = \frac{P[C_1 \cap H_1]}{P[H_1]} \le \frac{P[C_1 \cap H_1]}{P[\bar{C}_1]} = \frac{P[H_1 \mid P[C_1]}{P[\bar{C}_1]}$$

$$\le \frac{4^{-k+1}\gamma(m,1)}{1-\gamma(m,1)},$$

which proves .

Given that $\gamma(m, 1)$ is so small, for large m, Algorithm RP actually exhibits the following behavior in practice: it generates a random value $n \in \{2, ..., m\}$; if n is odd and composite, then the very first iteration of the Miller–Rabin test will detect this with overwhelming probability, and no more iterations of the test are performed on this n; otherwise, if n is prime, the algorithm will perform k – 1 more iterations of the Miller–Rabin test, "just to make sure."

TRIAL DIVISION

Significant efficiency gains can be achieved by testing if a given candidate n is divisible by any prime up to a given bound s, before we subject n to a Miller–Rabin test. This strategy makes sense, since for a small, "single precision" prime p, we can test if $p \mid n$ essentially in time $O(\text{len}(n))$, while a single iteration of the Miller–Rabin test takes time $O(\text{len}(n)3)$.

To be more precise, let us define the following algorithm:

Algorithm MRS. On input n, k, s, where $n, k, s \in \mathbb{Z}$, and $n > 1$, $k \geq 1$, and $s > 1$, do the following:

for each prime $p \leq s$ do

if $p \mid n$ then

if $p = n$ then return true else return false

repeat k times

$\alpha \xleftarrow{\not{c}} \mathbb{Z}_n^+$

if $\alpha \notin L_n'$ return false

return true

In an implementation of the above algorithm, one would most likely use the sieve of Eratosthenes to generate the small primes.

Note that MRS(n, k, 2) is equivalent to MR(n, k). Also, it is clear that the probability that MRS(n, k, s) makes a mistake is no more than the probability that MR(n, k) makes a mistake. Therefore, using MRS in place of MR will not increase the probability that the output of Algorithm RP is a composite—indeed, it is likely that this probability decreases significantly.

Let us now analyze the impact on the running time Algorithm RP. To do this, we need to estimate the probability $\sigma(m, s)$ that a randomly chosen integer between 2 and m is not divisible by any prime up to s. If m is sufficiently large with respect to s, the following heuristic argument can be made rigorous. The probability that a random integer is divisible by a prime p is about $1/p$, so the probability that it is not divisible by p is about $1 - 1/p$. Assuming that these events are essentially independent for different values of p (this is the heuristic part), we estimate

$$\sigma(m,s) \approx \prod_{p \leq s} (1 - 1/p).$$

Assuming for the time being that the approximation in is sufficiently accurate, then using Mertens' theorem, we may deduce that

$$\sigma(m, s) = O(1/\log s).$$

Later, when we make this argument more rigorous, we shall see thatn holds provided s is not too large relative to m, and in particular, if $s = O((\log m)^c)$ for some constant c.

The estimate gives us a bound on the probability that a random integer passes the trial division phase, and so must be subjected to Miller–Rabin; however, performing the trial division takes some time, so we also need to estimate the expected number $\kappa(m, s)$ of trial divisions performed on a random integer between 2 and m. Of course, in the worst case, we divide by all primes up to s, and so $\leq \kappa(m, s) \leq \pi(s) = O(s/\log s)$, but we can get a better bound, as follows. Let $p_1, p_2, ..., p_r$ be the primes up to s, and for $i = 1, ..., r$, let q_i be the probability that we perform at least i trial divisions. By Theorem, we have

$$k(m,s) = \sum_{i=1}^{r} q_i .$$

Moreover, $q_1 = 1$, and $q_i = \sigma(m, p_{i-1})$ for $i = 2, ..., r$. From this, it follows that

$$k(m,s) = 1 + \sum_{i=2}^{r} \sigma(m, p_{i-1}) = O\left(\sum_{p \leq s} 1/\log p\right)$$

As a simple consequence of Chebyshev's theorem, we obtain

$$\kappa(m, s) = O(s/(\log s)^2).$$

We now derive a bound on the running time of Algorithm RP, assuming that IsPrime(.) is implemented using MRS(., k, s). Let $\ell := \text{len}(m)$. Our argument follows the same lines as was used to derive the estimate. Let us define random variables measuring various quantities during the first iteration of the main loop in Algorithm RP: N_1 is the value of n; K_1 is the number of Miller–Rabin tests actually performed; Z_1 is the running time. Also, let C_1 be the event that N_1 is composite, and let D_1 be the event that N_1 passes the trial division check. Then we have

$$\begin{aligned} E[K_1] &= E[K_1 \mid C_1 \cap D_1]\, P[C_1 \cap D_1] + E[K_1 \mid C_1 \cap \bar{D}_1]\, P[C_1 \cap \bar{D}_1] \\ &\quad + E[K_1 \mid \bar{C}_1]\, P[\bar{C}_1] \\ &\leq 4/3 \,.\, P[C_1 \cap D_1] + 0 \,.\, P[C_1 \cap \bar{D}_1] + k \,.\, P[\bar{C}_1] \\ &\leq 4/3 \,.\, P[D_1] + k \,.\, P[\bar{C}_1]. \end{aligned}$$

By and Chebyshev's theorem, it follows that

$$E[K_1] = O(1/\text{len}(s) + k/\ell).$$

Let us write $Z_1 = Z_1' + Z_1''$, where Z_1' is the amount of time spent performing the Miller–Rabin test, and Z_1'' is the amount of time spent performing trial division. By, we have $E[Z_1'] = O(\ell^3/\text{len}(s) + k\ell^2)$. Further, assuming that each individual trial division step takes time $O(\ell)$, then by (10.10) we have $E[Z_1''] = O(\ell s/\text{len}(s)^2)$. Hence,

$$E[Z_1] = O(\ell^3/\text{len}(s) + k\ell^2 + \ell s/\text{len}(s)^2).$$

It follows that if Z is the total running time of Algorithm RP, then

$$E[Z] = O(\ell^4/\operatorname{len}(s) + k\ell^3 + \ell^2 s/\operatorname{len}(s)^2).$$

Clearly, we want to choose the parameter s so that the time spent performing trial division is dominated by the time spent performing the Miller–Rabin test. To this end, let us assume that $\ell \leq s \leq \ell^2$. Then we have

$$E[Z] = O(\ell^4/\operatorname{len}(\ell) + k\ell^3).$$

This estimate does not take into account the time to generate the small primes using the sieve of Eratosthenes.

These values might be pre-computed, in which case this time is zero, but even if we compute them on the fly, this takes time $O(s \operatorname{len}(\operatorname{len}(s)))$, which is dominated by the running time of the rest of the algorithm for the values of s under consideration.

Thus, by sieving up to a bound s, where $\ell \leq s \leq \ell^2$, we effectively reduce the running time by a factor proportional to $\operatorname{len}(\ell)$, which is a very real and noticeable improvement in practice.

As we already mentioned, the above analysis is heuristic, but the results are correct. We shall now discuss how this analysis can be made rigorous; however, we should remark that any such rigorous analysis is mainly of theoretical interest only — in any practical implementation, the optimal choice of the parameter s is best determined by experiment, with the analysis being used only as a rough guide.

Now, to make the analysis rigorous, we need prove that the estimate is sufficiently accurate. Proving such estimates takes us into the realm of "sieve theory." The larger m is with respect to s, the easier it is to prove such estimates. We shall prove only the simplest and most naive such estimate, but it is still good enough for our purposes.

Before stating any results, let us restate the problem slightly differently. For a given real number $y \geq 0$, let us call a positive integer "y-rough" if it is not divisible by any prime p up to y.

For all real numbers $x \geq 0$ and $y \geq 0$, let us define $R(x, y)$ to be the number of y-rough positive integers up to x. Thus, since $\sigma(m, s)$ is the probability that a random integer between 2 and m is s-rough, and 1 is by definition s-rough, we have $\sigma(m, s) = (R(m, s) - 1)/(m - 1)$.

Theorem: For all real $x \geq 0$ and $y \geq 0$, we have

$$\left| R(x,y) - x\prod_{p \leq y}(1 - 1/p) \right| \leq 2^{\pi(y)}.$$

Proof: To simplify the notation, we shall use the Mφbius function μ. Also, for a real number u, let us write $u = \lfloor u \rfloor + \{u\}$, where $0 \leq \{u\} < 1$. Let Q be the product of the primes up to the bound y.

Now, there are $\lfloor x \rfloor$ positive integers up to x, and of these, for each prime p dividing Q, precisely $\lfloor x/p \rfloor$ are divisible by p, for each pair p, p' of distinct primes dividing Q, precisely $\lfloor x/pp' \rfloor$ are divisible by pp', and so on. By inclusion/exclusion, we have

$$R(x, y) = \sum_{d|Q} \mu(d)\lfloor x/d \rfloor = \sum_{d|Q} \mu(d)(x/d) - \sum_{d|Q} \mu(d)\{x/d\}.$$

Moreover,

$$\sum_{d|Q} \mu(d)(x/d) = x\sum_{d|Q} \mu(d)/d = x\sum_{p \leq y}(1-1/p),$$

and

$$\left|\sum_{d|Q} \mu(d)\{x/d\}\right| \leq \sum_{d|Q} 1 = 2^{\pi(y)}.$$

That proves the theorem.

This theorem says something non-trivial only when y is quite small. Nevertheless, using Chebyshev's theorem on the density of primes, alorng with Mertens' theorem, it is not hard to see that this theorem implies that holds when $s = O((\log m)^c)$ for some constant c, which implies the estimate above, when $\ell \leq s \leq \ell^2$.

EXERCISE

1. Design an algorithm that takes as input a prime q and a bound m, and outputs a random prime p between 2 and m such that $p \equiv 1 \pmod{q}$. Clearly, we need to assume that m is suffciently large with respect to q. State how large m must be with respect to q, and under these assumptions, show that your algorithm runs in time $O(\ell^4/\operatorname{len}(\ell) + k\ell^3)$, and that its output is incorrect with probability $O(4^{-k}\ell)$. As usual, $\ell := \operatorname{len}(m)$.

2. Suppose that s is a function of m such that $s = O((\log m)^c)$ for some positive constant c. Show that $\sigma(m, s) = O(1/\log s)$.

3. Let f be a polynomial with integer coefficients. For real $x \geq 0$ and $y \geq 0$, define $R_f(x, y)$ to be the number of positive integers t up to x such that $f(t)$ is y-rough.

For each positive integer m, define $\omega_f(m)$ to be the number of integers $t \in \{0, ..., m-1\}$ such that $f(t) \equiv 0 \pmod{m}$. Show that

$$\left| R_f(x,y) - x\prod_{p\le y}(1-\omega_f(p)/p) \right| \le \prod_{p\le y}(1+\omega_f(p)).$$

GENERATING A RANDOM ℓ-BIT PRIME

In some applications, we want to generate a random prime of fixed size – a random 1024-bit prime, for example. More generally, let us consider the following problem: given an integer $\ell \geq 2$, generate a random '-bit prime, that is, a prime in the interval $[2^{\ell-1}, 2^{\ell})$.

Bertrand's postulate implies that there exists a constant $c > 0$ such that $\pi(2^{\ell}) - \pi(2^{\ell-1}) \geq c2^{\ell-1}/\ell$ for all $\ell \geq 2$.

Now let us modify Algorithm RP so that it takes as input an integer $\ell \geq 2$, and repeatedly generates a random n in the interval $\{2^{\ell-1}, ..., 2^{\ell} - 1\}$ until IsPrime(n) returns true. Let us call this variant Algorithm RP'. Further, let us implement IsPrime(.) as MR(., k), for some auxiliary parameter k, and define $\gamma'(\ell, k)$ to be the probability that the output of Algorithm RP' – with this implementation of IsPrime – is composite.

So we have

$$\gamma'(\ell,k) \le 4^{-k}\frac{2^{\ell-1}}{\pi(2^{\ell})-\pi(2^{\ell-1})} = O(4^{-k}\ell);$$

moreover, if the output of Algorithm *RP'* is prime, then every ℓ-bit prime is equally likely, and the expected running time is $O(\ell^4 + k\ell^3)$. By doing some trial division, this can be reduced to $O(\ell^4/\text{len}(\ell) + k\ell^3)$.

The function $\gamma'(\ell, k)$ has been studied a good deal; for example, the following explicit bound is known:

Theorem: Fo]r all $\ell \geq 2$, we have

$$\gamma'(\ell,1) \le \ell^2 4^{2-\sqrt{\ell}}.$$